T0178820

Spoken Language Processing

Spoken Language Processing

Edited by
Joseph Mariani

First published in France in 2002 by Hermes Science/Lavoisier entitled *Traitement automatique du langage parlé 1 et 2* © LAVOISIER, 2002

First published in Great Britain and the United States in 2009 by ISTE Ltd and John Wiley & Sons, Inc.

ISTE Ltd
27-37 St George's Road
London SW19 4EU
UK

www.iste.co.uk

John Wiley & Sons, Inc.
111 River Street
Hoboken, NJ 07030
USA

www.wiley.com

© ISTE Ltd, 2009

Library of Congress Cataloging-in-Publication Data

Traitement automatique du langage parlé 1 et 2. English
 Spoken language processing / edited by Joseph Mariani.
   p. cm.
 Includes bibliographical references and index.
 ISBN 978-1-84821-031-8
 1. Automatic speech recognition. 2. Speech processing systems. I. Mariani, Joseph. II. Title.
 TK7895.S65T7213 2008
 006.4'54--dc22

2008036758

British Library Cataloguing-in-Publication Data
A CIP record for this book is available from the British Library
ISBN: 978-1-84821-031-8

Printed and bound in Great Britain by CPI Antony Rowe, Chippenham, Wiltshire.

FSC
Mixed Sources
Product group from well-managed
forests and other controlled sources

Cert no. SGS-COC-2953
www.fsc.org
© 1996 Forest Stewardship Council

Table of Contents

**Chapter 11. Multimodal Speech: Two or Three senses are
Better than One**   377
Jean-Luc SCHWARTZ, Pierre ESCUDIER and Pascal TEISSIER

Chapter 12. Speech and Human-Computer Communication   417
Wolfgang MINKER & Françoise NÉEL

Preface

This book, entitled *Spoken Language Processing*, addresses all the aspects covering the automatic processing of spoken language: how to automate its production and perception, how to synthesize and understand it. It calls for existing know-how in the field of signal processing, pattern recognition, stochastic modeling, computational linguistics, human factors, but also relies on knowledge specific to spoken language.

The automatic processing of spoken language covers activities related to the analysis of speech, including variable rate coding to store or transmit it, to its synthesis, especially from text, to its recognition and understanding, should it be for a transcription, possibly followed by an automatic indexation, or for human-machine dialog or human-human machine-assisted interaction. It also includes speaker and spoken language recognition. These tasks may take place in a noisy environment, which makes the problem even more difficult.

The activities in the field of automatic spoken language processing started after the Second World War with the works on the *Vocoder* and *Voder* at Bell Labs by Dudley and colleagues, and were made possible by the availability of electronic devices. Initial research work on basic recognition systems was carried out with very limited computing resources in the 1950s. The computer facilities that became available to researchers in the 1970s made it possible to achieve initial progress within laboratories, and microprocessors then led to the early commercialization of the first voice recognition and speech synthesis systems at an affordable price. The steady progress in the speed of computers and in the storage capacity accompanied the scientific advances in the field.

Research investigations in the 1970s, including those carried out in the large DARPA "Speech Understanding Systems" (SUS) program in the USA, suffered from a lack of availability of speech data and of means and methods for evaluating the performance of different approaches and systems. The establishment by DARPA, as part of its following program launched in 1984, of a national language resources center, the Linguistic Data Consortium (LDC), and of a system assessment center, within the National Institute of Standards and Technology (NIST, formerly NBS), brought this area of research into maturity. The evaluation campaigns in the area of speech recognition, launched in 1987, made it possible to compare the different approaches that had coexisted up to then, based on "Artificial Intelligence" methods or on stochastic modeling methods using large amounts of data for training, with a clear advantage to the latter. This led progressively to a quasi-generalization of stochastic approaches in most laboratories in the world. The progress made by researchers has constantly accompanied the increasing difficulty of the tasks which were handled, starting from the recognition of sentences read aloud, with a limited vocabulary of 1,000 words, either speaker-dependent or speaker-independent, to the dictation of newspaper articles for vocabularies of 5,000, 20,000 and 64,000 words, and then to the transcription of radio or television broadcast news, with unlimited size vocabularies. These evaluations were opened to the international community in 1992. They first focused on the American English language, but early initiatives were also carried out on the French, German or British English languages in a French or European context. Other campaigns were subsequently held on speaker recognition, language identification or speech synthesis in various contexts, allowing for a better understanding of the pros and cons of an approach, and for measuring the status of technology and the progress achieved or still to be achieved. They led to the conclusion that a sufficient level of maturation has been reached for putting the technology on the market, in the field of voice dictation systems for example. However, it also identified the difficulty of other more challenging problems, such as those related to the recognition of conversational speech, justifying the need to keep on supporting fundamental research in this area.

This book consists of two parts: a first part discusses the analysis and synthesis of speech and a second part speech recognition and understanding. The first part starts with a brief introduction of the principles of speech production, followed by a broad overview of the methods for analyzing speech: linear prediction, short-term Fourier transform, time-representations, wavelets, cepstrum, etc. The main methods for speech coding are then developed for the telephone bandwidth, such as the CELP coder, or, for broadband communication, such as "transform coding" and quantization methods. The audio-visual coding of speech is also introduced. The various operations to be carried out in a text-to-speech synthesis system are then presented regarding the linguistic processes (grapheme-to-phoneme transcription, syntactic and prosodic analysis) and the acoustic processes, using rule-based approaches or approaches based on the concatenation of variable length acoustic units. The different types of speech signal modeling – articulatory, formant-based, auto-regressive, harmonic-noise or PSOLA-like – are then described. The evaluation of speech synthesis systems is a topic of specific attention in this chapter. The extension of speech synthesis to talking faces animation is the subject of the next chapter, with a presentation of the application fields, of the interest of a bimodal approach and of models used to synthesize and animate the face. Finally, computational auditory scene analysis opens prospects in the signal processing of speech, especially in noisy environments.

The second part of the book focuses on speech recognition. The principles of speech recognition are first presented. Hidden Markov models are introduced, as well as their use for the acoustic modeling of speech. The Viterbi algorithm is depicted, before introducing language modeling and the way to estimate probabilities. It is followed by a presentation of recognition systems, based on those principles and on the integration of those methodologies, and of lexical and acoustic-phonetic knowledge. The applicative aspects are highlighted, such as efficiency, portability and confidence measures, before describing three types of recognition systems: for text dictation, for audio documents indexing and for oral dialog. Research in language identification aims at recognizing which language is spoken, using acoustic, phonetic, phonotactic or prosodic information. The characteristics of languages are introduced and the way humans or machines can achieve that task is depicted, with a large presentation of the present performances of such systems. Speaker recognition addresses the recognition and verification of the identity of a person based on his voice. After an introduction on what characterizes a voice, the different types and designs of systems are presented, as well as their theoretical background. The way to evaluate the performances of speaker recognition systems and the applications of this technology are a specific topic of interest. The use of speech or speaker recognition systems in noisy environments raises especially difficult problems to solve, but they must be taken into account in any operational use of such systems. Various methods are available, either by pre-processing the signal, during the parameterization phase, by using specific distances or by adaptation methods. The Lombard effect, which causes a change in the production of the voice signal itself due to the noisy environment surrounding the speaker, benefits from a special attention. Along with recognition based solely on the acoustic signal, bi-modal recognition combines two acquisition channels: auditory and visual. The value added by bimodal processing in a noisy environment is emphasized and architectures for the audiovisual merging of audio and visual speech recognition are presented. Finally, applications of automatic spoken language processing systems, generally for human-machine communication and particularly in telecommunications, are described. Many applications of speech coding, recognition or synthesis exist in many fields, and the market is growing rapidly. However, there are still technological and psychological barriers that require more work on modeling human factors and ergonomics, in order to make those systems widely accepted.

The reader, undergraduate or graduate student, engineer or researcher will find in this book many contributions of leading French experts of international renown who share the same enthusiasm for this exciting field: the processing by machines of a capacity which used to be specific to humans: language.

Finally, as editor, I would like to warmly thank Anna and Frédéric Bimbot for the excellent work they achieved in translating the book *Traitement automatique du langage parlé*, on which this book is based.

Joseph Mariani
November 2008

Chapter 1

Speech Analysis

1.1. Introduction

1.1.1. *Source-filter model*

Speech, the acoustic manifestation of language, is probably the main means of communication between human beings. The invention of telecommunications and the development of digital information processing have therefore entailed vast amounts of research aimed at understanding the mechanisms of speech communication.

Speech can be approached from different angles. In this chapter, we will consider speech as a signal, a one-dimensional function, which depends on the time variable (as in [BOI 87, OPP 89, PAR 86, RAB 75, RAB 77]). The acoustic speech signal is obtained at a given point in space by a sensor (microphone) and converted into electrical values. These values are denoted $s(t)$ and they represent a real-valued function of real variable t, analogous to the variation of the acoustic pressure. Even if the acoustic form of the speech signal is the most widespread (it is the only signal transmitted over the telephone), other types of analysis also exist, based on alternative physiological signals (for instance, the electroglottographic signal, the palatographic signal, the airflow), or related to other modalities (for example, the image of the face or the gestures of the articulators). The field of speech analysis covers the set of methods aiming at the extraction of information on and from this signal, in various applications, such as:

Chapter written by Christophe D'ALESSANDRO.

– speech coding: the compression of information carried by the acoustic signal, in order to save data storage or to reduce transmission rate;

– speech recognition and understanding, speaker and spoken language recognition;

– speech synthesis or automatic speech generation, from an arbitrary text;

– speech signal processing, which covers many applications, such as auditory aid, denoising, speech encrypting, echo cancellation, post-processing for audiovisual applications;

– phonetic and linguistic analysis, speech therapy, voice monitoring in professional situations (for instance, singers, speakers, teachers, managers, etc.).

Two ways of approaching signal analysis can be distinguished: the model-based approach and the representation-based approach. When a voice signal model (or a voice production model or a voice perception model) is assumed, the goal of the analysis step is to identify the parameters of that model. Thus, many analysis methods, referred to as *parametric methods*, are based on the source-filter model of speech production; for example, the linear prediction method. On the other hand, when no particular hypothesis is made on the signal, mathematical representations equivalent to its time representation can be defined, so that new information can be drawn from the coefficients of the representation. An example of a non-parametric method is the short-term Fourier transform (STFT). Finally, there are some hybrid methods (sometimes referred to as semi-parametric). These consist of estimating some parameters from non-parametric representations. The sinusoidal and cepstral representations are examples of semi-parametric representation.

This chapter is centered on the linear acoustic source-filter speech production model. It presents the most common speech signal analysis techniques, together with a few illustrations. The reader is assumed to be familiar with the fundamentals of digital signal processing, such as discrete-time signals, Fourier transform, Laplace transform, Z-transforms and digital filters.

1.1.2. *Speech sounds*

The human speech apparatus can be broken down into three functional parts [HAR 76]: 1) the lungs and trachea, 2) the larynx and 3) the vocal tract. The abdomen and thorax muscles are the engine of the breathing process. Compressed by the muscular system, the lungs act as bellows and supply some air under pressure which travels through the trachea (subglottic pressure). The airflow thus expired is then modulated by the movements of the larynx and those of the vocal tract.

The larynx is composed of the set of muscles, articulated cartilage, ligaments and mucous membranes located between the trachea on one side, and the pharyngeal cavity on the other side. The cartilage, ligaments and muscles in the larynx can set the vocal cords in motion, the opening of which is called the *glottis*. When the vocal cords lie apart from each other, the air can circulate freely through the glottis and no sound is produced. When both membranes are close to each other, they can join and modulate the subglottic airflow and pressure, thus generating isolated pulses or vibrations. The fundamental frequency of these vibrations governs the pitch of the voice signal (F_0).

The vocal tract can be subdivided into three cavities: the pharynx (from the larynx to the velum and the back of the tongue), the oral tract (from the pharynx to the lips) and the nasal cavity. When it is open, the velum is able to divert some air from the pharynx to the nasal cavity. The geometrical configuration of the vocal tract depends on the organs responsible for the articulation: jaws, lips, tongue.

Each language uses a certain subset of sounds, among those that the speech apparatus can produce [MAL 74]. The smallest distinctive sound units used in a given language are called *phonemes*. The phoneme is the smallest spoken unit which, when substituted with another one, changes the linguistic content of an utterance. For instance, changing the initial /p/ sound of "pig" (/pɪg/) into /b / yields a different word: "big" (/bɪg/). Therefore, the phonemes /p/ and /b/ can be distinguished from each other.

A set of phonemes, which can be used for the description of various languages [WEL 97], is given in Table 1.1 (described both by the International Phonetic Alphabet, IPA, and the computer readable Speech Assessment Methodologies Phonetic Alphabet, SAMPA). The first subdivision that is observed relates to the excitation mode and to the vocal tract stability: the distinction between vowels and consonants. Vowels correspond to a periodic vibration of the vocal cords and to a stable configuration of the vocal tract. Depending on whether the nasal branch is open or not (as a result of the lowering of the velum), vowels have either a nasal or an oral character. Semivowels are produced when the periodic glottal excitation occurs simultaneously with a fast movement of the vocal tract, between two vocalic positions.

Consonants correspond to fast constriction movements of the articulatory organs, i.e. generally to rather unstable sounds, which evolve over time. For fricatives, a strong constriction of the vocal tract causes a friction noise. If the vocal cords vibrate at the same time, the fricative consonant is then voiced. Otherwise, if the vocal folds let the air pass through without producing any sound, the fricative is unvoiced. Plosives are obtained by a complete obstruction of the vocal tract, followed by a release phase. If produced together with the vibration of the vocal cords, the plosive is voiced, otherwise it is unvoiced. If the nasal branch is opened during the mouth closure, the produced sound is a nasal consonant. Semivowels are considered voiced consonants, resulting from a fast movement which briefly passes through the articulatory position of a vowel. Finally, liquid consonants are produced as the combination of a voiced excitation and fast articulatory movements, mainly from the tongue.

SAMPA		IPA		Unicode		label and exemplification
symbol	ASCII			hex	dec.	
						Vowels
A	65	ɑ	script a	0251	593	open back unrounded, Cardinal 5, Eng. *start*
{	123	æ	ae ligature	00E6	230	near-open front unrounded, Eng. *trap*
6	54	ɐ	turned a	0250	592	open schwa, Ger. *besser*
Q	81	ɒ	turned script a	0252	594	open back rounded, Eng. *lot*
E	69	ɛ	epsilon	025B	603	open-mid front unrounded, Fr. *même*
@	64	ə	turned e	0259	601	schwa, Eng. *banana*
3	51	ɜ	rev. epsilon	025C	604	long mid central, Eng. *nurse*
I	73	ɪ	small cap I	026A	618	lax close front unrounded, Eng. *kit*
O	79	ɔ	turned c	0254	596	open-mid back rounded, Eng. *thought*
2	50	ø	o-slash	00F8	248	close-mid front rounded, Fr. *deux*
9	57	œ	oe ligature	0153	339	open-mid front rounded, Fr. *neuf*
&	38	Œ	s.c. OE ligature	0276	630	open front rounded, Swedish *skörd*
U	85	ʊ	upsilon	028A	650	lax close back rounded, Eng. *foot*
}	125	ʉ	barred u	0289	649	close central rounded, Swedish *sju*
V	86	ʌ	turned v	028C	652	open-mid back unrounded, Eng. *strut*
Y	89	ʏ	small cap Y	028F	655	lax [y], Ger. *hübsch*

Consonants						
B	66	β	beta	03B2	946	Voiced bilabial fricative, Sp. *cabo*
C	67	ç	c-cedilla	00E7	231	voiceless palatal fricative, Ger. *ich*
D	68	ð	eth	00F0	240	Voiced dental fricative, Eng. *then*
G	71	ɣ	gamma	0263	611	Voiced velar fricative, Sp. *fuego*
L	76	ʎ	turned y	028E	654	Palatal lateral, It. *famiglia*
J	74	ɲ	left-tail n	0272	626	Palatal nasal, Sp. *año*
N	78	ŋ	eng	014B	331	velar nasal, Eng. *thing*
R	82	ʁ	inv. s.c. R	0281	641	Voiced uvular fricative. or trill, Fr. *roi*
S	83	ʃ	esh	0283	643	voiceless palatoalveolar fricative, Eng. *ship*
T	84	θ	theta	03B8	952	voiceless dental fricative, Eng. *thin*
H	72	ɥ	turned h	0265	613	labial-palatal semivowel, Fr. *huit*
Z	90	ʒ	ezh (yogh)	0292	658	vd. palatoalveolar fric., Eng. *measure*
?	63	ʔ	dotless ?	0294	660	glottal stop, Ger. *Verein*, also Danish *stød*

Table 1.1. *Computer-readable Speech Assessment Methodologies Phonetic Alphabet, SAMPA, and its correspondence in the International Phonetic Alphabet, IPA, with examples in 6 different languages [WEL 97]*

In speech production, sound sources appear to be relatively localized; they excite the acoustic cavities in which the resulting air disturbances propagate and then radiate to the outer acoustic field. This relative independence of the sources with the transformations that they undergo is the basis for the acoustic theory of speech production [FAN 60, FLA 72, STE 99]. This theory considers source terms, on the one hand, which are generally assumed to be non-linear, and a linear filter on the other hand, which acts upon and transforms the source signal. This source-filter decomposition reflects the terminology commonly used in phonetics, which describes the speech sounds in terms of "phonation" (source) and "articulation" (filter). The source and filter acoustic contributions can be studied separately, as they can be considered to be decoupled from each other, in a first approximation. From the point of view of physics, this model is an approximation, the main advantage of which is its simplicity. It can be considered as valid at frequencies below 4 or 5 kHz, i.e. those frequencies for which the propagation in the vocal tract consists of one-dimensional plane waves. For signal processing purposes, the acoustic model can be described as a linear system, by neglecting the source-filter interaction:

$$s(t) = e(t) * v(t) * l(t) = [p(t) + r(t)] * v(t) * l(t) \qquad [1.1]$$

$$= \left[\sum_{i=-\infty}^{+\infty} \delta(t - iT_0) * u_g(t) + r(t) \right] * v(t) * l(t) \qquad [1.2]$$

$$S(\omega) = E(\omega) \times V(\omega) \times L(\omega) = [P(\omega) + R(\omega)] \times V(\omega) \times L(\omega) \qquad [1.3]$$

$$= \left[\left(\sum_{i=-\infty}^{+\infty} \delta(\omega - iF_0) \right) \left| U_g(\omega) \right| e^{j\theta_{u_g}(\omega)} + \left| R(\omega) \right| e^{j\theta_r(\omega)} \right]$$
$$\times \left| V(\omega) \right| e^{j\theta_v(\omega)} \times \left| L(\omega) \right| e^{j\theta_l(\omega)} \qquad [1.4]$$

where $s(t)$ is the speech signal, $v(t)$ the impulse response of the vocal tract, $e(t)$ the vocal excitation source, $l(t)$ the impulse response of the lip radiation component, $p(t)$ the periodic part of the excitation, $r(t)$ the non-periodic part of the excitation, $u_g(t)$ the glottal airflow wave, T_0 the fundamental period, $r(t)$ the noise part of the excitation, δ the Dirac distribution, and where $S(\omega)$, $V(\omega)$, $E(\omega)$, $L(\omega)$, $P(\omega)$, $R(\omega)$, $U_g(\omega)$ denote the Fourier transforms of $s(t)$, $v(t)$, $e(t)$, $l(t)$, $p(t)$, $r(t)$, $u_g(t)$ respectively. $F_0 = 1/T_0$ is the voicing fundamental frequency. The various terms of the source-filter model are now going to be studied in more details.

1.1.3. Sources

The source component $e(t)$, $E(\omega)$ is a signal composed a periodic part (vibrations of the vocal cords, characterized by F_0 and the glottal airflow waveform) and a noise part. The various phonemes use both types of source excitation either separately or simultaneously.

1.1.3.1. Glottal airflow wave

The study of glottal activity (phonation) is particularly important in speech science. Physical models of the glottis functioning, in terms of mass-spring systems have been investigated [FLA 72]. Several types of physiological signals can be used to conduct studies on the glottal activity (for example, electroglottography, fast photography, see [TIT 94]). From the acoustic point of view, the glottal airflow wave, which represents the airflow traveling through the glottis as a function of time, is preferred to the pressure wave. It is indeed easier to measure the glottal airflow rather than the glottal pressure, from physiological data. Moreover, the pseudo-periodic voicing source $p(t)$ can be broken down into two parts: a pulse train, which represents the periodic part of the excitation and a low-pass filter, with an impulse response u_g, which corresponds to the (frequency-domain and time-domain) shape of the glottal airflow wave.

The time-domain shape of the glottal airflow wave (or, more precisely, of its derivative) generally governs the behavior of the time-domain signal for vowels and voiced signals [ROS 71]. Time-domain models of the glottal airflow have several properties in common: they are periodical, always non-negative (no incoming airflow), they are continuous functions of the time variable, derivable everywhere except, in some cases, at the closing instant. An example of such a time-domain model is the Klatt model [KLA 90], which calls for 4 parameters (the fundamental frequency F_0, the voicing amplitude AV, the opening ratio O_q and the frequency T_L of a spectral attenuation filter). When there is no attenuation, the KGLOTT88 model writes:

$$U_g(t) = \begin{cases} at^2 - bt^3 & for \ \ 0 \le t \le O_q T_0 \\ 0 & for \ \ O_q T_0 \le t \le T_0 \end{cases} \quad with \quad \begin{vmatrix} a = \dfrac{27}{4} \dfrac{AV}{O_q^2 T_0} \\ b = \dfrac{27}{4} \dfrac{AV}{O_q^3 T_0^2} \end{vmatrix} \qquad [1.5]$$

when $T_L \ne 0$, $U_g(t)$ is filtered by an additional low-pass filter, with an attenuation at 3,000 Hz equal to T_L dB.

The LF model [FAN 85] represents the derivative of the glottal airflow with 5 parameters (fundamental period T_0, amplitude at the minimum of the derivative or at the maximum of the wave Ee, instant of maximum excitation Te, instant of maximum airflow wave Tp, time constant for the return phase Ta):

$$U_g'(t) = \begin{cases} E_e e^{a(t-T_e)} \dfrac{\sin(\pi t / T_p)}{\sin(\pi T_e / T_p)} & for \ \ 0 \le t \le T_e \\ -\dfrac{E_e}{\varepsilon T_a}(e^{-\varepsilon(t-T_e)} - e^{-\varepsilon(T_0-T_e)}) & for \ \ T_e \le t \le T_0 \end{cases} \qquad [1.6]$$

In this equation, parameter ε is defined by an implicit equation:

$$\varepsilon T_a = 1 - e^{-\varepsilon(T_0-T_e)} \qquad [1.7]$$

All time-domain models (see Figure 1.1) have at least three main parameters: the voicing amplitude, which governs the time-domain amplitude of the wave, the voicing period, and the opening duration, i.e. the fraction of the period during which the wave is non-zero. In fact, the glottal wave represents the airflow traveling through the glottis. This flow is zero when the vocal chords are closed. It is positive when they are open. A fourth parameter is introduced in some models to account for the speed at which the glottis closes. This closing speed is related to the high frequency part of the speech spectrum.

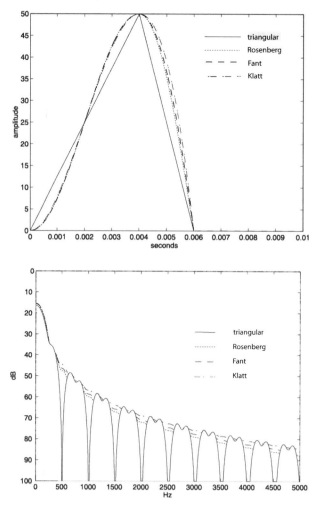

Figure 1.1. *Models of the glottal airflow waveform in the time domain: triangular model, Rosenberg model, KGLOT88, LF and the corresponding spectra*

The general shape of the glottal airflow spectrum is one of a low-pass filter. Fant [FAN 60] uses four poles on the negative real axis:

$$U_g(s) = \frac{U_{g0}}{\prod_{r=1}^{4}(1 - s/s_r)} \qquad [1.8]$$

with $s_{r_1} \approx s_{r_2} = 2\pi \times 100$ Hz, and $s_{r_3} = 2\pi \times 2{,}000$ Hz, $s_{r_4} = 2\pi \times 4{,}000$ Hz. This is a spectral model with six parameters (F_0, U_{g0} and four poles), among which two are fixed (s_{r_3} and s_{r_4}). This simple form is used in [MAR 76] in the digital domain, as a second-order low-pass filter, with a double real pole in K:

$$U_g(z) = \frac{U_{g0}}{(1 - Kz^{-1})^2} \qquad [1.9]$$

Two poles are sufficient in this case, as the numerical model is only valid up to approximately 4,000 Hz. Such a filter depends on three parameters: gain U_{g0}, which corresponds to the voicing amplitude, fundamental frequency F_0 and a frequency parameter K, which replaces both s_{r_1} and s_{r_2}. The spectrum shows an asymptotic slope of −12 dB/octave when the frequency increases. Parameter K controls the filter's cut-off frequency. When the frequency tends towards zero, $|U_g(0)| \sim U_{g0}$. Therefore, the spectral slope is zero in the neighborhood of zero, and −12 dB/octave, for frequencies above a given bound (determined by K). When the focus is put on the derivative of the glottal airflow, the two asymptotes have slopes of +6 dB/octave and −6 dB/octave respectively. This explains the existence of a maximum in the speech spectrum at low frequencies, stemming from the glottal source.

Another way to calculate the glottal airflow spectrum is to start with time-domain models. For the Klatt model, for example, the following expression is obtained for the Laplace transform L, when there is no additional spectral attenuation:

$$L(n_g')(s) = \frac{27}{4}\frac{1}{s}\left(e^{-s} + \frac{2(1 + 2e^{-s})}{s} - \frac{6(1 - e^{-s})}{s^2}\right) \qquad [1.10]$$

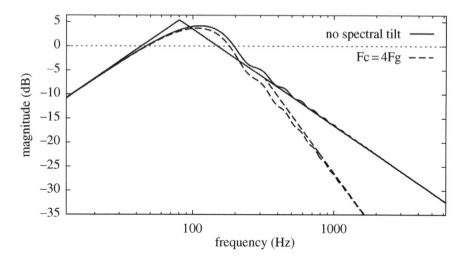

Figure 1.2. *Schematic spectral representation of the glottal airflow waveform. Solid line: abrupt closure of the vocal cords (minimum spectral slope). Dashed line: dampened closure. The cut-off frequency owed to this dampening is equal to 4 times the spectral maximum F_g*

It can be shown that this is a low-pass spectrum. The derivative of the glottal airflow shows a spectral maximum located at:

$$f_g = \frac{\sqrt{3}}{\pi} \frac{1}{O_q T_0}$$

[1.11]

This sheds light on the links between time-domain and frequency-domain parameters: the opening ratio (i.e. the ratio between the opening duration of the glottis and the overall glottal period) governs the spectral peak frequency. The time-domain amplitude rules the frequency-domain amplitude. The closing speed of the vocal cords relates directly to the spectral attenuation in the high frequencies, which shows a minimum slope of -12 dB/octave.

1.1.3.2. *Noise sources*

The periodic vibration of the vocal cords is not the only sound source in speech. Noise sources are involved in the production of several phonemes. Two types of noise can be observed: transient noise and continuous noise. When a plosive is produced, the holding phase (total obstruction of the vocal tract) is followed by a release phase. A transient noise is then produced by the pressure and airflow impulse generated by the opening of the obstruction. The source is located in the vocal tract, at the point where the obstruction and release take place. The impulse is a wide-band noise which slightly varies with the plosive.

For continuous noise (fricatives), the sound originates from turbulences in the fast airflow at the level of the constriction. Shadle [SHA 90] distinguishes noise caused by the lining and noise caused by obstacles, depending on the incidence angle of the air stream on the constriction. In both cases, the turbulences produce a source of random acoustic pressure downstream of the constriction. The power spectrum of this signal is approximately flat in the range of 0 – 4,000 Hz, and then decreases with frequency.

When the constriction is located at the glottis, the resulting noise (aspiration noise) shows a wide-band spectral maximum around 2,000 Hz. When the constriction is in the vocal tract, the resulting noise (frication noise) also shows a roughly flat spectrum, either slowly decreasing or with a wide maximum somewhere between 4 kHz and 9 kHz. The position of this maximum depends on the fricative. The excitation source for continuous noise can thus be considered as a white Gaussian noise filtered by a low-pass filter or by a wide band-pass filter (several kHz wide).

In continuous speech, it is interesting to separate the periodic and non-periodic contributions of the excitation. For this purpose, either the sinusoidal representation [SER 90] or the short-term Fourier spectrum [DAL 98, YEG 98] can be used. The principle is to subtract from the source signal its harmonic component, in order to obtain the non-periodic component. Such a separation process is illustrated in Figure 1.3.

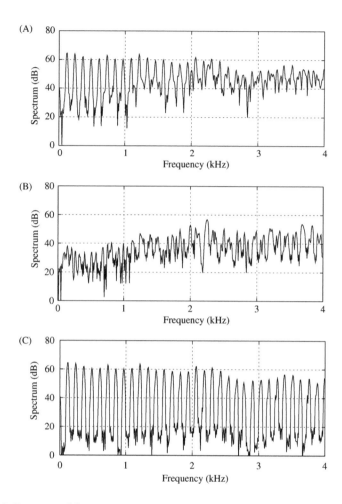

Figure 1.3. *Spectrum of the excitation source for a vowel. (A) the complete spectrum; (B) the non-periodic part; (C) the periodic part*

1.1.4. *Vocal tract*

The vocal tract is an acoustic cavity. In the source-filter model, it plays the role of a filter, i.e. a passive system which is independent from the source. Its function consists of transforming the source signal, by means of resonances and anti-resonances. The maxima of the vocal tract's spectral gain are called *spectral formants*, or more simply *formants*. Formants can generally be assimilated to the spectral maxima which can be observed on the speech spectrum, as the source spectrum is globally monotonous for voiced speech. However, depending on the source spectrum, formants and resonances may turn out to be shifted. Furthermore, in some cases, a source formant can be present. Formants are also observed in unvoiced speech segments, at least those that correspond to cavities located in front of the constriction, and thus excited by the noise source.

1.1.4.1. *Multi-tube model*

The vocal tract is an acoustic duct with a complex shape. At a first level of approximation, its acoustic behavior may be understood to be one of an acoustic tube. Hypotheses must be made to calculate the propagation of an acoustic wave through this tube:

– the tube is cylindrical, with a constant area section A;

– the tube walls are rigid (i.e. no vibration terms at the walls);

– the propagation mode is (mono-dimensional) plane waves. This assumption is satisfied if the transverse dimension of the tube is small, compared to the considered wavelengths, which correspond in practice to frequencies below 4,000 Hz for a typical vocal tract (i.e. a length of 17.6 cm and a section of 8 cm^2 for the neutral vowel);

– the process is adiabatic (i.e. no loss by thermal conduction);

– the hypothesis of small movements is made (i.e. second-order terms can be neglected).

Let A denote the (constant) section of the tube, x the abscissa along the tube, t the time, $p(x, t)$ the pressure, $u(x, t)$ the speed of the air particles, $U(x, t)$ the volume velocity, ρ the density, L the tube length and C the speed of sound in the air (approximately 340 m/s). The equations governing the propagation of a plane wave in a tube (Webster equations) are:

$$\frac{1}{C^2}\frac{\partial^2 p}{\partial t^2} = \frac{\partial^2 p}{\partial x^2} \quad \text{and} \quad \frac{1}{C^2}\frac{\partial^2 u}{\partial t^2} = \frac{\partial^2 u}{\partial x^2} \tag{1.12}$$

This result is obtained by studying an infinitesimal variation of the pressure, the air particle speed and the density: $p(x, t) = p_0 + \partial p(x, t)$, $u(x, t) = u_0 + \partial u(x, t)$, $\rho(x, t) = \rho_0 + \partial \rho(x, t)$, in conjunction with two fundamental laws of physics:

1) the conservation of mass entering a slice of the tube comprised between x and $x+dx$: $A\partial x \partial \rho = \rho A \partial u \partial t$. By neglecting the second-order term ($\partial \rho \partial u \partial t$), by using the ideal gas law and the fact that the process is adiabatic, ($p/\rho = C^2$), this equation can be rewritten $\partial p/C^2 \partial t = \rho_0 \partial u/\partial x$;

2) Newton's second law applied to the air in the slice of tube yields: $A\partial p = \rho A\partial x(\partial u/\partial t)$, thus $\partial p/\partial x = \rho_0\partial u/\partial t$.

The solutions of these equations are formed by any linear combination of functions $f(t)$ and $g(t)$ of a single variable, twice continuously derivable, written as a forward wave and a backward wave which propagate at the speed of sound:

$$f^+(x,t) = f\left(t - \frac{x}{C}\right) \quad \text{and} \quad f^-(x,t) = g\left(t + \frac{x}{C}\right) \qquad [1.13]$$

and thus the pressure in the tube can be written:

$$p(x,t) = f\left(t - \frac{x}{C}\right) + g\left(t + \frac{x}{C}\right) \qquad [1.14]$$

It is easy to verify that function p satisfies equation [1.12]. Moreover, functions f and g satisfy:

$$\frac{\partial f(t - x/C)}{\partial t} = -c\frac{\partial f(t - x/C)}{\partial x} \quad \text{and} \quad \frac{\partial g(t + x/C)}{\partial t} = c\frac{\partial g(t + x/C)}{\partial x} \qquad [1.15]$$

which, when combined for example with Newton's second law, yields the following expression for the volume velocity (the tube having a constant section A):

$$U(x,t) = \frac{A}{\rho C}\left[f\left(t - \frac{x}{C}\right) - g\left(t + \frac{x}{C}\right)\right] \qquad [1.16]$$

It must be noted that if the pressure is the sum of a *forward function* and a *backward function*, the volume velocity is the difference between these two functions. The expression $Z_c = \rho C/A$ is the ratio between the pressure and the volume velocity, which is called the *characteristic acoustic impedance of the tube*. In general, the acoustic impedance is defined in the frequency domain. Here, the term "impedance" is used in the time domain, as the ratio between the forward and backward parts of the pressure and the volume velocity. The following electroacoustical analogies are often used: "acoustic pressure" for "voltage"; "acoustic volume velocity" for "intensity".

The vocal tract can be considered as the concatenation of cylindrical tubes, each of them having a constant area section A, and all tubes being of the same length. Let Δ denote the length of each tube. The vocal tract is considered as being composed of p sections, numbered from 1 to p, starting from the lips and going towards the glottis. For each section n, the forward and backward waves (respectively from the glottis to the lips and from the lips to the glottis) are denoted f_n and b_n. These waves are defined at the section input, from $n+1$ to n (on the left of the section, if the glottis is on the left). Let $R_n = \rho C/A_n$ denote the acoustic impedance of the section, which depends only on its area section.

Each section can then be considered as a quadripole with two inputs f_{n+1} and b_{n+1}, two outputs f_n and b_n and a transfer matrix T_{n+1}:

$$\begin{bmatrix} f_n \\ b_n \end{bmatrix} = T_{n+1} \begin{bmatrix} f_{n+1} \\ b_{n+1} \end{bmatrix} \qquad\qquad [1.17]$$

For a given section, the transfer matrix can be broken down into two terms. Both the interface with the previous section (1) and the behavior of the waves within the section (2) must be taken into account:

1) At the level of the discontinuity between sections n and $n+1$, the following relations hold, on the left and on the right, for the pressure and the volume velocity:

$$\begin{cases} P_{n+1} = R_{n+1}(f_{n+1} + b_{n+1}) \\ U_{n+1} = f_{n+1} - b_{n+1} \end{cases} \quad \text{and} \quad \begin{cases} P_n = R_n(f_n + b_n) \\ U_n = f_n - b_n \end{cases} \qquad [1.18]$$

as the pressure and the volume velocity are both continuous at the junction, we have $R_{n+1}\,(f_{n+1}+b_{n+1}) = R_n\,(f_n+b_n)$ and $f_{n+1}-b_{n+1} = f_n-b_n$, which enables the transfer matrix at the interface to be calculated as:

$$\begin{bmatrix} f_n \\ b_n \end{bmatrix} = \frac{1}{2R_n} \begin{bmatrix} R_{n+1} + R_n & R_{n+1} - R_n \\ R_{n+1} - R_n & R_{n+1} + R_n \end{bmatrix} \begin{bmatrix} f_{n+1} \\ b_{n+1} \end{bmatrix} \qquad [1.19]$$

After defining acoustic reflection coefficient k, the transfer matrix $T_{n+1}^{(1)}$ at the interface is:

$$T_{n+1}^{(1)} = \frac{1}{k+1} \begin{bmatrix} 1 & -k \\ -k & 1 \end{bmatrix} \quad \text{with} \quad k = \frac{R_n - R_{n+1}}{R_n + R_{n+1}} = \frac{A_{n+1} - A_n}{A_{n+1} + A_n} \qquad [1.20]$$

2) Within the tube of section $n+1$, the waves are simply submitted to propagation delays, thus:

$$f_n(t) = f_{n+1}\left(t - \frac{\Delta}{C}\right) \quad \text{and} \quad b_n(t) = b_{n+1}\left(t + \frac{\Delta}{C}\right) \qquad [1.21]$$

The phase delays and advances of the wave are all dependent on the same quantity Δ/C. The signal can thus be sampled with a sampling period equal to $Fs = C/(2\Delta)$ which corresponds to a wave traveling back and forth in a section. Therefore, the z-transform of equations [1.21] can be considered as a delay (respectively an advance) of Δ/C corresponding to a factor $z^{-1/2}$ (respectively $z^{1/2}$).

$$F_n(z) = F_{n+1}(z)z^{-1/2} \quad \text{and} \quad B_n(z) = B_{n+1}(z)z^{1/2} \qquad [1.22]$$

from which the transfer matrix $T_{n+1}^{(2)}$ corresponding to the propagation in section $n + 1$ can be deduced.

In the z-transform domain, the total transfer matrix T_{n+1} for section $n+1$ is the product of $T_{n+1}^{(1)}$ and $T_{n+1}^{(2)}$:

$$T_{n+1} = \frac{1}{k+1}\begin{bmatrix} 1 & -k \\ -k & 1 \end{bmatrix}\begin{bmatrix} z^{-1/2} & 0 \\ 0 & z^{1/2} \end{bmatrix} = \frac{z^{-1/2}}{k+1}\begin{bmatrix} 1 & -kz \\ -k & z \end{bmatrix} \qquad [1.23]$$

The overall volume velocity transfer matrix for the p tubes (from the glottis to the lips) is finally obtained as the product of the matrices for each tube:

$$\begin{bmatrix} f_0 \\ b_0 \end{bmatrix} = T\begin{bmatrix} f_p \\ b_p \end{bmatrix} \quad \text{with} \quad T = \prod_{i=1}^{p} T^i \qquad [1.24]$$

The properties of the volume velocity transfer function for the tube (from the glottis to the lips) can be derived from this result, defined as $A_u = (f_0-b_0)/(f_p-b_p)$. For this purpose, the lip termination has to be calculated, i.e. the interface between the last tube and the outside of the mouth. Let (f_l,b_l) denote the volume velocity waves at the level of the outer interface and (f_0,b_0) the waves at the inner interface. Outside of the mouth, the backward wave b_l is zero. Therefore, b_0 and f_0 are linearly dependent and a reflection coefficient at the lips can be defined as $k_l = b_0/f_0$. Then, transfer function A_u can be calculated by inverting T, according to the coefficients of matrix T and the reflection coefficient at lips k_l:

$$A_u = \frac{\det(T)(1-k_l)}{T_{21} + T_{22} - k_l(T_{11} + T_{12})} \qquad [1.25]$$

It can be verified that the determinant of T does not depend on z, as this is also not the case for the determinant of each elementary tube. As the coefficients of the transfer matrix are the products of a polynomial expression of z and a constant multiplied by $z^{-1/2}$ for each section, the transfer function of the vocal tract is therefore an all-pole function with a zero for $z=0$ (which accounts for the propagation delay in the vocal tract).

1.1.4.2. All-pole filter model

During the production of oral vowels, the vocal tract can be viewed as an acoustic tube of a complex shape. Its transfer function is composed of poles only, thus behaving as an acoustic filter with resonances only. These resonances correspond to the formants of the spectrum, which, for a sampled signal with limited bandwidth, are of a finite number N. In average, for a uniform tube, the formants are spread every kHz; as a consequence, a signal sampled at $F=1/T$ kHz (i.e. with a bandwidth of $F/2$ kHz), will contain approximately $F/2$ formants and $N=F$ poles will compose the transfer function of the vocal tract from which the signal originates:

$$V(z) = \frac{U_l}{U_g}(z) = \frac{K_1 z^{-N/2}}{\prod_{i=1}^{N}(1 - \hat{z}_i z^{-1})(1 - \hat{z}_i^* z^{-1})} \qquad [1.26]$$

Developing the expression for the conjugate complex poles $\hat{z}_i, \hat{z}_i^* = \exp[-\pi B_i T \pm 2i\pi f_i T]$ yields:

$$V(z) = \frac{K_1 z^{-N/2}}{\prod_{i=1}^{N}[1 - 2\exp(\pi B_i T)\cos(2\pi f_i T)z^{-1} + \exp(-2\pi B_i T)z^{-2}]} \qquad [1.27]$$

where B_i denotes the formant's bandwidth at -6 dB on each side of its maximum and f_i its center frequency.

To take into account the coupling with the nasal cavities (for nasal vowels and consonants) or with the cavities at the back of the excitation source (the subglottic cavity during the open glottis part of the vocalic cycle or the cavities upstream the constriction for plosives and fricatives), it is necessary to incorporate in the transfer function a finite number of zeros $\bar{z}_j, \bar{z}_j^*$ (for a band-limited signal).

$$V(z) = \frac{U_l}{U_g}(z) = \frac{K_2 \prod_{j=1}^{M}(1 - \bar{z}_j z^{-1})(1 - \bar{z}_j^* z^{-1})}{\prod_{i=1}^{N}(1 - \hat{z}_i z^{-1})(1 - \hat{z}_i^* z^{-1})} \qquad [1.28]$$

Any zero in the transfer function can be approximated by a set of poles, as $1 - az^{-1} = 1/\sum_{n=0}^{\infty} a^n z^{-n}$. Therefore, an all-pole model with a sufficiently large number of poles is often preferred in practice to a full pole-zero model.

1.1.5. *Lip-radiation*

The last term in the linear model corresponds to the conversion of the airflow wave at the lips into a pressure wave radiated at a given distance from the head. At a first level of approximation, the radiation effect can be assimilated to a differentiation: at the lips, the radiated pressure is the derivative of the airflow. The pressure recorded with the microphone is analogous to the one radiated at the lips, except for an attenuation factor, depending on its distance to the lips. The time-domain derivation corresponds to a spectral emphasis, i.e. a first-order high-pass filtering. The fact that the production model is linear can be exploited to condense the radiation term at the very level of the source. For this purpose, the derivative of the source is considered rather than the source itself. In the spectral domain, the consequence is to increase the slope of the spectrum by approximately +6 dB/octave, which corresponds to a time-domain derivation and, in the sampled domain, to the following transfer function:

$$L(z) = \frac{P}{U_l}(z) \approx 1 - K_d z^{-1} \qquad\qquad [1.29]$$

with $K_d \approx 1$.

1.2. Linear prediction

Linear prediction (or LPC for *Linear Predictive Coding*) is a parametric model of the speech signal [ATA 71, MAR 76]. Based on the source-filter model, an analysis scheme can be defined, relying on a small number of parameters and techniques for estimating these parameters.

1.2.1. *Source-filter model and linear prediction*

The source-filter model of equation [1.4] can be further simplified by grouping in a single filter the contributions of the glottis, the vocal tract and the lip-radiation term, while keeping a flat-spectrum term for the excitation. For voiced speech, $P(z)$ is a periodic train of pulses and for unvoiced speech, $N(z)$ is a white noise.

$$S(z) = P(z)\,U_g(z)\,V(z)\,L(z) = P(z)\,H(z) \qquad \text{voiced speech} \qquad [1.30]$$

$$S(z) = R(z)\,V(z)\,L(z) = N(z)\,H(z) \qquad \text{unvoiced speech} \qquad [1.31]$$

Considering the lip-radiation spectral model in equation [1.29] and the glottal airflow model in equation [1.9], both terms can be grouped into the flat spectrum source E, with unit gain (the gain factor G is introduced to take into account the amplitude of the signal). Filter H is referred to as the *synthesis filter*. An additional simplification consists of considering the filter H as an all-pole filter. The acoustic theory indicates that the filter V, associated with the vocal tract, is an all-pole filter only for non-nasal sounds whereas is contains both poles and zeros for nasal sounds. However, it is possible to approximate a pole/zero transfer function with an all-pole filter, by increasing the number of poles, which means that, in practice, an all-pole approximation of the transfer function is acceptable. The inverse filter of the synthesis filter is an all-zero filter, referred to as the *analysis filter* and denoted A. This filter has a transfer function that is written as an Mth-order polynomial, where M is the number of poles in the transfer function of the synthesis filter H:

$$S(z) = G\,E(z)\,H(z) \qquad\qquad H(z)\text{: synthesis filter} \qquad [1.32]$$

$$= \frac{G\,E(z)}{A(z)} \qquad \text{with} \quad A(z) = \sum_{i=0}^{M} a_i z^{-i} \ : \text{analysis filter} \qquad [1.33]$$

Linear prediction is based on the correlation between successive samples in the speech signal. The knowledge of p samples until the instant $n{-}1$ allows some prediction of the upcoming sample, denoted $\hat{s}_n$, with the help of a prediction filter, the transfer function of which is denoted $F(z)$:

$$s_n \approx \hat{s}_n = \alpha_1 s_{n-1} + \alpha_2 s_{n-2} + \cdots + \alpha_p s_{n-p} = \sum_{i=1}^{p} \alpha_i s_{n-i} \qquad [1.34]$$

$$\hat{S}(z) = S(z)(\alpha_1 z^{-1} + \alpha_2 z^{-2} + \cdots + \alpha_p z^{-p}) = S(z)\left(\sum_{i=1}^{P} \alpha_i z^{-i} \right) \qquad [1.35]$$

$$\hat{S}(z) = S(z)F(z) \qquad [1.36]$$

The prediction error ε_n between the predicted and actual signals is thus written:

$$\varepsilon_n = s_n - \hat{s}_n = s_n - \left(\sum_{i=1}^{p} \alpha_i s_{n-i} \right) \qquad [1.37]$$

$$\mathcal{E}(z) = S(z) - \hat{S}(z) = S(z) \left(1 - \sum_{i=1}^{P} \alpha_i z^{-i} \right) \qquad [1.38]$$

Linear prediction of speech thus closely relates with the linear acoustic production model: the source-filter production model and the linear prediction model can be identified with each other. The residual error ε_n can then be interpreted as the source of excitation e and the inverse filter A is associated with the prediction filter (by setting $M = p$).

$$\varepsilon_n + \sum_{i=1}^{p} \alpha_i s_{n-i} = G\, e(n) - \sum_{i=1}^{p} a_i s_{n-i} \qquad [1.39]$$

The identification of filter A assumes a flat spectrum residual, which corresponds to a white noise or a single pulse excitation. The modeling of the excitation source in the framework of linear prediction can therefore be achieved by a pulse generator and a white noise generator, piloted by a voiced/unvoiced decision. The estimation of the prediction coefficients is obtained by minimizing the prediction error. Let ε_n^2 denote the square prediction error and E the total square error over a given time interval, between n_0 and n_1:

$$\varepsilon_n^{\,2} = [s_n - \sum_{i=1}^{p} \alpha_i s_{n-i}]^2 \quad \text{and} \quad E = \sum_{n=n_0}^{n_1} \varepsilon_n^2 \qquad [1.40]$$

The expression of coefficients α_k that minimizes the prediction error E over a frame is obtained by zeroing the partial derivatives of E with respect to the α_k coefficients, i.e., for $k = 1, 2, \ldots, p$:

$$\frac{\partial E}{\partial \alpha_k} = 0 \quad \text{i.e.} \quad 2 \sum_{n=n_0}^{n_1} s_{n-k} \left[s_n - \sum_{i=1}^{p} \alpha_i s_{n-i} \right] = 0 \qquad [1.41]$$

Finally, this leads to the following system of equations:

$$\sum_{n=n_0}^{n_1} s_{n-k} s_n = \sum_{i=1}^{p} \alpha_i \sum_{n=n_0}^{n_1} s_{n-k} s_{n-i} \qquad 1 \le k \le p \qquad [1.42]$$

and, if new coefficients c_{ki} are defined, the system becomes:

$$c_{k0} = \sum_{i=1}^{p} \alpha_i c_{ki} \quad 1 \leq k \leq p \quad \text{with} \quad c_{ki} = \sum_{n=n_0}^{n_1} s_{n-i} s_{n-k} \tag{1.43}$$

Several fast methods for computing the prediction coefficients have been proposed. The two main approaches are the *autocorrelation* method and the *covariance* method. Both methods differ by the choice of interval $[n_0, n_1]$ on which total square error E is calculated. In the case of the covariance method, it is assumed that the signal is known only for a given interval of N samples exactly. No hypothesis is made concerning the behavior of the signal outside this interval. On the other hand, the autocorrelation method considers the whole range $-\infty, +\infty$ for calculating the total error. The coefficients are thus written:

$$c_{ki} = \sum_{n=p}^{N-1} s_{n-i} s_{n-k} \qquad \text{covariance} \tag{1.44}$$

$$c_{ki} = \sum_{n=-\infty}^{+\infty} s_{n-i} s_{n-k} \qquad \text{autocorrelation} \tag{1.45}$$

The covariance method is generally employed for the analysis or rather short signals (for instance, one voicing period, or one closed glottis phase). In the case of the covariance method, matrix $[c_{ki}]$ is symmetric. The prediction coefficients are calculated with a fast algorithm [MAR 76], which will not be detailed here.

1.2.2. *Autocorrelation method: algorithm*

For this method, signal s is considered as stationary. The limits for calculating the total error are $-\infty, +\infty$. However, only a finite number of samples are taken into account in practice, by zeroing the signal outside an interval $[0, N-1]$, i.e. by applying a time window to the signal. Total quadratic error E and coefficients c_{ki} become:

$$E = \sum_{n=-\infty}^{+\infty} \varepsilon_n^2 \quad \text{and} \quad c_{ki} = \sum_{n=-\infty}^{+\infty} s_{n-i} s_{n-k} = \sum_{n=-\infty}^{+\infty} s_n s_{n+|k-i|} \tag{1.46}$$

Those are the autocorrelation coefficients of the signal, hence the name of the method. The roles of k and i are symmetric and the correlation coefficients only depend on the difference between k and i.

The samples of the signal s_n (resp. $s_{n+|k-i|}$) are non-zero only for $n \in [0, N-1]$ ($n+|k-i| \in [0, N-1]$ respectively). Therefore, by rearranging the terms in the sum, it can be written for $k = 0, \ldots, p$:

$$c_{ki} = \sum_{n=0}^{N-|k-i|-1} s_n s_{n+|k-i|} = r(|k-i|) \quad \text{with} \quad r(k) = \sum_{n=0}^{N-k-1} s_n s_{n+k} \qquad [1.47]$$

The p equation system to be solved is thus (see [1.43]):

$$\sum_{i=1}^{p} a_i r(|k-i|) = 0 \qquad 1 \le k \le p \qquad [1.48]$$

Moreover, one equation follows from the definition of the error E:

$$E = \sum_{n=-\infty}^{+\infty} \sum_{i=0}^{p} a_i s_{n-i} \sum_{j=0}^{p} a_j s_{n-j} = \sum_{n=-\infty}^{+\infty} s_n \sum_{i=0}^{p} a_i s_{n-i} = \sum_{i=0}^{p} a_i r_i \qquad [1.49]$$

as a consequence of the above set of equations [1.48]. An efficient method to solve this system is the recursive method used in the Levinson algorithm.

Under its matrix form, this system is written:

$$\begin{bmatrix} r_0 & r_1 & r_2 & r_3 & \cdots & r_p \\ r_1 & r_0 & r_1 & r_2 & \cdots & r_{p-1} \\ r_2 & r_1 & r_0 & r_1 & \cdots & r_{p-2} \\ r_3 & r_2 & r_1 & r_0 & \cdots & r_{p-3} \\ \cdots & \cdots & \cdots & \cdots & \cdots & \cdots \\ r_p & r_{p-1} & r_{p-2} & r_{p-3} & \cdots & r_0 \end{bmatrix} \begin{bmatrix} 1 \\ a_1 \\ a_2 \\ a_3 \\ \cdots \\ a_p \end{bmatrix} = \begin{bmatrix} E \\ 0 \\ 0 \\ 0 \\ \cdots \\ 0 \end{bmatrix} \qquad [1.50]$$

The matrix is symmetric and it is a Toeplitz matrix. In order to solve this system, a recursive solution on prediction order n is searched for. At each step n, a set of

$n+1$ prediction coefficients is calculated: $a_0^n, a_1^n, a_2^n, ..., a_n^n$. The process is repeated up to the desired prediction order p, at which stage: $a_0^p = a_0$, $a_1^p = a_1$, $a_2^p = a_2, ..., a_p^p = a_p$. If we assume that the system has been solved at step $n-1$, the coefficients and the error at step n of the recursion are obtained as:

$$
\begin{bmatrix} a_0^n \\ a_1^n \\ a_2^n \\ \cdots \\ a_{n-1}^n \\ a_n^n \end{bmatrix}
=
\begin{bmatrix} a_0^{n-1} \\ a_1^{n-1} \\ a_2^{n-1} \\ \cdots \\ a_{n-1}^{n-1} \\ 0 \end{bmatrix}
+ k_n
\begin{bmatrix} 0 \\ a_{n-1}^{n-1} \\ a_{n-2}^{n-1} \\ \cdots \\ a_1^{n-1} \\ a_0^{n-1} \end{bmatrix}
\qquad [1.51]
$$

$$
\begin{bmatrix} E_n \\ 0 \\ 0 \\ \cdots \\ 0 \\ 0 \end{bmatrix}
=
\begin{bmatrix} E_{n-1} \\ 0 \\ 0 \\ \cdots \\ 0 \\ q \end{bmatrix}
+ k_n
\begin{bmatrix} q \\ 0 \\ 0 \\ \cdots \\ 0 \\ E_{n-1} \end{bmatrix}
\qquad [1.52]
$$

i.e. $a_i^n = a_i^{n-1} + k_n a_{n-i}^{n-1}$, where it can be easily shown from equations [1.50], [1.51] and [1.52] that:

$$
k_n = -\frac{1}{E_{n-1}} \sum_{i=0}^{n-1} a_i^{n-1} r_{n-i} \quad \text{and} \quad E_n = E_{n-1}(1 - k_n^2) \qquad [1.53]
$$

As a whole, the algorithm for calculating the prediction coefficients is (coefficients k_i are called reflection coefficients):

1) $E_0 = r_0$

2) step n: $k_n = -\dfrac{1}{E_{n-1}} \sum_{i=0}^{n-1} a_i^{n-1} r_{n-i}$

3) $a_n^n = k_n$ and $a_0^n = 1$

4) $a_i^n = a_i^{n-1} + k_n a_{n-i}^{n-1}$ for $1 \le i \le n\text{-}1$

5) $E_n = (1 - k_n^2) E_{n-1}$

These equations are solved recursively, until the solution for order p is reached.

In many applications, one of the goals is to identify the filter associated with the vocal tract, for instance to extract the formants [MCC 74]. Let us consider vowel signals, the spectra of which are shown in Figures 1.4, 1.5 and 1.6 (these spectra were calculated with a short-term Fourier transform (STFT) and are represented on a logarithmic scale). The linear prediction analysis of these vowels yields filters which correspond to the prediction model which could have produced them. Therefore, the magnitude of the transfer function of these filters can be viewed as the spectral envelope of the corresponding vowels.

Linear prediction thus estimates the filter part of the source-filter model. To estimate the source, the speech signal can be filtered by the inverse of the analysis filter. The residual signal subsequently obtained represents the derivative of the source signal, as the lip-radiation term is included in the filter (according to equation [1.30]). The residual signal must thus be integrated in order to obtain an estimation of the actual source, which is represented in Figure 1.7, both in the frequency and time domains.

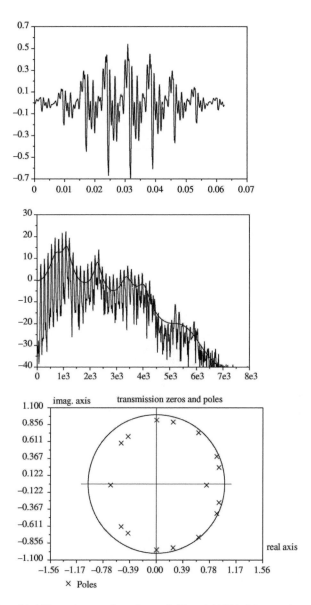

Figure 1.4. *Vowel /a/. Hamming windowed signal (F_e = 16 kHz). Magnitude spectrum on a logarithmic scale and gain of the LPC model transfer function (autocorrelation method). Complex poles of the LPC model (16 coefficients)*

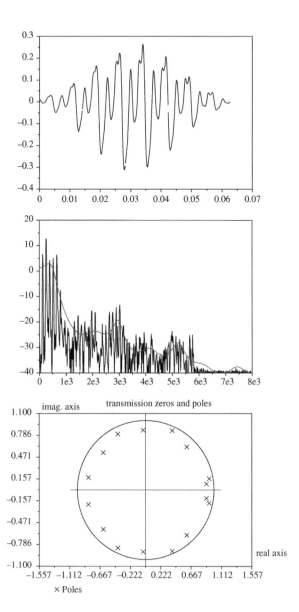

Figure 1.5. *Vowel /u/. Hamming windowed signal (F_e = 16 kHz). Magnitude spectrum on a logarithmic scale and gain of the LPC model transfer function (autocorrelation method). Complex poles of the LPC model (16 coefficients)*

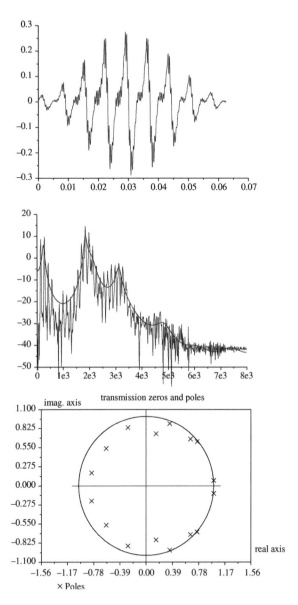

Figure 1.6. *Vowel /i/. Hamming windowed signal (Fe = 16 kHz). Magnitude spectrum on a logarithmic scale and gain of the LPC model transfer function (autocorrelation method). Complex poles of the LPC model (16 coefficients)*

1.2.3. *Lattice filter*

We are now going to show that reflection coefficients k_i obtained by the autocorrelation method correspond to the reflection coefficients of a multi-tube acoustic model of the vocal tract. For this purpose, new coefficients b_i^n must be introduced, which are defined at each step of the recursion as:

$$b_i^n = a_{n-i}^n \qquad i = 0, 1, \cdots, n \tag{1.54}$$

The $\{b_i^p\}$ coefficients, where p is the prediction order, can be used to *postdict* the signal, i.e. to predict the preceding sample of the signal. Let's form the estimate $\hat{s}_{n-p}$:

$$\hat{s}_{n-p} = -b_0 s_n - b_1 s_{n-1} - \cdots - b_{p-1} s_{n-p+1}$$

$$= \sum_{i=0}^{p-1} -b_i s_{n-i} = \sum_{i=0}^{p-1} \alpha_{p-i} s_{n-i} \tag{1.55}$$

A postdiction, or backward error, ε_n^- can be defined as:

$$\varepsilon_n^- = s_{n-p} - \hat{s}_{n-p} = \sum_{i=0}^{p} b_i s_{n-i} \quad \text{with} \quad b_p = 1 \tag{1.56}$$

The total forward prediction error E (of equation [1.40]) is denoted E^+, while the total backward prediction error is denoted E^-. In a same manner as in the previous development, it can be shown that, for the autocorrelation method, we have $E^- = E^+$. Subsequently, the backward prediction coefficients b_i obtained via the minimization of the total backward error are identical to the a_i coefficients, and the Levinson algorithm can be rewritten as:

$$a_i^n = a_i^{n-1} + k_n b_{i-1}^{n-1} \text{ and } b_i^n = b_{i-1}^{n-1} + k_n a_i^{n-1} \text{ with } \begin{cases} a_n^{n-1} = 0 \\ b_n^{n-1} = 0 \end{cases} \tag{1.57}$$

If we consider the forward and backward prediction errors for a same instant (at order n):

$$\varepsilon_i^{n+} = \sum_{j=0}^{n} a_j s_{i-j} \quad \text{and} \quad \varepsilon_i^{n-} = \sum_{j=0}^{n} b_j s_{i-j} \tag{1.58}$$

and then equations [1.57] yield:

$$\varepsilon_i^{n+} = \varepsilon_i^{(n-1)+} + k_n \varepsilon_{i-1}^{(n-1)-} \quad \text{and} \quad \varepsilon_i^{n-} = \varepsilon_{i-1}^{(n-1)-} + k_n \varepsilon_i^{(n-1)+} \tag{1.59}$$

The z-transforms of these equations provide:

$$\begin{bmatrix} E^{n+}(z) \\ E^{n-}(z) \end{bmatrix} = \begin{bmatrix} 1 & k_n z^{-1} \\ k_n & z^{-1} \end{bmatrix} \begin{bmatrix} E^{(n-1)+}(z) \\ E^{(n-1)-}(z) \end{bmatrix} \tag{1.60}$$

with, for $n = 0$: $\varepsilon_i^{0-} = \varepsilon_i^{0+} = s_i$.

To complete the analogy between linear prediction and multi-tube acoustic model, a slightly different definition of the backward prediction coefficients must be resorted to: $b_i^n = a_{n-i+1}^n$ for $i = 1, 2, ..., n+1$. The total backward error has the same expression and the Levinson algorithm is written:

$$a_i^n = a_i^{n-1} + k_n b_i^{n-1} \quad \text{and} \quad b_i^n = b_{i-1}^{n-1} + k_n a_{i-1}^{n-1} \quad \text{with} \quad \begin{cases} a_n^{n-1} = 0 \\ b_0^{n-1} = 0 \end{cases} \tag{1.61}$$

from which the error recursion matrix can be deduced:

$$\begin{bmatrix} E^{n+}(z) \\ E^{n-}(z) \end{bmatrix} = \begin{bmatrix} 1 & k_n \\ k_n z^{-1} & z^{-1} \end{bmatrix} \begin{bmatrix} E^{(n-1)+}(z) \\ E^{(n-1)-}(z) \end{bmatrix} \tag{1.62}$$

for $n = 0$, $\varepsilon_i^{0+} = s_i$ and $\varepsilon_i^{0-} = s_{i-1}$, i.e. $E^{0+}(z) = S(z)$ and $E^{0-}(z) = z^{-1}S(z)$. The inverse matrix from equation [1.62] is:

$$\frac{1}{1-k_n^2} \begin{bmatrix} 1 & -k_n z \\ -k_n & z \end{bmatrix} \tag{1.63}$$

Except for a multiplicative factor, this is the matrix of equation [1.23], obtained for a section of the multi-tube vocal tract model. This justifies the naming of the k_n coefficients as *reflection coefficients*. This is the inverse matrix, as the linear prediction algorithm provides the analysis filter. On the contrary, the matrix for an elementary section of the multi-tube acoustic model corresponds to the synthesis filter, i.e. the inverse of the analysis filter. Note that this definition of backward prediction coefficients introduces a shift of one sample between the forward error and the backward error, which in fact corresponds to the physical situation of the multi-tube model, in which the backward wave comes back only after a delay due to the propagation time in the tube section. On the contrary, if the definition of [1.54] is used, there is no shift between forward and backward errors.

Equation [1.62] allows for the analysis and synthesis of speech by linear prediction, with a lattice filter structure. In fact, for each step in the recursion, crossed terms are used that result from the previous step. A remarkable property of lattice filters is that the prediction coefficients are not directly used in the filtering algorithm. Only the signal and the reflection coefficients intervene. Moreover, it can be shown [MAR 76, PAR 86] that the reflection coefficients resulting from the autocorrelation method can be directly calculated using the following formula:

$$k_n = \frac{\sum_{i=0}^{N-1} \varepsilon_i^{(n-1)+} \varepsilon_i^{(n-1)-}}{\sqrt{\sum_{i=0}^{N-1} \left(\varepsilon_i^{(n-1)+}\right)^2 \sum_{i=0}^{N-1} \left(\varepsilon_i^{(n-1)-}\right)^2}} \qquad [1.64]$$

These coefficients are sometimes called PARCOR coefficients (for PARtial error CORrelation). The use of equation [1.64] is thus an alternate way to calculate the analysis and synthesis filters, which is equivalent to the autocorrelation method, but without calculating explicitly the prediction coefficient. Other lattice filter structures have been proposed. In the Burg method, the calculation of the reflection coefficients is based on the minimization (in the least squares sense) of the sum of the forward and backward errors. The error term to minimize is:

$$E^N = \sum_{i=0}^{N-1} \left[\left(\varepsilon_i^{n+}\right)^2 + \left(\varepsilon_i^{n-}\right)^2 \right] \qquad [1.65]$$

By writing that $\partial E^n / \partial k_n = 0$, in order to find the optimal k_n coefficients, we obtain:

$$k_n = \frac{2\sum_{i=0}^{N-1} \varepsilon_i^{(n-1)+} \varepsilon_i^{(n-1)-}}{\sum_{i=0}^{N-1} \left(\varepsilon_i^{(n-1)+}\right)^2 \sum_{i=0}^{N-1} \left(\varepsilon_i^{(n-1)-}\right)^2} \qquad [1.66]$$

These coefficients no longer correspond to the autocorrelation method, but they possess good stability properties, as it can be shown that $-1 \leq k_n \leq 1$. Adaptive versions of the Burg algorithm also exist [MAK 75, MAK 81].

1.2.4. Models of the excitation

In addition to the filter part of the linear prediction model, the source part has to be estimated. One of the terms concerning the source is the synthesis gain G. There is no unique solution to this problem and additional hypotheses must be made. A commonly accepted hypothesis is to set the total signal energy equal to that of the impulse response of the synthesis filter. Let us denote as $h(n)$ the impulse response and $r_h(k)$ the corresponding autocorrelation coefficients. Thus:

$$h(n) = G\delta(n) + \sum_{i=1}^{p} \alpha_i h(n-i) \quad \text{and} \quad r_h(k) = \sum_{i=1}^{p} \alpha_i r_h(k-i) \qquad [1.67]$$

Indeed, for $k > 0$, the autocorrelation coefficients are infinite sums of terms such as:

$$h(n-k)h(n) = G\delta(n)h(n-k) + h(n-k)\sum_{i=1}^{p} \alpha_i h(n-i) \qquad [1.68]$$

and the terms $\delta(n)h(n-k)$ are always zero, for $k \neq 0$. Equaling the total energies is equivalent to equaling the 0th order autocorrelations. Thanks to recurrence equation [1.67], the autocorrelation coefficients of the signal and of the impulse response can be identified with each other: $r_h(i) = r(i)$, for $i = 0, 1, ..., p$. For $n = 0$, $h(0) = G$; therefore, reusing equation [1.67] yields:

$$r_h(0) = Gh(0) + \sum_{i=1}^{p} \alpha_i r_h(i) \quad \text{therefore:} \quad G^2 = r(0) - \sum_{i=1}^{p} \alpha_i r(i) \qquad [1.69]$$

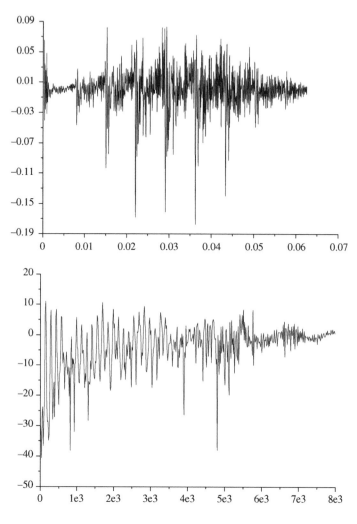

Figure 1.7. *Vowel /i/. Residual signal and its magnitude spectrum on a logarithmic scale*

In the conventional linear prediction model, the excitation is either voiced or unvoiced, for each analysis frame. In the case of a voiced signal, the excitation is a periodic pulse train at the fundamental period (see Figure 1.9), and for an unvoiced signal, the excitation is a Gaussian white noise (see Figure 1.8). The mixture of these two sources is not allowed, which is a definite drawback for voiced sounds for which a noise component is also present in the excitation.

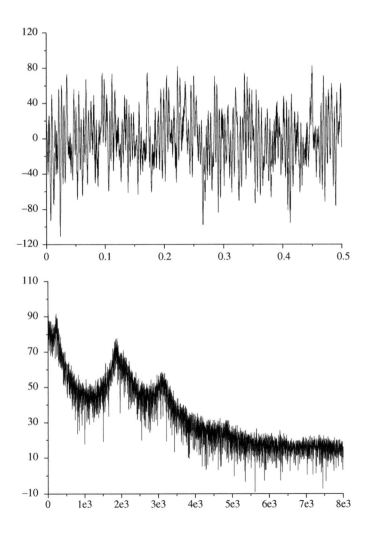

Figure 1.8. *Vowel /i/. LPC synthesis with a noise source. Top: source; bottom: spectrum*

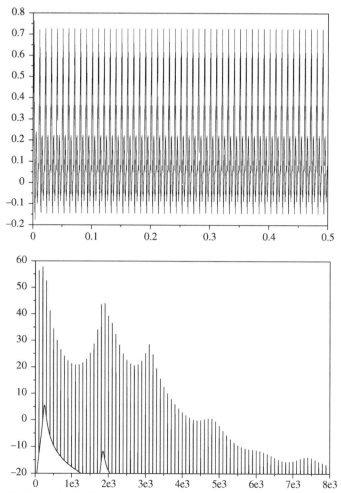

Figure 1.9. *Vowel /i/. LPC synthesis with a pulse train. Top: source; bottom: spectrum*

Multi-pulse linear prediction [ATA 82] offers a solution to the problem of mixed excitation by considering the source as a collection of pulses, for each frame, without discriminating between voiced and unvoiced cases. The relation between the source and the production model is difficult to establish. Nevertheless, the coding quality can be transparent. The choice of positions and amplitudes for the excitation pulses is an iterative procedure. Each new pulse is chosen so as to minimize the square error between the original and reconstructed signals. An error threshold is set to decide when to stop the procedure. The validation of this approach is experimental (listening test) and the average number of pulses for a 10 ms frame typically ranges between 8 and 12.

Other methods of linear prediction use a time-domain model of the glottis airflow wave [FUJ 87].

1.3. Short-term Fourier transform

The speech signal can also be analyzed without any reference to a production model. We are thus going to set aside now the source-filter model and we will consider representations and analysis schemes that do not make any hypothesis on the way speech is produced (non-parametric representations). The nature of the speech signal leads to the use of time-frequency descriptions. In fact, for a voice signal to bear a code and a linguistic content, it is necessary for its spectral content to evolve over time.

1.3.1. *Spectrogram*

A prototypical time-frequency-energy representation is the spectrogram. Until quite recently, the Sonagraph, one of the first commercialized analogical spectrographs, has been an essential piece of furniture in any research laboratory in acoustics or phonetics. A considerable level of expertise has therefore been developed for the interpretation of spectrograms [CAL 89, LEH 67]. Nowadays, the spectrogram is calculated with a short-term Fourier transform. Starting from the time-domain signal $x(\tau)$, a short-term signal is defined as $x(t, \tau) = x(\tau) w(t - \tau)$, which represents, for a given t, the signal x observed through the analysis window w, centered on t, and called *analysis frame*. The short-term Fourier transform is the Fourier transform of signal $x(t, \tau)$, denoted $\tilde{x}(t,v)$:

$$\tilde{x}(t,v) = \int_{-\infty}^{+\infty} x(t,\tau)e^{-2i\pi v\tau}d\tau = \int_{-\infty}^{+\infty} x(\tau)w(t-\tau)e^{-2i\pi v\tau}d\tau \qquad [1.70]$$

The time-frequency representation of $x(\tau)$ by $\tilde{x}(t,v)$ can be interpreted as a linear filtering operation. After identifying a convolution in the expression of $\tilde{x}(t,v)$, written as a function of t (which corresponds to a sliding analysis window), the analysis formula can be rewritten as $\tilde{x}^{w}(t,v) = w(t) * x(t)\exp(-2i\pi vt)$, which provides an interpretation of the short-term Fourier transform as a linear filtering of the signal $x(t)$ by the filter with impulse response $w(t)$, modulated by $e^{-2i\pi vt}$: $w(t)$ is thus called the analysis filter. For each analysis instant, the spectrogram, denoted SX, is the square of the modulus of the short-term Fourier analysis:

$$SX(t,v) = \left| \int_{-\infty}^{+\infty} x(\tau)w(t-\tau)e^{-2i\pi v\tau}\,d\tau \right|^2 \qquad [1.71]$$

$$= \left| \int_{-\infty}^{+\infty} \tilde{x}(f)\tilde{w}(f-v)e^{-2i\pi ft}\,df \right|^2 \qquad [1.72]$$

In order to represent the 3D of the analysis, time is represented along the x-axis, frequency along the y-axis, and intensity (in logarithmical scale) as the gray level of the plot (darker shades corresponding to higher intensities). The duration and spectral characteristics of the temporal window govern the effective bandwidth and duration of the analysis window. The filter bandwidth (which governs the desired frequency accuracy) is inversely proportional to the time precision that can be expected. This leads to two broad categories of analysis: 1) the narrow-band (typically 45 kHz) which smoothes the fast time-domain variations of the signal but which is able to reveal the fine structure in the frequency domain (each harmonic of a periodic speech sound can be accurately distinguished); 2) the wide-band analysis (typically 300 Hz) which, conversely, enables the visualization of time-domain events, but with a low resolution in the frequency domain.

1.3.2. Interpretation in terms of filter bank

The spectrogram has been presented in the previous section as a visual analysis tool. The short-term Fourier transform can also be used to represent and modify digital signals, and not only to visualize them. Let us consider a digital signal $x(n)$, and the short-term signal centered on instant n, $x(n, m) = w(n-m)x(m)$. The short-term Fourier transform (STFT) of x is defined as:

$$\tilde{x}(n,\omega) = \sum_{m=-\infty}^{\infty} x(n,m)e^{-i\omega m} = \sum_{m=-\infty}^{\infty} w(n-m)x(m)e^{-i\omega m} \qquad [1.73]$$

Note that this infinite summation is in reality limited to the N samples for which the analysis window w is non-zero. The signal can be reconstructed by the inverse Fourier transform, i.e. the inverse expression of equation [1.73]:

$$x(n,m) = w(n-m)x(m) = \frac{1}{2\pi} \int_{\omega=-\pi}^{\pi} \tilde{x}(n,\omega)e^{i\omega m}\,d\omega \qquad [1.74]$$

i.e., if the window is normalized so that $w(0) = 1$, and taking $m = n$:

$$x(n) = \frac{1}{2\pi} \int_{\omega=-\pi}^{\pi} \tilde{x}(n,\omega) e^{i\omega n} d\omega \qquad\qquad [1.75]$$

Equation [1.73] can be interpreted as a discrete convolution $w(n) * x(n)$ exp($-i\omega n$), i.e. as a low-pass filtering of the demodulated signal for frequency ω. Synthesis formula [1.75] then corresponds to a continuous summation (integration) of a set of oscillators controlled in time and frequency, by the output of the low-pass analysis filters. It is also relevant to consider this process as the summation of the outputs $f_\omega(n)$ of a set of band-pass filters with impulse response $w(n)e^{i\omega n}$ obtained when frequency ω is fixed:

$$x(n) = \frac{1}{2\pi} \int_{\omega=-\pi}^{\pi} f_\omega(n) d\omega \quad \text{with} \qquad\qquad [1.76]$$

$$f_\omega(n) = \tilde{x}(n,\omega) e^{i\omega n} = \left(w(n) * x(n) e^{-i\omega n} \right) e^{i\omega n} = w(t) e^{i\omega n} * x(n) \qquad [1.77]$$

From a practical point of view, a finite number of filters is sufficient. Let us consider N band-pass filters f_k, chosen with uniformly spaced central frequencies, so that $\omega_k = 2\pi k/N$ for $k = 0, 1, ..., N-1$. If the number of filters is sufficient (i.e. $N \geq K$, where K is the duration of the analysis window w), the signal can be reconstructed as:

$$x(n) = \frac{1}{N} \sum_{k=0}^{N-1} f_k(n) = \frac{1}{N} \sum_{k=0}^{N-1} \tilde{x}(n,\omega_k) e^{i\omega_k n} \qquad\qquad [1.78]$$

$$= \frac{1}{N} \sum_{k=0}^{N-1} e^{i\omega_k n} \sum_{m=-\infty}^{\infty} x(n,m) e^{-i\omega_k m} \qquad\qquad [1.79]$$

A variant of the filter bank method is the phase vocoder. For this type of STFT, the short-term spectral samples are represented by their amplitudes and phases, which is particularly convenient for coding and modification purposes, and has enjoyed wide success [FLA 72, POR 81a, POR 81b].

1.3.3. Block-wise interpretation

The interpretation of the STFT in terms of filter bank is a sampling based interpretation (or frequential blocks). Another way to interpret the STFT is to consider a block-wise time-domain analysis [ALL 77]. Let $\tilde{x}(n,\omega_k)$ denote the discrete Fourier transform of the sequence $x(n,m) = w(n-m)x(m)$. Applying an inverse discrete Fourier transform, a relation analogous to [1.74] can be written:

$$x(n,m) = w(n-m)x(m) = \frac{1}{N}\sum_{k=0}^{N-1}\tilde{x}(n,\omega_k)e^{i\omega_k m} \qquad [1.80]$$

The signal $x(n,m)$ can be considered as a time-domain block, which can be sampled at the rate of R samples, i.e. $w(sR-m)x(m)$. From this series of short-term signals, signal x can be retrieved by summing with respect to s (except for an constant window term):

$$\sum_{p=-\infty}^{\infty} w(pR-m)x(m) = x(m)\sum_{p=-\infty}^{\infty} w(pR-m) \qquad [1.81]$$

Each block, for a given s, yields a short-term Fourier spectrum. The overlapping summation (in s) of all inverse transforms for each discrete short-term spectrum again gives the initial signal at instant m, corresponding to the following block-wise synthesis formula:

$$x(m) = \frac{\sum_{p=-\infty}^{\infty} x(pR,m)}{\sum_{p=-\infty}^{\infty} w(pR-m)} = \frac{\sum_{p=-\infty}^{\infty} 1/N \sum_{k=0}^{N-1} \tilde{x}(pR,\omega_k)e^{i\omega_k m}}{\sum_{p=-\infty}^{\infty} w(pR-m)} \qquad [1.82]$$

This interpretation illustrates that the signal can be represented as a double sum, over time and frequency, of temporally and frequentially localized components. The term OLA (for Overlap-Add) is frequently used to designate the block-wise synthesis formula. A variant of the block-wise method [POR 81a, POR 81b] calls for an interpolating filter f, which has low-pass characteristics, for the re-synthesis. The synthesis formula then becomes [CRO 80]:

$$x(m) = \sum_{p=-\infty}^{\infty} f(pR-m)\, x(pR,m) \qquad [1.83]$$

An advantage of this method is the possibility that it offers to modify the short-term signals before re-synthesis.

1.3.4. Modification and reconstruction

One of the most essential uses of STFT is signal modifications, for instance, its filtering, interpolation, compression, denoising, etc. In general, a modification step is inserted between the analysis and synthesis operations. Thus, a problem arises concerning the validity of these modifications. Some specific constraints must be satisfied for a time-frequency function $f(n,k)$ to be the STFT $\tilde{x}(n,\omega_k)$ of a signal

$x(m)$. For instance, a particular modification of $\tilde{x}(n,\omega_k)$ does not necessarily lead to a valid signal using an inverse STFT. However, there are projection methods for modifying the signal in the STFT space and appropriately re-synthesizing a modified time-domain signal. One of these methods has been proposed in [GRI 84]. Let $\tilde{y}(sR,\omega)$ denote the modified STFT of a signal. The approach consists of searching for the closest signal $x(n)$ to this modified STFT, in the least square sense. Let d designate this distance in the spectral domain:

$$d = \sum_{s=-\infty}^{\infty} \frac{1}{2\pi} \int_{\omega=-\pi}^{\pi} [\tilde{y}(sR,\omega) - \tilde{x}(sR,\omega)]^2 d\omega \qquad [1.84]$$

In the time domain, the Parseval theorem enables us to write:

$$d = \sum_{s=-\infty}^{\infty} \sum_{k=-\infty}^{\infty} [y(sR,k) - x(sR,k)]^2 \qquad [1.85]$$

In order to minimize this distance, its partial derivative with respect to $x(k)$ is set to zero. The inverse discrete Fourier transform of the time-domain samples of the modified STFT is calculated and, as $x(sR,k) = w(sR-k)x(k)$, a simple calculus gives the following result:

$$x(k) = \frac{\sum_{s=-\infty}^{\infty} w(sR-k)y(sR,k)}{\sum_{s=-\infty}^{\infty} w^2(sR-k)} \qquad [1.86]$$

This synthesis formula turns out to be very similar to the block-wise synthesis formula in equation [1.82].

1.4. A few other representations

1.4.1. *Bilinear time-frequency representations*

The spectrogram is a fundamental tool for the acoustic and phonetic analysis of speech. The link between the impulse response and the magnitude of the analysis filter for the spectrogram leads to a time-frequency "uncertainty" which prevents good resolution simultaneously in the time domain and in the frequency domain. If the analysis window is assumed to be centered in 0, its effective duration and its effective bandwidth can be defined as:

$$\Delta t = \sqrt{\frac{\int_{-\infty}^{+\infty} t^2 |w(t)|^2 \, dt}{\int_{-\infty}^{+\infty} |w(t)|^2 \, dt}} \quad \text{and} \quad \Delta v = \sqrt{\frac{\int_{-\infty}^{+\infty} v^2 |\tilde{w}(v)|^2 \, dv}{\int_{-\infty}^{+\infty} |\tilde{w}(v)|^2 \, dv}} \qquad [1.87]$$

It can be shown [FLA 93, GAB 46] that the time-frequency uncertainty expresses as an inequality: $\Delta t . \Delta v \geq 1/4\pi \approx 0.08$, the equality being reached for a Gaussian window (i.e. proportional to e^{-at^2}, with a real). This lack of resolution shows up in various manifestations: the time-domain and frequency-domain supports of the signals are not respected; rapid frequency modulations are underestimated; when a signal contains abrupt changes, it is difficult to adjust the optimal bandwidth. In order to circumvent the intrinsic lack of accuracy of the spectrogram in the time-frequency domain, some more general representations have been investigated. The expression for the spectrogram of equation [1.72] can be rewritten:

$$SX(t,v) = \int_{-\infty}^{+\infty} \int_{-\infty}^{+\infty} x(a) \, x^*(b) \, w^*(t-a) \, w(t-a) \, e^{-2i\pi v(a-b)} \, da \, db \qquad [1.88]$$

$$= \int_{-\infty}^{+\infty} \int_{-\infty}^{+\infty} x\left(u+\frac{\tau}{2}\right) x^*\left(u-\frac{\tau}{2}\right) w^*\left(t-u-\frac{\tau}{2}\right) w\left(t-u+\frac{\tau}{2}\right) e^{-2i\pi v\tau} \, du \, d\tau \qquad [1.89]$$

thanks to the substitution rule: a = u+τ/2, b = u–τ/2. Thus, the analysis window appears as a time-frequency analysis "kernel", and a general class of representations can be formulated as follows:

$$C(t,v;\phi) = \int_{-\infty}^{+\infty} \int_{-\infty}^{+\infty} \phi(a-t,\tau) x(u+\frac{\tau}{2}) x^*(u-\frac{\tau}{2}) e^{-2i\pi f\tau} \, du \, d\tau \qquad [1.90]$$

The Wigner-Ville distribution is obtained when $\varphi(t,\tau) = \delta(t)$:

$$W(t,v) = \int x^*(t-\tau/2) \, x(t+\tau/2) e^{-2i\pi v\tau} \, d\tau \qquad [1.91]$$

An argument in favor of this representation is its simultaneous resolution in time and frequency, which goes beyond the limitations of the spectrogram. The Wigner-Ville representation can be considered as an energy distribution in the time-frequency plane, since the summation of the distribution over time (respectively, over frequency) enables us to retrieve the energy density in the frequency domain (respectively, the instantaneous power). However, this interpretation is debatable because of the existence of significant interference terms between the frequency components, which show up on the time-frequency plots. The representation is not linear (the representation of the sum of two signals is not the sum of the representations). Moreover, negative values can be locally obtained on the distribution. The undesirable properties can be lessened by using the analytical signal instead of the real signal, and with the help of smoothing techniques. A particularly interesting type of smoothing is based on a time-frequency separable kernel φ, resulting from the product of a time-domain window v and a spectral window w, yielding the smoothed pseudo-Wigner-Ville transform:

$$PW(t,v) = \int_{-\infty}^{+\infty} \int_{-\infty}^{+\infty} w(\tau)\, v(u-t)\, x^*(u-\tau/2)\, x\; (u+\tau/2)\, e^{-2i\pi v\tau}\, du\, d\tau \qquad [1.92]$$

Another type of kernel is the conic kernel [LOU 93]: $\varphi(t,\tau) = g(\tau)rect(t/\tau)$ where $rect(t/\tau)$ is equal to 1 for $|t/\tau| \le 1/2$ and 0 otherwise.

Bilinear distributions have been proposed as an analysis tool for speech by many authors [CHE 84, COH 86, COH 89, VEL 89, WOK 87]. In general, these studies use much finer time scales and frequency scales than the spectrogram, short durations (of about one phoneme) and/or narrow frequency bands (of about a formant's bandwidth). Under these conditions, these representations show resolution levels greater than the spectrogram. However, none of these methods have so far reached a level of expertise comparable to that of the spectrogram, which, despite its lower accuracy in certain situations, remains more usual and more conventional, from a practical point of view.

1.4.2. Wavelets

The wavelet transform is a "time-scale" method [COM 89, MEY 90]. For the continuous transform, the signal is represented as a sum of functions obtained as the translations (at instants τ) and the time-domain dilations (by factors a) of a prototype function o (c_0 being a constant):

$$x(t) = \frac{1}{c_0} \int_{-\infty}^{+\infty} \int_{-\infty}^{+\infty} P_{t,a} \frac{1}{\sqrt{a}}\, o(\frac{t-\tau}{a}) \frac{d\tau\, da}{a^2} \qquad [1.93]$$

with:

$$P_{t,a} = \frac{1}{\sqrt{a}} \int_{-\infty}^{+\infty} o^*(\frac{t-\tau}{a}) x(t) dt = \sqrt{a} \int_{-\infty}^{+\infty} \tilde{o}(av)\tilde{x}(v) e^{2i\pi v\tau} dv \qquad [1.94]$$

These are methods with a constant relative resolution in the frequency domain, (v/v_c= *constant*, where v is the bandwidth of the filters and v_c their central frequency), similarly to the acoustic analysis using a third of one-third octave band filters. The use of scales instead of frequencies in a linear transform enables us, to a certain extent, to overcome the limitation in the resolution due to the time-frequency uncertainty. Indeed, if a signal portion contains a large bandwidth process, this process will be observed on several analysis scales; in other words, it will be represented accurately both in time and frequency, on different scales.

The wavelet transform can be interpreted as being twofold, like the STFT. It appears as a linear filtering, the characteristics of which are determined by parameter a, for the central frequencies and by the shape of the wavelet, for the filter. In addition, the wavelet transform can be interpreted as the decomposition of the signal on a family of functions derived from $o(t)$, and the synthesis operation as a weighted summation of wavelets. Thus, several wavelet-based methods are extensions of the STFT, for the visualization and the time-scale modification of the signal. Considering several analysis bands per octave, the plot of amplitudes and/or phases of the wavelet coefficients yield a spectrograph-like visual representation. The logarithmic scale for the frequencies resembles the auditory scale, which is a positive point in favor of this representation.

Moreover, for analysis purposes, the properties of the wavelet transform have been used to study the fundamental frequencies and the voicing periods. Two properties have been exploited: 1) it is a zero-phase linear representation; 2) if the signal, or one of its derivatives, shows discontinuities, the modulus of the transform points towards these discontinuities. Using a wavelet obtained as the derivative of a low-pass filter impulse response, it is possible to localize the discontinuities in the signal [KAD 92, VUN 99]. The periodic part of the source is characterized by the voicing fundamental frequency. In many applications, it is useful to detect voiced periods, by localizing the instants of glottal closure [ANA 75]. The wavelet transform is particularly well adapted to this type of analysis, as shown in Figure 1.10.

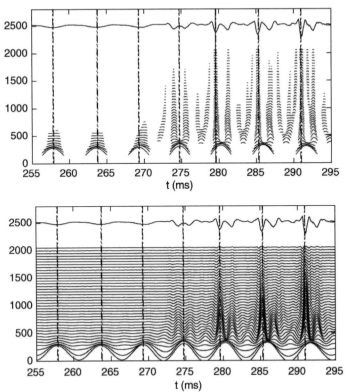

Figure 1.10. *Wavelet-based analysis of the transition between a voiced fricative and a vowel. 35 filters are used in the band 100 Hz – 2,100 Hz. This type of analysis enables the detection of glottal closure instants*

The discrete wavelet transform has a structure similar to the sub-band coding scheme using quadrature mirror filters [EST 77, MAL 89]. Many applications in speech coding have been proposed, but they will not be discussed in this chapter.

1.4.3. *Cepstrum*

Cepstral analysis is another signal analysis method particularly well suited to the speech signal [CHI 77, NOL 67, OPP 69]. By considering a (stable) discrete signal $s(n)$, and its z-transform $S(z)$, the complex cepstrum of this signal is another signal denoted $\hat{s}(n)$, such that its z-transform is the logarithm of $S(z)$:

$$\hat{S}(z) = \log[S(z)] \quad \text{with} \quad \log[S(z)] = \log|S(z)| + i\Phi(S(z)) \qquad [1.95]$$

where the logarithm function log [] denotes the complex logarithm defined as log[z]=log |z|+iφ(z). Note that the phase of the logarithm of the complex cepstrum is ambiguous (it is defined except for an integer multiple of 2π). The complex cepstrum can thus be calculated by inverting the z-transform. As the signal is stable, its Fourier transform exists, and the z-transform can be inverted:

$$\hat{s}(n) = \frac{1}{2\pi} \int_{-\pi}^{\pi} \log\left[S(e^{i\omega})\right] e^{i\omega n} d\omega = \frac{1}{2\pi} \int_{-\pi}^{\pi} \left(\log\left|S(e^{i\omega})\right| + i\Phi(S(e^{i\omega}))\right) e^{i\omega n} d\omega$$

[1.96]

It is important to note that, because of the Hermitian symmetry of the spectrum for a real signal, the complex cepstrum of a real signal is in fact real (and not complex, as its name would indicate).

$$\hat{s}(n) = \frac{1}{\pi} \int_{0}^{\pi} \log\left|S(e^{i\omega})\right| e^{i\omega n} d\omega$$

[1.97]

The real cepstrum, denoted $\breve{s}(n)$, is defined as the inverse Fourier transform of the log-module of the Fourier transform of the signal, i.e. as the real part of the complex cepstrum $\hat{s}(n)$:

$$\breve{s}(n) = \frac{1}{2\pi} \int_{-\pi}^{\pi} \log\left|S(e^{i\omega})\right| e^{i\omega n} d\omega \quad \text{i.e.} \quad \breve{s}(n) = \frac{\hat{s}(n) + \hat{s}^{*}(-n)}{2}$$

[1.98]

An advantage of the complex cepstrum is its inversibility. It is possible to retrieve the signal from its complex cepstrum, using the property: exp(log[S(z)]) = S(z). On the contrary, the real cepstrum is not invertible, since it is based only on the modulus of the Fourier transform: exp(log |S(z)| =|S(z)|.

The cepstrum can be used for the deconvolution between the source and the filter. In fact, as the logarithm function transforms a product into a sum, equation [1.32] can be rewritten as (assuming $G = 1$):

$$\hat{S}(z) = \hat{E}(z) + \hat{H}(z) \quad \text{therefore} \quad \hat{s}(n) = \hat{e}(n) + \hat{h}(n)$$

[1.99]

If both terms occupy distinct domains in the cepstral space, their separation may be achievable. For this purpose, it is worth studying the cepstrum of speech signals. For voiced speech, the source can be decomposed into a glottal airflow wave convolved with a periodic pulse train. The cepstrum for a periodic pulse train is written:

$$\hat{p}(n) = \sum_{i=-\infty}^{\infty} \beta_i \delta(n - iP_0) \quad \text{if} \quad p(n) = \sum_{i=-\infty}^{\infty} \alpha_i \delta(n - iP_0) \qquad [1.100]$$

Therefore, these pulses appear in the speech cepstrum as separated apart from the fundamental period. If the excitation pulses are incorporated in the source part, the filter part of the production model contains the vocal tract, the lip radiation and the glottal airflow wave. In the most general case, the filter can be considered as a pole-zero transfer function, with factors on both sides of the unit circle:

$$V(z) = K \frac{\prod_{j=1}^{M}(1 - a_j z^{-1}) \prod_{j=1}^{N}(1 - b_j z)}{\prod_{i=1}^{O}(1 - c_i z^{-1}) \prod_{i=1}^{P}(1 - d_i z)} \quad \text{with} \quad \begin{cases} |a_j| < 1 \\ |b_j| < 1 \\ |c_i| < 1 \\ |d_i| < 1 \end{cases} \qquad [1.101]$$

Using the Taylor series $\log(1 - a) = -\sum_{n=1}^{\infty} a^n / n$ for $|a| < 1$, and assuming $K > 0$, the complex cepstrum becomes:

$$\hat{v}(n) = \begin{cases} \log(K) & n = 0 \\ \sum_{i=1}^{Q} \frac{c_i^n}{n} - \sum_{j=1}^{M} \frac{a_j^n}{n} & n > 0 \\ \sum_{j=1}^{N} \frac{b_j^{-n}}{n} - \sum_{i=1}^{P} \frac{d_i^{-n}}{n} & n < 0 \end{cases} \qquad [1.102]$$

Therefore, the cepstrum of the filter component of the speech production model decreases rapidly (in a^n/n) when n increases. In general, this property enables the separation of the filter and the source in the speech signal, using equation [1.99]. The contributions pertaining to the vocal tract and the source can be separated from each other by applying a simple time-domain window on the cepstrum, in the neighborhood of zero, in order to isolate the vocal tract contribution. Typically, a well marked peak indicates the voicing period when the signal is voiced. Figure 1.11 shows a cepstrum representation obtained for a vowel and illustrates the source/filter separation operation.

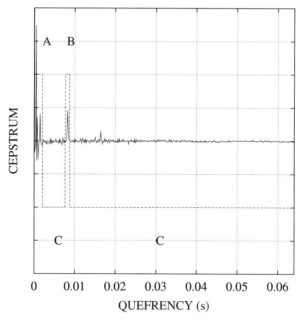

Figure 1.11. *Cepstrum of the vowel /a/. Region A corresponds to the vocal tract (first cepstrum coefficients). Region B corresponds to the voicing peak (fundamental period). Region C corresponds to the non-periodic components in the signal*

1.4.4. *Sinusoidal and harmonic representations*

For voiced speech, the excitation e is periodic. The signal can be considered as the sum of sinusoidal components. Two main methods have developed this type of representation, in two different ways: sinusoidal coding and harmonic coding.

The sinusoidal representation [MCA 86, QUA 86] generalizes the approach of decomposing the signal as a sum of sinusoidal segments. It is based on the source-filter model. The speech signal is written as an excitation $e(t)$ passed through a filter with an impulse response $h(t, \tau)$, which evolves over time.

$$s(t) = \int_0^t h(t, t - \tau)e(\tau)d\tau \qquad [1.103]$$

Excitation $e(t)$ is expressed as a sum of sinusoids, and the vocal tract part is represented via its transfer function $H(t, v)$:

$$e(t) = \sum_{l=1}^{L(t)} a_l(t) \cos\left(\int_{t_l}^{t} \omega_l(\tau) d\tau + \varphi_l\right) \quad \text{and} \quad H(t,v) = M(t,v) e^{i\phi(t,v)} \quad [1.104]$$

It must be noted that the number $L(t)$ of sinusoidal segments in the excitation varies with time, as well as amplitudes a_l and frequencies ω_l. The initial phases φ_l depend on the starting time of sinusoid t_l. The signal resulting from the full model is written:

$$s(t) = \sum_{l=1}^{L(t)} A_l(t) \cos(\Psi_l(t)), \quad \begin{cases} A_l(t) = a_l(t) M(t, v_l(t)) \\ \Psi_l(t) = \int_{t_l}^{t} \omega_l(\tau) d\tau + \varphi_l + \phi(t, v_l(t)) \end{cases} \quad [1.105]$$

This analysis scheme is based on the notion of sinusoidal *track*, referring to the components of the sum in the synthesis formula [1.105]. The number $L(t)$ of tracks varies with time: each track is active during a given lapse of time and this has to be determined by a tracking algorithm. In the case of discrete signals, equation [1.105] reduces to:

$$s(n) = \sum_{l=1}^{L} A_l(n) \cos(\omega_l n + \theta_l) \quad [1.106]$$

It is thus necessary to estimate the number of components, their amplitudes, phases and frequencies. The analysis step is based on the STFT. Let $\tilde{x}_r(k)$ denote the kth sample in the discrete Fourier transform, for frame number r (the analysis window, of length N, is assumed to be non-zero between 0 and $N-1$):

$$\tilde{x}_r(k) = \sum_{n=0}^{N-1} w(n) x(n + rH) e^{-i\omega_k n} \quad \text{for } r = 0, 1, \ldots \quad [1.107]$$

where $w(n)$ is a time-domain analysis window, H, the position of the analysis window, and N the frame duration. From this complex short-term spectrum, the amplitude and phase spectra for $\tilde{x}_r(k)$ can be calculated. For each frame, the spectral peaks are obtained by searching for all local maxima on the amplitude spectrum and then eliminating those whose amplitude is below a given threshold. The position of the peaks provides frequencies ω_l and amplitudes A_l of the sinusoidal components. Phases θ_l of these components are calculated as the phases of the STFT for frequency ω_l. For each frame, a set of L spectral peaks is thus obtained. The synthesis signal is calculated by an overlap-add of the short-term signals from equation [1.106]. In this case, the sinusoidal tracks are not explicit. It is much more advantageous to follow the sinusoidal tracks explicitly and then to interpolate the synthesis parameter along these tracks. This approach therefore requires us to track the peaks over time, i.e. to match in amplitude, phase and frequency the peaks estimated for a given frame with those of the following frame. In fact, the number of peaks may vary across frames and to solve this problem, an algorithm has to be put into operation for detecting the "birth", "continuation" and "death" of sinusoidal components across frames. Let q be the number of peaks detected for frame $r+1$, with frequencies g_1, g_2, ..., g_q. These peaks have to be matched with those of frame r, with frequencies f_1, f_2, ..., f_p. Each track f_i searches its peak in frame $r+1$, by choosing the one with the closest frequency. A threshold on the difference $|g_i - f_i|$ is used to decide whether the match is admissible or not. Three situations may arise:

1) a match is found – there is continuation of that track from the current frame r to the next one $r+1$;

2) no match is found – the track is considered to have vanished when moving from frame r to frame $r+1$: the amplitude of that peak is set to zero in frame $r+1$;

3) some peaks in frame $r+1$ have not found any match in frame r and are thus considered as belonging to a new track – this is considered as the start of the track in frame r, with amplitude 0.

After applying the matching algorithm briefly sketched above, all peaks for a given frame r correspond to a peak in frame $r+1$, some of them with a zero amplitude (just started or about to vanish). Synthesis between instants r and $r+1$ can then be achieved using equation [1.06]. However, as the parameters change at each frame, an interpolation procedure is required in order to avoid abrupt changes from one frame to the next. The amplitudes are linearly interpolated between their values A_r and A_{r+1} taken at the frame borders, i.e.:

$$\hat{A}^r(n) = A^r + (A^{r+1} - A^r)\left(\frac{n}{H}\right) \qquad\qquad [1.108]$$

Phases and frequencies are interpolated using a cubic function:

$$\hat{\theta}^r(n) = \theta^r + \omega^r n + v n^2 + \gamma n^3 \qquad\qquad [1.109]$$

where coefficients v and γ are determined according to ω_r, ω_{r+1}, θ_r and θ_{r+1}, via a procedure which will not be described here. The synthesis formula for frame r is thus:

$$\hat{s}^r(n) = \sum_{l=1}^{L'} \hat{A}_l^r(n)\cos(\theta_l^r n) \qquad\qquad [1.110]$$

The formula for sinusoidal synthesis can be viewed as a reduced form of the STFT synthesis, which uses only the spectral peaks of a narrow-band STFT. For voiced speech, the sinusoidal tracks correspond to the harmonics of the signal. For unvoiced speech, the sinusoidal representation remains valid as long as many tracks are used, since the sinusoidal components then form an approximation of the STFT spectrum. The sinusoidal representation delivers a signal which is perceptually identical to the original and can therefore be used as a coding scheme. Modifications to the fundamental frequency and duration are also possible, by separating on each sinusoidal segment the contributions from the source and from the filter.

Another example of sinusoidal decomposition of voiced signal is *harmonic coding* [ALM 83, TRA 88]. By generalizing the spectral behavior of a periodic pulse train to quasi-periodic pulse train, the generalized harmonics can be defined as the spectral components of $e(t)$ at rank k:

$$e(t) = \sum_k A(t) e^{ik\Phi(t)} \qquad\qquad [1.111]$$

Note that, in this case, the components are harmonic, i.e. they are multiples of the same fundamental pseudo-frequency $\Phi(t)$. The amplitudes and phases vary slowly with time. To each generalized harmonic in the excitation corresponds a generalized harmonic in the voiced signal, via the application of $h(t,\tau)$. The parameter estimation algorithm resorts to a method which is similar to the one presented for sinusoidal coding, but it requires the prior knowledge of the fundamental frequency. The complete coder uses harmonic coding on the one hand and transmits a residual signal on the other hand, in order to correct the errors on voiced frames and to handle unvoiced frames.

1.5. Conclusion

In this brief overview of the analysis methods for speech signals, we have tried to approach, or at least to mention, the most important methods. It can be observed that the development of methods for speech analysis dates back about 50 years. The acoustic spectrograph, conceived in the 1940s, is the ancestor of non-parametric methods, such as the STFT in the late 1970s and early 1980s and the time-frequency methods and wavelet approach in the 1980s and 1990s. From another angle, the linear acoustic theory of speech production, which originated in the 1950s and 1960s has been the starting point for parametric analysis methods. Many of these techniques are derived from the source-filter model: linear prediction and cepstrum analysis in the 1960s and 1970s, sinusoidal representations and periodic/non-periodic decompositions in the 1980s and 1990s. Undoubtedly, the future will bring further improvements to the source-filter model, probably thanks to the simulation of fluid mechanics equations, as many works have shown that the source-filter coupling is sometimes important.

For a more comprehensive overview on speech analysis methods, it would be necessary to take into account other speech signals (visual signals, movement sensors, for instance), analysis schemes derived from perception and auditory models, multi-sensor analysis such as microphone arrays, and of course, composite methods, which combine several methods at a time.

At last, speech analysis is not limited to speech as a linguistic object. More and more research is focused on non-linguistic aspects of speech [LIE 95], such as emotions, attitudes, the distance between speakers, etc. Many works in voice analysis concern musical voices (singers), professional situations (teachers, speakers, actors, chairpersons, etc.) and voice pathologies.

1.6. References

[ALL 77] ALLEN J.B., "Short-term spectral analysis, synthesis, and modification by discrete Fourier transform", *IEEE Trans. on ASSP*, vol. ASSP-25, 1977.

[ALM 83] ALMEIDA L.B., TRIBOLET J.M., "Nonstationary modeling of voiced speech", *IEEE trans. on ASSP*, vol. ASSP-31, 1983.

[ANA 75] ANANTHAPADMANABHA T.V., YEGNANARAYANA B., "Epoch extraction of voiced speech", *IEEE Trans. on ASSP*, vol. 33, no. 6, p. 562-570, 1975.

[ATA 71] ATAL B.S., HANAUER B.S., "Speech analysis and synthesis by linear prediction of the speech wave", *J. Acoust. Soc. Am.*, vol. 50, no. 2, p. 637-655, 1971.

[ATA 82] ATAL B.S., REMDE J.R., "A new model of LPC excitation for producing natural-sounding speech at low bit rates", *Proceedings of IEEE-ICASSP-82*, 1982.

[BOI 87] BOITE R., KUNT M., *Traitement de la parole*, Dunod, Paris, 1987.

[CAL 89] CALLIOPE, *La parole et son traitement automatique*, Masson, Paris, 1989.

[CHE 84] CHESTER D., TAYLOR F.J., "The Wigner distribution in speech processing applications", *Journal of the Franklin Institute,* Pergamon Press, vol. 318, no. 6, p. 415-430, 1984.

[CHI 77] CHILDERS D.G., SKINNER D.P., KEMERAIT R.C., "The cepstrum: a guide to processing", *Proceedings of the IEEE*, vol. 65, no. 10, p. 1428-1443, 1977.

[COH 86] COHEN L., PICKOVER C.A., "A comparison of joint time-frequency distributions for speech signals", *IEEE Inter. Symp. on Circuits and Systems*, p. 42-45, 1986.

[COH 89] COHEN L., "Time-frequency distributions – a review", *Proceedings of the IEEE*, vol. 77, no. 7, p. 941-981, 1989.

[COM 89] COMBES J.M., GROSSMAN A., TCHMITCHIAN P. (ed.), *Wavelets, Time-frequency Methods and Phase Space*, Springer Verlag, Berlin, 1989.

[CRO 80] CROCHIERE R.E., "A weighted overlap-add method of short-time Fourier analysis/synthesis", *IEEE Trans. on ASSP*, vol. ASSP-28, no. 1, 1980.

[DAL 98] D'ALESSANDRO C., DARSINOS V., YEGNANARAYANA B., "Effectiveness of a periodic and aperiodic decomposition method for analysis of voice sources", *IEEE Trans. on Speech and Audio Processing*, vol. 6, no. 1, p. 12-23, 1998.

[EST 77] ESTEBAN D., GALAND C., "Application of quadrature mirror filters to split band voice coding schemes", *Proceedings of IEEE-ICASSP-77*, 1977.

[FAN 60] FANT G., *Acoustic Theory of Speech Production*, Mouton, La Hague-Paris, 1960.

[FAN 85] FANT G., LILJENCRANTS J., LIN Q., "A four-parameter model of glottal flow", *Speech Transmission Laboratory – QPSR*, vol. 85, no. 2, p. 1-13, 1985.

[FLA 72] FLANAGAN J.L., *Speech Analysis, Synthesis and Perception*, Springer Verlag, Berlin, 1972.

[FLA 93] FLANDRIN P., *Temps-Fréquence*, Hermes, Paris, 1993.

[FUJ 87] FUJISAKI H., LJUNGQVIST M., "Estimation of voice source and vocal tract parameters based on ARMA analysis and a model for the glottal source waveform", *Proceedings of IEEE-ICASSP-87*, 1987.

[GAB 46] GABOR D., "Theory of communication", *Journal of the IEE*, 93, London, p. 429-441, 1946.

[GRI 84] GRIFFIN D.W., LIM J.S., "Signal estimation from modified short-time Fourier transform", *IEEE trans. on ASSP*, vol. ASSP-32, no. 2, 1984.

[HAR 76] HARDCASTLE W.J., *Physiology of Speech Production*, Academic Press, London, 1976.

[KAD 92] KADAMBE S., BOUDREAUX-BARTELS G.F., "Application of the wavelet transform for pitch detection of speech signals", *IEEE Trans. on IT*, vol. 38, no. 2, p. 917-924, 1992.

[KLA 90] KLATT D., KLATT L., "Analysis, synthesis, and perception of voice quality variations among female and male talkers", *J. Acoust. Soc. Am.*, 87, p. 820-857.

[LEH 67] LEHISTE I. (ed.), *Readings in Acoustics Phonetics*, MIT Press, Boston, 1967.

[LIE 95] LIÉNARD J.S., "From speech variability to pattern processing: a non-reductive view of speech processing", in C. Sorin *et al., Levels in Speech Communication: Relations and Interactions*, Elsevier, p. 137-148, 1995.

[LOU 93] LOUGHIN P.J., ATLAS L.E., PITTON J.W., "Advanced time-frequency representations for speech processing", M. Cooke, S. Beet and M. Crawford (ed.), *Visual Representation of Speech Signals*, John Wiley & Sons, 1993.

[MAK 81] MAKHOUL J.I., COSELL L.K., "Adaptative lattice analysis of speech", *IEEE Transaction on Circuit and System*, vol. 28, no. 6, 1981.

[MAK 75] MAKHOUL J.I., "Linear prediction: a tutorial review", *Proceedings of IEEE*, p. 561580, 1975.

[MAL 74] MALMBERG B., *Manuel de phonétique générale*, Editions Picard, Paris, 1974.

[MAL 89] MALLAT S.G., "A theory for multiresolution signal decomposition: the wavelet representation", *IEEE Trans. on PAMI*, vol. 31, p. 674-693, 1989.

[MAR 76] MARKEL J.D., GRAY A.H., *Linear Prediction of Speech*, Springer Verlag, Berlin, 1976.

[MCA 86] MCAULAY R.J., QUATIERI T.F., "Speech analysis/synthesis based on a sinusoidal representation", *IEEE Trans. on ASSP*, vol. 34, no. 4, 1986.

[MCC 74] MCCANDLESS S.S., "An algorithm for automatic formant extraction using linear prediction spectra", *IEEE Trans. on ASSP*, vol. 22, no. 2, 1974.

[MEY 90] MEYER Y., *Ondelettes et opérateurs*, Hermann, Paris, 1990.

[NOL 67] NOLL A.M., "Cepstrum Pitch Determination", *J. Acoust. Soc. Am.*, vol. 41, no. 2, p. 293-309, 1967.

[OPP 69] OPPENHEIM A.V., "Speech analysis-synthesis system based on homomorphic filtering", *J. Acoust. Soc. Am.*, vol. 45, no. 2, p. 458-46, 1969.

[OPP 89] OPPENHEIM A.V., SCHAFER R.W., *Discrete-time Signal Processing*, Prentice Hall, Englewood Cliff, New Jersey, 1989.

[PAR 86] PARSONS T., *Voice and Speech Processing*, McGraw-Hill, 1987.

[POR 81a] PORTNOFF M.R., "Short-time Fourier analysis of sampled speech", *IEEE Trans. on ASSP*, vol. 29, no. 3, 1981.

[POR 81b] PORTNOFF M.R., "Time-frequency modification of speech based on short-time Fourier analysis", *IEEE Trans. on ASSP*, vol. 29, no. 3, 1981.

[QUA 86] QUATIERI T.F., MCAULAY R.J., "Speech transformations based on a sinusoidal representation", *IEEE Trans. on ASSP*, vol. 34, no. 6, 1986.

[RAB 77] RABINER L.R., SCHAFER R.W., *Digital Processing of Speech Signals*, Prentice Hall, Inc., Englewood Cliffs, New Jersey, 1977.

[RAB 75] RABINER L.R., GOLD B., *Theory and Application of Digital Signal Processing*, Prentice Hall, Englewood Cliffs, New Jersey, 1975.

[ROS 71] ROSENBERG A.E., "Effect of glottal pulse shape on the quality of natural vowels", *J. Acoust. Soc. Am.*, vol. 49, no. 2, 1971.

[SCH 85] SCHROEDER M.R., ATAL B.S., "Code-excited linear prediction (CELP): high-quality speech at very low bit rates", *Proceedings of IEEE-ICASSP-85*, p. 937-940, 1985.

[SER 90] SERRA X., SMITH J., "Spectral modeling synthesis: a sound analysis/synthesis system based on a deterministic plus stochastic decomposition", *Computer Music Journal*, vol. 14, no. 4, 1990.

[SHA 90] SHADLE C.H., "Articulatory-acoustic relationships in fricative consonants", in W.J. Hardcastle and A. Marchal (ed.), *Speech Production and Speech Modelling*, p. 187-209, Kluwer, 1990.

[STE 99] STEVENS K.N., *Acoustic Phonetics*, MIT Press, 1999.

[TIT 94] TITZE I., *Principle of Voice Production*, Prentice Hall, Englewood Cliffs, 1994

[TRA 88] TRANCOSO I.M., ALMEIDA L.B., Rodrigues J.S., Marques J.S., Tribolet J.M., "Harmonic coding – state of the art and future trends", *Speech Communication*, vol. 7, no. 2, 1988.

[VEL 89] VELEZ E.F., ABSHER R.G., "Transient analysis of speech signals using the Wigner time-frequency representation", *Proceedings of ICASSP'89*, Glasgow, p. 2242-2245, 1989.

[VUN 99] VU NGOC TUAN, D'ALESSANDRO C., "Robust glottal closure detection using the wavelet transform", *Proceedings of the European Conference on Speech Technology, Euro-Speech'99*, p. 2805-2808, 1999.

[WEL 97] WELLS, J.C., "SAMPA computer readable phonetic alphabet" in Gibbon D., Moore R. and Winski R. (eds.), 1997. *Handbook of Standards and Resources for Spoken Language Systems*, Berlin and New York: Mouton de Gruyter, Part IV, section B, 1997.

[WOK 87] WOKUREK W., HLAWATSCH F., KUBIN G., "Wigner distribution analysis of speech signals", in V. Cappellini and A.G. Constantinides (eds.), *Digital Signal Processing-87*, Elsevier Science Publishers, 1987.

[YEG 98] YEGNANARAYANA B., D'ALESSANDRO C., DARSINOS V., "An iterative algorithm for decomposition of speech signals into periodic and aperiodic components", *IEEE Trans. on Speech and Audio Processing*, vol. 6, no. 1, p. 1-11, 1998.

Chapter 2

Principles of Speech Coding

2.1. Introduction

35 years ago, speech coding was a domain which was only just emerging. Today, it has become a scientific research topic and an industrial engineering sector of the utmost importance. The main factor for this fabulous growth lies in the recent development of modern telecommunications systems, which has brought about a boom in the deployment of digital signal transmission in many applications. As digital systems outperform analog technologies in many respects, digital transmission and storage of speech signals have become an absolute necessity, inducing incredible developments in the area of speech coding, over recent years.

Even though the deployment of digital networks now enables the transmission of sounds, images and all sorts of data, voice communication – i.e., the telephone – remains the most essential means of communication between people. This is why the development of telecommunications systems has, for its primary goal, to meet the demands in the area of telephone communication, as the recent development of mobile phones illustrates. In this sector, speech coding technologies have played a key role.

Speech coding is nothing more than a set of techniques for compressing – i.e. reducing the bit-rate of – the speech signal, while being able to reconstruct it with a sufficiently high quality, so as to efficiently use transmission channels and/or to lessen the memory required to store the compressed speech. However, we might wonder whether it is still relevant nowadays to compress speech, given that very

---

Chapter written by Gang FENG and Laurent GIRIN.

high bit-rate digital transmissions are available through optical fibers and that the memory capacity of semiconductors almost doubles every 18 months. The answer is yes. On the one hand, optical fiber links cannot be made available everywhere. In the case of mobile telephones, the radio wave will remain the only relevant medium for a long time to come. As each user can only be allocated a limited bandwidth, the bit-rate constraint is still relevant. On the other hand, in many applications, telephone quality is not adapted and a much better voice quality is required, i.e. a wider bandwidth and therefore an increase in the bit-rate. Compression techniques will thus continue to be in demand and speech coding is a domain with great prospects in the future.

Over the past 35 years, engineers and researchers have gathered great knowledge and good know-how in the area of speech coding. They have achieved great progress in accompanying the development of telecommunications, computer science and micro-electronics. Whereas the principles of the initial coding systems (64 kbit/s PCM) are quite straightforward to understand by any student in engineering sciences, the development of a modern coding system requires an extended knowledge of mathematics and signal processing, together with some significant practical experience. At this stage, it is important to underline the strongly experimental dimension of speech coding. A speech coding system does not amount to a mathematical model of the voice. Speech coding is considered as the art of distributing a limited number of bits over many parameters, so as to obtain an optimal quality for the coded speech. Even though some theoretical tools in signal processing can guide us in this task, there is no direct method that enables the prior determination of an optimal and universal scheme among a very large number of possible configurations. In this area, experience still plays a definite role.

The experimental dimension of speech coding can also be explained by a deeper reason: it is difficult to measure the quality of a coded speech signal objectively. Indeed, any compression technique can introduce more or less significant distortions, owing to the limited binary resources. These distortions generally result in a perceptible degradation of the voice quality, which is not always easy to describe. Moreover, the link between these distortions and the subjective speech quality is not always well established. Thus, not all visible distortions on the speech signal are audible, and the location of the distortions that can be heard in the signal is not always obvious. This very subjective property of the human ear increases considerably the difficulties met in the design of a speech coding system, since the minimization of the perceivable degradations is far from being easy to achieve.

2.1.1. *Main characteristics of a speech coder*

When dealing with a speech coding system, any specialist and any user must take into account the three most important properties of the coder: its bit-rate, its quality and its complexity.

The bit-rate, measured in terms of the number of bits transmitted per second, corresponds directly to the binary resources that need to be dedicated to reproduce the speech signal. In most cases, it is a synonym of the *compression factor*. The lower the bit-rate, the higher the compression factor and the fewer resources required for transmission or storage. It is therefore natural to target the lowest bit-rate. However, a low bit-rate has a cost: this usually results in a poorer coded speech quality or a high coding complexity.

The coding quality refers generally to the perceived quality of the coded speech. As any compression system induces more or less significant degradations, the quality of a coder is a way to measure these degradations. When they are sufficiently light, to the point that they can barely be perceived, the coder is said to be transparent (or quasi-transparent). Otherwise, the degradations can be perceived as the addition of a background noise, more or less uncorrelated to the speech signal. More generally, degradations can often manifest themselves as any type of voice deformation (change of timbre, metallic sounding, granular sensation, etc.).

As indicated above, it is difficult to measure the degradations caused by a coding system objectively. The well-known "signal-to-noise ratio" (SNR), which measures the average coding noise power, provides only a very rough indication, since different types of noise can yield the same average power, whereas they are not perceived as comparable by the ear. Even though researchers invest efforts in the quest for an objective measure, which would reflect what the ear hears, the evaluation of a speech coder is usually a subjective exercise. The MOS (mean opinion score) is a conventional measure which consists of presenting to a large panel of listeners a sufficiently large corpus of speech coded by the tested coder, mixed with a reference corpus of calibrated quality. The listeners qualify the overall quality of each sentence presented according to a five degree scale (excellent, good, fair, poor, bad). The average value over the entire set of speakers provides the MOS score for the coder[1]. While this evaluation indeed indicates the global quality of a coder, especially with respect to a reference, it has two major drawbacks: a heavy set up and a non-guaranteed reproducibility.

---

1 The subjective evaluation of coded speech quality is a vast research topic. In addition to the MOS, other measures exist. The reader can for instance refer to [DVO 88], [ITU 96a] and [ITU 96b].

The complexity of a coder reflects the degree of sophistication of the underlying algorithm. As these algorithms are usually designed with the aim to be implemented on a digital signal processor (DSP), the complexity is generally measured in terms of computational load per time unit or MIPS (millions of instructions per second), and in terms of memory requirements. Note that, as the DSP performances increase continuously, the coders tend to be more and more complex.

In any case, these three dimensions of a coder are tightly linked and it is not realistic to try to design a low bit-rate, excellent quality, low complexity speech coder. A compromise must be reached. For instance, researchers and engineers will try to conceive a new coder with a lower complexity for a fixed bit-rate and quality. If complexity is predominant, quality will have to be slightly sacrificed, so as to be able to implement the algorithm on a given DSP. Similarly, a user will have to be lenient on the compression factor, if high constraints are set in priority on quality and complexity.

In addition to these three fundamental properties of a speech coder, it is necessary to specify the bandwidth at which it operates. Currently, most coders operate at the telephone bandwidth (300-3,400 Hz), essentially for the sector of conventional telephone systems. For several years now, research has strengthened on wideband coders (50-7,000 Hz) or even FM bandwidth quality (20-15,000 Hz), in order to meet new demands, for the future generations of digital telephone or video conferencing applications. A wider bandwidth permits a considerably better coded speech quality, at the expense of a significant increase in coding rate. We will see further on that the nature of wideband speech and the specific requirements concerning speech quality, result in significantly different considerations for wideband coders, as opposed to telephone coders.

Another important specification for a speech coder is its intrinsic algorithmic delay, i.e. the inherent calculation delay of each algorithm below which speech reconstruction is not possible. This delay depends solely on the compression algorithm. As a general rule, the lower the coding bit-rate, the higher the delay is bound to be, because of the look-ahead that may be necessary to reach such a bit-rate. For some applications, where interactive communication is needed, a too high delay is undesirable. This is why a coder embedded in such applications must have a reasonable delay (below 20 ms for interactive communication[2], for instance). It is worth noting that the delay constraint has a strong influence on the choice of algorithms.

---

2 All other delays (recording, transmission, etc.) are added to this figure, and the total should not exceed 150 ms.

A coder can be described using other specifications, such as its robustness to transmission errors, its behavior after several successive coding-decoding operations, etc., which are not detailed here. The reader is invited to refer to specialized papers (for example, [COX 97]).

2.1.2. *Key components of a speech coder*

How does a speech coder work? What are the main principles that govern the design of a compression algorithm? It is well known that sampling a band limited signal can be achieved without any loss of information, as long as the Shannon rule is respected. On the other hand, when a signal is digitized, i.e. when the samples are represented on a linear scale with a finite number of bits, an irreversible quantization error is introduced, which manifests itself as the presence of noise over the signal. This simple linear quantization procedure can be considered as a form of coding, as it results in a representation of the signal with a limited number of bits. However, this process is not *stricto sensu* a compression, as it only consists of suppressing information in the signal when it is digitized.

So, what does the term "compression" mean? Compression means that a restricted number of bits are used to represent a given quantity of information, which would require a higher bit-rate if no compression was taking place. In other words, before compression, the signal contains a certain level of redundancy and the compression is precisely achieved in order to remove some of this redundancy. Ideally, compression should not cause any loss of information to the signal[3].

This summarizes the first coding principle: any redundancy contained in a signal should not be transmitted. The benefit of this principle is obvious: as only limited binary resources are available, all bits are doomed to carry as much information as possible, and certainly not any redundancy!

But where is this redundancy, and how can we eliminate it? Let us consider an example. For transmitting a sine signal, it is easy to understand that it is not necessary to transmit each and every sample. The frequency, the phase and the amplitude are sufficient to describe such a signal (we assume that no noise is present). This means that the sine signal contains a lot of redundancy. Not every signal is as redundant, but the principle remains: a redundant signal is predictable, i.e. given a number of past samples, it is possible to estimate the value of the current

---

3 Loosely speaking, the expression "compression system" is used to refer to a coding system, as it allows the reduction of the transmission rate without over-deteriorating the coded speech quality. As far as coding schemes with information loss are concerned, they are referred to as "coding with loss" in order to differentiate them from the strict meaning of "compression" which corresponds to "loss-less coding".

sample. Suppressing redundancy consists merely of decreasing the signal's predictability. It can thus be easily understood that a white noise is impossible to compress, as it contains no redundancy which makes it hopeless to try to predict the value of a sample from its past.

The predictability of a signal reveals itself by the presence of correlations between samples. From this point of view, correlation means redundancy. In the frequency domain, signal correlation corresponds to a non-flat spectrum. Spectrum flatness can thus be used to measure redundancy quantitatively. Let $\Gamma_X(e^{j\theta})$ denote the power spectral density (PSD) of a digital signal $X(n)$. The spectrum flatness can be defined as the following coefficient:

$$\xi_X = \frac{\exp\left(\frac{1}{2\pi}\int_{-\pi}^{\pi}\ln\Gamma_X(e^{j\theta})d\theta\right)}{\frac{1}{2\pi}\int_{-\pi}^{\pi}\Gamma_X(e^{j\theta})d\theta} \qquad [2.1]$$

This spectrum flatness coefficient has the following properties: the value of ξ_X is necessarily between 0 and 1. $\xi_X = 1$ corresponds to a completely flat spectrum and $\xi_X = 0$ if some part of the spectrum is equal to 0. The closer ξ_X is to 1, the flatter the spectrum, while the closer ξ_X is to 0, the more contrasted the spectrum. When some part of the spectrum is zero, the degree of contrast is considered as maximum[4], hence the zero value of ξ_X.

It has been shown that the redundancy caused by the correlation between samples is a simple function of ξ_X, which can be expressed as follows (in number of bits per sample):

$$-\frac{1}{2}\log_2 \xi_X$$

In summary, if the spectrum is flat, $\xi_X = 1$ and there is no redundancy (this is the case for white noise). On the other hand, if the spectrum is strongly contrasted, ξ_X is rather small and redundancy can be quite high.

Referring to these notions helps to understand where the main source of redundancy is located in the speech signal. The main feature of speech spectra for voiced sounds is the presence of formants (resulting from resonances in the vocal system), i.e. quite contrasted spectra. This means that the level of redundancy is high and that compression is possible.

---

4 It is easier to understand this property when the spectrum is thought of on a logarithmic scale along the y-axis (PSD in dB, for instance).

In order to remove redundancy, a natural idea arises: flattening the spectrum by any possible means[5]. This can be obtained easily with filters. Let us take a concrete example. Suppose that $X(n)$ is a signal with a non-flat spectrum (we will also assume it is a stationary signal for the time being), which therefore contains some redundancy. The direct transmission of the raw signal samples is not very efficient because of the redundancy. To remove the redundancy, signal $X(n)$ can be passed through a filter, the transfer function of which is chosen so that output signal $Y(n)$ has a flat spectrum, i.e. with little remaining redundancy (such a filter is called a "whitening filter"). As a consequence, the transmission of signal $Y(n)$ will be more efficient. Concerning the recovery of $X(n)$ from $Y(n)$, it can be obtained by passing $Y(n)$ through another filter, the transfer function of which is the inverse of the first one, used for whitening the signal. Note that, in this example with a stationary signal, the transmission of the whitening filter coefficients is almost costless in terms of the transmission rate, since it can be done once for all.

This very simple example illustrates the underlying idea of a compression system: removing all redundancies. In the case of speech signals, the situation is very similar. The main difference lies in the fact that speech is not a stationary signal. However, as speech can be considered as locally stationary over a few tens of milliseconds, the above approach remains valid, provided the filter coefficients are regularly refreshed.

It should be mentioned that the signal correlation is not the only source of redundancy, even though it is usually the predominant factor. The amplitude distribution in the signal is also a source of redundancy, even when there is no correlation at all. A signal with a non-uniform amplitude distribution contains more redundancy than that of a signal with a uniform distribution. If the same number of bits is used with the same uniform quantizer for both signals, some efficiency is lost in the first case, because of the redundancy. As it is quite rare to come across a signal with a totally uniform amplitude distribution, it is fundamental to remove the redundancy resulting from the non-uniform amplitude distribution, for a more efficient compression.

A second principle in coding can be formulated as follows: only what the ear hears should be transmitted. Let us illustrate this principle with an example. Psychoacoustic studies [ZWI 81] show that the ear is not able to distinguish between two distinct sequences of white noise, as long as they follow the same statistical properties. This suggests that the complete transmission of such a sequence can be replaced by the transmission of parameters which describe the statistical properties of the signal, thus reducing considerably the coding rate. These studies reveal other

---

5 Other strategies also exist to remove redundancy stemming from correlations in the signal, for example, transform coding.

phenomena which can be exploited to reduce the transmission bit-rate. For example, frequency masking: i.e., a particular sound with a given level, which is perfectly audible on its own, can become inaudible when associated with another louder sound, with a neighboring frequency. A more accurate description of this effect will be provided to the reader further in this chapter (section 2.3.1) and it will be seen how it can be used to save on the transmission for some of the signal components.

Imagine now that everything has been done in order to remove redundancy and inaudible components from the signal. The last step is to transmit the signal components or the parameters which bear the useful information. As the binary resources are limited, these must be distributed over the useful signals and parameters. At this stage, some loss of information cannot be avoided. The quantization effects, due to the limited number of bits, will induce distortions of the coded speech. These will be discussed later.

To sum up, the main components involved in a speech coder are:

– a number of processing steps (filtering, transformation, etc.) aimed at removing the redundancy in the signal, and, if need be, its inaudible components. After these first steps, the new signal and the parameters must bear the relevant information required for being able to reconstruct the speech signal;

– a procedure (generally a quantization step) intended to distribute the binary resources over the signal and parameters, so as to minimize the distortions thus incurred.

Note that only a few coders show a structure in which these two blocks are clearly distinct. Most of the time, they are combined and occasionally subdivided into several stages.

In the next sections of this chapter, we will present several coding techniques which are commonly used and have been acknowledged for their efficiency. Note that our goal is neither to present an exhaustive coverage of all existing methods, nor to provide sharp technical details concerning the coders. Our objective is rather to present *how* these coders work and to explain *why* these approaches are successful. Repeatedly, the reader will be provided with references pointing to more specific technical details[6]. In addition, a new track will be presented at the end of this chapter: the joint exploitation, the acoustic speech signal and the image of the speaker, in order to achieve an even more efficient compression.

---

6 For instance, the reader is invited to refer to [GER 92a, JAY 84, JAY 92, MOR 95, NOL 93], for general works and articles on speech coding.

2.2. Telephone-bandwidth speech coders

The most widespread speech coding systems are those which operate over the telephone bandwidth (from 300 to 3,400 Hz). This type of speech is said to be of *toll quality*. Note that most worldwide telephone systems use this frequency band, which explains why the corresponding coding techniques are so popular.

For telephone speech, the reference coding system is the well-known PCM (pulse code modulation) at 64 kbit/s. The signal is sampled at 8 kHz and quantized on 8 bits. This system is considered to be the simplest coding scheme. Its main property lies in a non-uniform quantization: a logarithmic law is used to keep the signal-to-noise ratio approximately constant over a wide range of input amplitude values, so as to yield a quality almost equivalent to a 12 bit uniform quantizer. The resulting distortion is moderate: with an average SNR of 38 dB, this coder is considered virtually transparent. It was standardized by the CCITT[7] in 1972, as the G.711 standard [CCI 84a].

In the PCM system, redundancy owed to correlations is not exploited. The goal of removing this redundancy has yielded a new type of coding scheme: ADPCM (*adaptive differential PCM*) which operates at 32 kbit/s and provides a speech quality comparable to PCM at 64 kbit/s. This corresponds to a bit-rate reduction by a factor of 2. In terms of transmission rate per sample, it means that the ADPCM coder functions at 4 bits per sample. The ADPCM coder is based on differential coding (a principle similar to predictive coding)[8] which enables redundancy reduction using a predictive filter. A feature of this coder is that no filter-related information is transmitted. The filter coefficients are estimated in the coder and in the decoder, from the coded speech, using a rather elaborate gradient algorithm [BEN 86].

The 32 kbit/s ADPCM coder was standardized by CCITT in 1984 as the G.721 standard [CCI 84b]. Extensions at 40, 24 and 16 kbit/s were proposed in 1988 (G.726 standard [CCI 88a]). Let us underline that nowadays ADPCM is still quite widespread worldwide.

---

7 CCITT stands for *"Comité Consultatif International Télégraphique et Téléphonique"*: the international body in charge of the standardization of various telecommunication devices, including speech coders. Today, this role is in the hands of the ITU (International Telecommunication Union).

8 For correlated signals, the difference between two successive samples is statistically smaller than the value of the samples themselves. Therefore, for efficient coding, the difference is transmitted instead of the absolute value: this is differential coding. However, this difference can also be calculated between the current sample and a linear combination of several previous samples. This corresponds to predictive coding.

In the 1980s, speech coding experienced a very fast growth and revolutionary progress has been achieved. The most striking event was certainly the conception of CELP (*code excited linear prediction*) coding in 1985 [SCH 85]. The principle of CELP is to introduce, in LPC-based coders, an analysis-by-synthesis procedure and a vector quantization step. This enables an even more efficient allocation of binary resources. Thanks to this very powerful technique, the compression factor reached a value of 4 as soon as the early 1990s. And above all, this technique considerably stimulated research in this direction and many new solutions were subsequently proposed to further improve the speech quality and lower the coder complexity. To date, a wide range of CELP coders have been designed and among them, one that is particularly efficient is the LDCELP (*low-delay CELP*), standardized in 1992 by CCITT for coding speech at 16 kbit/s (2 bits per sample), known as the G.728 standard [CCI 92, CHE 92].

Three years later (in 1995), the ITU put forward a new recommendation for speech coding at 8 kbit/s: the G.729 standard [ITU 96c]. This corresponds to a compression factor of 8, i.e. an average bit-rate of 1 bit per sample! Like its predecessors, this new coder was based on CELP but had a particular structure (CS-ACELP for *conjugate-structure algebraic CELP* [SAL 98]), which will be detailed further below.

In addition to these coders, standardized on a world level to fulfill telephone demands, several other coders have been designed and proposed as continental or governmental standards. Let us cite for example the mobile phone speech coder GSM (13.2 kbit/s RPE-LTP [KRO 86, HEL 89]) and the FS1016 American standard (CELP at 4.8 kbit/s [CAM 91]).

It is generally acknowledged that it should be possible to reach a compression factor of 16, i.e. coding speech at 4 kbit/s (half a bit per sample) while maintaining a telephone-like quality. This is a challenge for researchers and engineers: research in this area is very active and the upcoming ITU standardization for speech coding at 4 kbit/s is currently awaited.

Even though there is no theoretical limit that could prevent a further bit-rate reduction, it is difficult to imagine today a speech coder operating below 2 kbit/s, while offering a telephone quality, a reasonable complexity and an admissible delay. Of course, there are several coders which operate below 4 kbit/s, but they do not provide telephone quality. With such a low bit-rate, a number of distortions become clearly audible. In this context, the objective is to provide coded speech which is quite natural and sufficiently intelligible. For this purpose, some researchers are investigating solutions requiring only a few hundred bits per second.

2.2.1. *From predictive coding to CELP*

CELP coders are based on predictive coding, i.e. redundancy reduction exploiting the signal's predictability. This goal can be achieved with a prediction filter, which calculates an estimate of the current sample as a linear combination of several past samples. This is *linear predictive coding* (LPC). The prediction error, i.e. the difference between the actual and the predicted samples, is often called the *residual* signal. It contains very little correlation and therefore very little redundancy, which is ideal for transmission. Note that the residual signal obtained by linear prediction is precisely the result of a spectral flattening operation using the whitening filter described above.

For coding purposes, a first option is to quantify the residual signal on a sample-by-sample basis. This approach is more efficient than a direct quantization of the signal itself, because the residual signal has much less energy than the original signal. The ratio between these two energies varies according to the signal and its predictor, but it leads towards a theoretical asymptotic value which is exactly equal to the signal's spectral flatness coefficient ξ_X defined above. This property is important, since a smaller residual average power implies a smaller quantization error average power, i.e. a higher SNR. As a consequence, a *compression gain* is observed after coding as compared to a direct quantification. Assuming that the residual signal follows the same amplitude distribution as the original signal (which would be the case for a Gaussian signal and an ideal predictor of infinite order), the maximal compression gain is written:

$$G_{p\,\text{max}} = \frac{1}{\xi_X} \qquad\qquad [2.2]$$

Note that the maximal compression gain is a simple function of the redundancy contained in the signal. The compression gain thus makes it possible to code speech with improved quality for a given bit-rate or to lower the bit-rate while keeping the quality constant.

The main idea in CELP coding comes precisely from the nature of the residual signal which closely resembles a white noise, in spite of its variable power caused by the non-stationarity of speech. This is where the idea of replacing the residual signal by a number of blocks[9] of white noise stems from, so that it becomes useless to transmit the residual itself. As obviously a single segment of white noise is not sufficient to replace a whole residual signal, it is important to find out how many

---

9 It is not realistic to replace a long block of residual signal because of the non-stationarity of speech. A reasonable duration is of a few milliseconds.

(small) blocks of white noise are necessary to reproduce a good quality speech signal.

Schroeder and Atal's work has demonstrated the feasibility of such a scheme [SCH 85]. They have shown that it is indeed possible to reproduce speech with a relatively small number of white noise segments. They proposed a speech coder based on these principles and this is how the first CELP coder started.

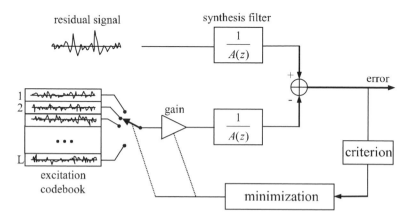

Figure 2.1. *Principles of a CELP coder. The residual signal is replaced by one of the L elements from the excitation codebook, resulting from an analysis-by-synthesis procedure which minimizes the difference between the reproduced signal and the original one*

Figure 2.1 illustrates the principles of such a coder. It is important to stress that, if the residual signal is passed through the synthesis filter (which is the inverse of the predictive analysis filter), the original signal is perfectly reconstructed. When the residual is replaced by a block of white noise, one among the L blocks within the excitation codebook is selected and its amplitude adjusted, so as to minimize the difference between the coded speech and the original. The optimal excitation is thus obtained via an analysis-by-synthesis procedure. Finally, the index corresponding to the optimal codebook entry and the (quantized) gain factor are transmitted and the speech signal can be (approximately) reconstructed with the same synthesis process in the decoder (in which the codebook is also stored). The advantage in terms of transmission rate of the CELP coder is due to the transmission of an index instead of the residual signal itself.

For the CELP coder to be operational, two remaining problems must be solved: 1) the definition of a criterion so as to measure the difference between the coded speech and the original signal and 2) the procedure for determining the optimal excitation and the corresponding gain factor. For the first point, the ultimate criterion is the human ear. However, in the lack of a suitable ear model (and also given the complexity constraint), simpler objective criteria are used, such as the mean square error. However, the direct use of this criterion leads to a relatively flat spectrum for the resulting error. As the error is precisely the difference between the coded speech and the original, i.e. the coding distortion, the latter also shows a flat spectrum. Such a property is not favorable because the distortion is not evenly masked by the signal. A better solution is to filter, using a *perceptual filter*, the difference before minimization (see Figure 2.2). In this way, the spectral shape of the coding distortion becomes less perceptible. The perceptual filter is dependent on the signal and it can be calculated from the LPC analysis filter. This point will be detailed in section 2.2.2.

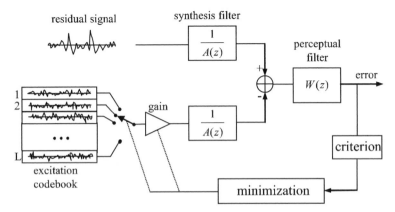

Figure 2.2. *Introduction of a perceptual filter in a CELP coder in order to modify the spectral shape of the coding distortion, so as to improve its masking by the signal*

In order to determine the optimal excitation, let us first note that, in Figure 2.2, the synthesis filter and the perceptual filter can be combined into a single filter: the weighted synthesis filter. Let $h(n)$ denote the impulse response of this filter[10]. A matrix $\mathbf{H}$ can be constructed, with time-shifted versions of $h(n)$. For any index k in the codebook, error ε_k to be minimized is written:

$$\varepsilon_k = \mathbf{H}\mathbf{r} - G_k \mathbf{H}\mathbf{e}_k \qquad [2.3]$$

Here, $\mathbf{r}$ is the residual signal vector, $\mathbf{e}_k$ is the kth entry in the excitation codebook and G_k is the gain factor. For each $\mathbf{e}_k$, an optimal gain $G_{k_{\text{opt}}}$ is calculated so as to minimize the mean square error $J_k = \varepsilon_k^T \varepsilon_k$. In fact, $G_{k_{\text{opt}}}$ must be such that ε_k is orthogonal to $\mathbf{H}\mathbf{e}_k$, i.e.:

10 Theoretically, $h(n)$ is an infinite impulse response. However, in practice, it is possible to truncate it at a reasonably short length.

$$G_{k_{opt}} = \frac{\mathbf{r}^T \mathbf{H}^T \mathbf{H} \mathbf{e}_k}{\mathbf{e}_k^T \mathbf{H}^T \mathbf{H} \mathbf{e}_k} = \frac{\mathbf{r}^T \mathbf{\Gamma} \mathbf{e}_k}{\mathbf{e}_k^T \mathbf{\Gamma} \mathbf{e}_k}$$ [2.4]

where $\mathbf{\Gamma} = \mathbf{H}^T \mathbf{H}$ is the correlation matrix of $h(n)$.

Using this expression for $G_{k_{opt}}$, the quadratic error is written:

$$J_k = \mathbf{r}^T \mathbf{\Gamma} \mathbf{r} - \frac{(\mathbf{r}^T \mathbf{\Gamma} \mathbf{e}_k)^2}{\mathbf{e}_k^T \mathbf{\Gamma} \mathbf{e}_k}$$ [2.5]

As a consequence, the minimization of J_k is equivalent to searching for index k which maximizes:

$$T_k = \frac{(\mathbf{r}^T \mathbf{\Gamma} \mathbf{e}_k)^2}{\mathbf{e}_k^T \mathbf{\Gamma} \mathbf{e}_k}$$ [2.6]

This term corresponds to the inter-correlation between **Hr** and **He**$_k$. The maximization of this term can be interpreted as the selection of **e**$_k$ so that **He**$_k$ resembles **Hr** as much as possible.

In order to determine the optimal excitation, each codebook entry must be tried. It is easy to imagine the heaviness of the corresponding computational load needed for this exhaustive search, in particular for large codebooks. This problem has somehow hindered the development of CELP coders for some time after it was conceived. The advent of CELP coders in the engineering world thus had to wait until fast search algorithms were designed.

Now that the most essential components of a CELP coder have been described, let us make a few observations. The search for an entry in a codebook, so as to approximate a target vector, can be considered as a vector quantization procedure. This process yields a more optimal result, in terms of quantization error, as compared to a sample-by-sample quantization of the residual signal (scalar quantization) [MAK 85]. However, the "analysis by synthesis" procedure involved in the coding scheme impacts the search process as compared to a direct vector quantization (VQ) of the residual. In the standard VQ case, the closest entry to the residual in the codebook is searched for as the optimal approximation of the residual itself. In the CELP optimization, the existence of the synthesis and perceptual filters modifies the result and the optimal entry is not necessarily the one that is most similar to the residual.

The dimensions of the excitation vectors (i.e., their length) and the size of the codebook (i.e. its number of entries) are key elements that impact the quality, the bit-rate and the complexity of the CELP coder. There is no theoretical method for determining these parameters optimally. The most efficient choice is driven by experimentation.

To illustrate this more clearly, let us consider a typical CELP coder, similar to the one proposed by Schroeder and Atal in 1985. Excitation vectors of dimension 40 samples (5 ms) will be considered and the codebook will be assumed to contain only 1,024 entries (10 bits). In other words, 10 bits are dedicated every 5 ms to code the index k, and we will assume that (typically) 5 more bits are used for the corresponding gain factor. Thus, we obtain a bit-rate of 3 kbit/s. If 40 bits are added every 20 ms to code the prediction coefficients, the total coding rate reaches about 5 kbit/s. And what about the quality? Experimentally, such a coder does not perform sufficiently well for reproducing telephone speech quality, but the coded signal is very natural and does not show strong distortions. However, the coder complexity is huge because of the exhaustive search of the optimal excitation in a 1,024 entry codebook. Today, it is still impossible to implement such a coder on a regular DSP in real-time.

Why is it so difficult to optimize a coder? Let us imagine that, in order to improve the coder quality, 3 more bits are added to code the excitation. This represents only 600 more bits/s, while the codebook size is multiplied by 8. The computational load soon becomes prohibitive while the coding quality is still not guaranteed. Conversely, if the codebook size is divided by 8, the transmission rate is just reduced by 600 bits/s which is not considerable. However, the coding quality will become unacceptable because the codebook is too small with respect to the dimension of the excitation vectors.

2.2.2. Improved CELP coders

Since CELP coders were first proposed, considerable progress has been made, putting the focus on complexity reduction. For this purpose, particular structures of the excitation codebook have been conceived, so that fast search techniques have been made possible for optimizing the coded excitation. At the same time, the other components in the coder have also been subject to intense research: quantization of prediction coefficients, long-term predictor, perceptual filter, etc. To sum up, nothing has been left to chance; everything has been carefully studied and optimized. These efforts have yielded considerable improvements of the coders.

Today, CELP coding has reached maturity. In this section, we present a selection of techniques which have significantly contributed to this progress[11].

2.2.2.1. *Excitation codebook*

The implementation of a CELP coder requires an exhaustive search of the optimal excitation in a codebook, which means a very large computational load. In order to lower the complexity of the coder, one approach is to limit the search to a relatively small subset within the codebook. This means that a process exists to identify *a priori* an area in the codebook where the optimal excitation is likely to be found. Unfortunately, this is impossible for codebooks composed of white noise blocks, as no simple relationship exists between different codebook entries. Therefore, fast search algorithms must rely on a structured codebook. Many studies have proceeded in this direction, based on different branches of vector quantization theory, which is considered today as a research field *per se*.

As an example of structured codebooks, one particularly interesting approach is the algebraic codebook, in which all non-zero values are generally forced to be equal to +1 or –1. Such codebook entries can be represented by a binary index in direct correspondence with the content of the vector. This enables the generation of the codebook entry from its index and saves storage resources in the coder and in the decoder, as it is not necessary to store the codebook. The main advantage of algebraic codebooks lies in the fact that there are fast algorithms for finding the optimal excitation.

A rich literature is available on structured codebooks and fast search algorithms. For more details, the reader is invited to refer to specialized papers, such as [ADO 87, BOU 95, DAV 86, KLE 90, LAM 88].

We illustrate the above principle on a concrete example: the CELP coder (G.729 standard). One of the features of this standard is a low complexity, in order to satisfy a number of targeted applications, in particular those concerning personal communication systems (PCS). Efforts have been made on this aspect, and the use of a structured dictionary has reached a good compromise.

In this type of coder, sequences of 40 samples (5 ms) in the residual signal are approximated by a vector which contains only 4 pulses with amplitudes equal to ±1 (the rest of the vector values being equal to zero). Whereas the number of non-zero pulses is very small, their locations are free, which leads to a rather wide variety of coded residuals. In fact, the 4 pulses can have $C_{40}^4 = 91,390$ different locations and

---

11 The following list is not exhaustive. The reader is invited to refer to [GER 92b] and [CHE 95] for more details concerning harmonic noise shaping and adaptive post-filtering, for example.

therefore 17 bits are necessary to code them. Additionally, 1 bit per pulse is needed to code its sign, yielding a dictionary which contains 91,630 × 16 distinct configurations (requiring 21 bits per entry altogether). This number is huge.

However, the introduction of two marginal constraints enables a significant reduction of this combinatory complexity:

1) the 4th pulse can only take one out of 16 predetermined locations within: 4, 5, 9, 10, 14, 15, …, 39, 40;

2) for the locations corresponding to the 3 remaining pulses, it is not possible for another pulse to take a location at a distance of 5 samples (or any multiple of 5) from an existing pulse.

With these 2 constraints, the total number of pulse locations is brought down to:

$$\frac{24 \times 16 \times 18}{1 \times 2 \times 3} \times 16 = 8,192$$

which is only 9% of the initial number of locations (91,930). Moreover, after a detailed study, it can be observed that more than 70% of the 91,930 locations can be slightly altered by shifting one or two pulses by one sample around their initial position, in order to satisfy the constraints. Therefore, a lot of efficiency is gained in terms of the transmission rate and of the complexity, while the loss of optimality due to these slight adjustments is negligible.

The particular structure of this codebook avoids a large part of the exhaustive search for determining the optimal excitation. It can be observed that, once the 4 locations are fixed, their sign can be readily determined, as the pulses have only two possible amplitude values (±1): of the $8,192 \times 16 = 2^{17}$ possible configurations in the dictionary, only $8,192 = 2^{13}$ of them need to be tested.

It is even possible to avoid testing these 8,192 configurations. As there are only 4 non-zero pulses in each entry, the test can be carried out in a progressive manner, by adding the pulses successively, so as to maximize criterion T_k of equation [2.6] gradually. If the matching score remains low with the first three pulses, it is reasonable to stop this track and save the 16 additional tests that would have been required to specify the location of the 4th pulse. The corresponding threshold can be determined experimentally and a significant computational load can be saved[12], while reaching a level of optimality comparable to the one that would have been achieved with an exhaustive search.

---

12 Experiments have shown that the optimization of the 4th pulse is necessary only one time out of eight on average.

We might wonder how an algebraic codebook composed only of signed binary vectors (and many zeros) can enable the reproduction of good quality coded speech, while these vectors do not *a priori* resemble the residual signal. Here are several explanatory elements.

First of all, the residual has a structure close to a white noise. If we consider a stationary white noise with unit power, any block of this signal which is long enough (N samples with $N \gg 1$) has an energy which tends towards a constant N. This slice of signal can be viewed as a vector in an N-dimensional space and its length tends towards $\sqrt{N}$. This means that the vectors corresponding to the different blocks of white noise tend to be distributed on a hyper-sphere in the N-dimensional space. The construction of an excitation codebook can be understood as the selection of a well-distributed finite subset of points on this hyper-sphere, for example, those located on the vertex of the hypercube inside the hyper-sphere. Note that these points can precisely have their coordinates composed of $+1$ and -1.

Statistics show that the residuals follow an amplitude distribution which is quite close to a Gaussian law. This means that small amplitude values are much more frequent than large ones. However, the latter contribute much more to the overall signal energy. This suggests that the residual signal can be approximated by vectors with only a few non-zero components, the rest of them being forced to zero. It is worthwhile noting that the multi-pulse LPC (MPLPC) coders [ATA 82], i.e. the predecessors of CELP coders, are based on this principle. Experience has indeed shown that one pulse every millisecond on average is sufficient to reproduce speech with good quality.

Let us finally note that a CELP coder usually contains some mechanism for approximating the residual signal from past excitation (see the long term prediction presented further down). The search for an optimal excitation in the codebook takes place only at the second phase, to complement the predicted excitation provided by the long term predictor. The entries in an algebraic codebook are well adapted for this purpose.

2.2.2.2. *Quantization of the prediction coefficients*

In a predictive coder, an analysis filter is used to reduce the speech signal predictability, thus suppressing the correlations in the residual signal (flat spectrum). Generally speaking, the analysis filter has the following form:

$$A(z) = 1 + \sum_{i=1}^{p} a_i z^{-i} \qquad\qquad [2.7]$$

Here, the a_i coefficients are the prediction coefficients, which can be obtained by minimizing the mean power of the residual signal. In most cases, these coefficients are calculated very efficiently with the Levinson-Durbin algorithm, from the short-term correlation coefficients of the signal. A typical value for the filter order p is 10, which yields a reasonable flat residual spectrum for telephone speech, for which the number of well-marked spectral peaks (or formants) is usually 4 or 5. Let us finally recall that, because speech is non-stationary, the prediction coefficients must be refreshed every 10 to 30 ms.

In speech coders for which the transmission of the prediction coefficients is necessary, these coefficients have to be quantized. It appears that the a_i coefficients are not adapted to quantization, as they are very sensitive to rounding approximations. Sometimes, a very small alteration of one coefficient can modify the behavior of the corresponding filter significantly, to the point that it can become unstable. To avoid this problem, more than a dozen bits per coefficient would be necessary, which is far too much.

Fortunately, equivalent representations of the a_i exist, which appear to be less sensitive to quantization. One of them is based on the fact that the polynomial $A(z)$ of degree p can be obtained uniquely from a polynomial $A_{p-1}(z)$ of degree p-1, as:

$$A(z) = A_{p-1}(z) + k_p z^{-P} A_{p-1}(z^{-1})$$ [2.8]

The polynomial $A_{p-1}(z)$ can itself be obtained from a polynomial of degree p–2 with the help of a coefficient k_{p-1}, etc. The p coefficients k_i thus obtained are often called reflection coefficients or PARCOR. They can be used uniquely to generate filter $A(z)$ and they form a representation of the prediction filter which is equivalent to the a_i coefficients.

The k_i coefficients have three remarkable properties as follows:

– for any stable filter, $|k_i| < 1$;

– higher order k_i coefficients (for example, k_9 and k_{10}) are much less important than lower orders k_i (for example, k_1 and k_2) for characterizing prediction filter $A(z)$;

– and last but not least, the k_i coefficients can be submitted to a rough quantization without causing too many spectral distortions.

For these reasons, a first generation of predictive coders worked with the k_i coefficients for transmission. A good spectral fidelity can be achieved with 36 to 40 bits for a vector of 10 coefficients.

In order to further explain the difference in behavior between the k_i and the a_i coefficients for quantization, it is worthwhile noting that polynomial $A(z)$ is completely determined by its roots, i.e. the poles of synthesis filter $1/A(z)$. The position of the poles with respect to the unit circle is particularly important in relation to the corresponding transfer function. These positions can be significantly affected by the quantization of the a_i coefficients, whereas they are much less sensitive to quantization of the k_i coefficients.

So as to reduce further the sensitivity to quantization, a non-linear transformation can be applied to the k_i coefficients [VIS 75]:

$$LAR_i = \log \frac{1+k_i}{1-k_i} \hspace{3cm} [2.9]$$

The sensitivity of the LAR (for *log area ratio*) coefficients is such that a modification of the poles by quantization has less impact when the poles are closer to the unit circle. Furthermore, the LAR coefficients are particularly well adapted to interpolation, which allows smoothing of the coefficients across successive frames.

In 1975, a new representation of the a_i coefficients was proposed, so as to make quantization even more efficient. The new coefficients were the *line spectrum pairs* (LSP) or *line spectrum frequencies* (LSF) [ITA 75]. They were based on the following property: by extending $A(z)$ at the order $p+1$ with the decomposition formula of [2.8], and assuming $k_{p+1}=\pm1$, the two following polynomials are obtained:

$$Q(z) = A(z) + z^{-(p+1)}A(z^{-1})$$
$$R(z) = A(z) - z^{-(p+1)}A(z^{-1}) \hspace{2.5cm} [2.10]$$

$Q(z)$ is symmetric and $R(z)$ is anti-symmetric. Each of them carries half of the information of $A(z)$. However, by construction, all the roots of these two polynomials are on the unit circle and can be written as $e^{j\omega_i}$. Thus one single parameter ω_i is sufficient to describe each root, and can be interpreted as a frequency. It can also be shown that, for any stable filter $A(z)$, the roots of $Q(z)$ and $R(z)$ are interleaved. Thus, we obtain p frequency values that describe $A(z)$. Moreover, the roots of $A(z)$ close to the unit circle correspond in general to two frequencies ω_i and ω_j, one from $Q(z)$ and one from $R(z)$, which fall on each side of the root's frequency, hence the name *line spectrum pair*. The closer the two lines are to each other, the closer the root of $A(z)$ is to the unit circle.

The strong physical meaning of coefficients ω_i and their relatively narrow distribution provide good quantization efficiency [SOO 84, SUG 86]. In the context of a scalar quantization, 34 bits are sufficient to represent 10 coefficients [ATA 89]. If the ordered structure of the LSF coefficients is exploited, it is possible to use differential coding techniques that reduce the number of bits to 30 [SOO 90]. Even more efficiency can be achieved with vector quantization. For instance, a *split-VQ* approach can bring the number of bits down to 24 [PAL 93]. If the correlation between successive frames is taken into account, it is possible to use some vector predictive coding to reduce the inter-frame redundancy. Such a scheme is used in the CELP G.729 coder, based on 18 bit LSP quantization, for each 10 ms frame [SAL 98].

Several more recent coding schemes can be found in the literature, to further reduce the coding rate [PAN 98, RAM 95, XIE 96, XYD 99]. Generally speaking, all of these sophisticated techniques bring down the coding bit-rate for the prediction coefficients from 2,000 bits/s (40 bits per 20 ms frames with 10 k_i coefficients) to 1,000 bits/s, or even less. This gain is advantageous for the design of new coders with lower and lower bit-rates.

2.2.2.3. Long-term predictor

The analysis filter suppresses the short-term correlation in the speech signal. However, this filter is not able to take into account the periodicity (or pseudo-periodicity) of the voiced sounds. This periodicity is still present in the residual signal, the resulting correlation is not properly exploited and the search in the excitation codebook is much less efficient.

It is therefore logical to remove all predictable information on the basis of the past excitation. In a CELP coder, this can be achieved by an initial search of the optimal excitation on the basis of the coded excitation of the previous frame. The search procedure is identical to the one previously described, with the goal to determine an optimal time-shift T and the corresponding gain factor β. T generally corresponds to the fundamental period of the speech signal, but this is not systematic as the periodicity may be less visible on the residual than it is on the original speech. Furthermore, T may correspond to two fundamental periods, if the fundamental period is short (for a female voice, for example). As the search for the excitation is a very expensive operation, the fundamental period can be estimated by simple means (correlation, for instance) and then the search can take place around the fundamental period only, so as to avoid the testing of all possible time-shifts on the previous excitation.

The determination of T and β makes it possible to reconstruct, after transmission, a rather good approximation of the residual signal. The fixed entry from the codebook comes into play in addition, so as to refine the coded excitation. We recall here that the search for this entry is done from the signal to which the contribution of the past excitation (time-shifted by T and multiplied by β) has been previously removed. This approach is equivalent to a removal of the signal's periodicity with a filter of the form $1 - \beta z^{-1}$. It is usually referred to as *long-term predictor* (LTP) [ATA 70]. The LTP scheme has been proven to be very efficient and it is used today in almost all CELP coders.

It must be stressed that, during the search from the previous excitation, the time-shift can only be an integer number of samples. However, the actual fundamental period may not necessarily correspond to an integer number of samples. Experiments have shown that, when using only integer shifts, the reproduction of periodicity is far from being perfect. That is why several solutions have been proposed, one of which consists of the handling of fractional delays [KLE 94, KRO 91]. To reach this aim, a first step consists of interpolating the past excitation in order to find an optimal excitation with a fractional time-shift. This, of course, increases considerably the coder complexity. But the corresponding quality seems to justify the approach. For example, in the G.729 CELP coder, fractional shifts of 1/3 sample are implemented [SAL 98].

The set of past excitations with various time-shifts can be grouped to form a codebook which resembles the fixed codebook, with the difference that this codebook varies from one frame to the next. This is why such a codebook is called an adaptive codebook [KLE 88]. The best shift corresponds to an index that is transmitted to the decoder.

2.2.2.4. *Perceptual weighting filter*

The perceptual filter is one of the most essential components in a CELP coder. Its function is to modify the spectral shape of the coding distortion, so that it becomes masked by the signal. In fact, the frequency masking effect in the human perception system suggests that the distortion spectrum can be relatively high in the areas where the signal is dominant (the formant areas, for instance) and has to be relatively small in the areas where the signal has little energy [ATA 79]. Obviously, such a shaping must follow the spectral variations of the signal.

In order to implement this principle, a perceptual filter of the following type can be used:

$$W(z) = \frac{A(z)}{A(z/\gamma)}$$

[2.11]

Parameter γ lies between 0 and 1. Its effect in equation [2.11] is to bring the roots of $A(z/\gamma)$ closer to the center of the unit circle, all the more as the value of γ is small.

It can be observed that transfer function $W(e^{j\theta})$ shows more or less deep minima in the signal's formant areas. The weighing role played by this filter during the minimization step, increases the coding distortions in these areas and decreases them elsewhere. Experiments have shown that a value of $\gamma = 0.85$ is a good compromise. It should be mentioned that the introduction of $W(z)$ also results in increasing the overall coding distortion, in terms of energy.

The efficiency of the perceptual filter can be further improved by using a more elaborate filter, such as:

$$W(z) = \frac{A(z/\gamma_1)}{A(z/\gamma_2)} \qquad [2.12]$$

The two parameters γ_1 and γ_2 offer a more flexible shaping of the coding distortion. However, fixed parameters cannot render all types of signal variations. Thus, in the most recent coders (such as the CELP G.729 [SAL 98]), these two parameters are adjusted according to the signal's spectrum, according to a heuristic algorithm. These techniques indeed offer a better performance, at the expense of a higher complexity of the coder. Fortunately, these parameters do not need to be known by the decoder.

2.2.3. Other coders for telephone speech

In addition to the CELP coders described in detail in this chapter, there are other coding schemes which provide very good performances for telephone bandwidth speech. Among these coders, we can mention the multi-pulse coders and the *regular pulse excited* (RPE) coders, which will not be described in this chapter, as they are very similar to CELP coders. However, it is worthwhile to present here a totally different coding structure: *multi-band excitation* (or MBE).

This coding scheme, initially introduced by Griffin and Lim [GRI 88] is essentially based on the harmonic structure of the speech spectrum. Indeed, voiced sound spectra show rather regular harmonic patterns (at least, in the telephone bandwidth), at integer multiples of the fundamental frequency. As a consequence, it is sufficient to transmit the latter, together with the amplitudes (modulus and phases) of the harmonics, in order to reproduce rather accurately the speech signal. In the decoder, the reconstruction of the speech signal is generally carried out in the time domain, by summing sine curves with various amplitudes and frequencies.

Note, however, that, even for voiced sounds, spectra contain frequency areas where the harmonic structure is not well pronounced. Moreover, these areas contribute in a major way to the naturalness of the speech signal. If they are not accurately reproduced, the reconstructed speech sounds very artificial, with a "metallic" timbre. To solve this problem, the MBE coding scheme considers several "sub-bands", which may contain one or several harmonics. If the harmonic structure is well pronounced in a sub-band, it is labeled as "voiced" and the reconstruction of this sub-band is obtained by summing sine curves. If the signal sub-band does not show a clear harmonic structure, or looks like a noise, it is considered to be "unvoiced" and the reconstruction is obtained by a noise generator. Thanks to this distinction, from which stems the name "*multi-band excitation*", a very high quality speech signal can be reproduced.

Even though the principles of an MBE coder can seem rather simple, its implementation requires relatively complex procedures. First of all, it is necessary to accurately estimate the fundamental frequency; otherwise, a small inaccuracy can induce significant errors for high frequency harmonics. Secondly, the MBE scheme requires the determination of the amplitude for each harmonic as well as the voiced or unvoiced nature of the sub-band. An analysis-by-synthesis procedure may seem to be a solution for determining all these parameters, like in the CELP coders, but complexity issues then arise. Note in fact that the number of parameters for a given frame varies according to the fundamental frequency, which complicates the allocation process for the binary resources. Finally, as some correlations generally exist between the harmonics, it is worthwhile exploiting them for an efficient coding.

A number of solutions have been developed in order to solve these technical problems. Today, several MBE coders exist with comparable performance to those of CELP coders [BRA 95, GAR 90]. Let us quote for example the standard developed by INMARSAT (*International Maritime Satellite*) which uses an MBE coder at 4.15 kbit/s for satellite communications used for navigation [INM 91]. Generally speaking, the MBE coders seem particularly well suited to noisy signals, as they are based on a relatively flexible model of the speech spectrum.

For telephone bandwidth speech coding, the vocoders offer as of today the best performance level, when the coding rate is below 3 kbit/s. For such bit-rates, the objective is to maintain intelligibility and naturalness, whereas transparency is not a priority. The most common vocoders in that context are based on linear prediction [ATA 71]. However, the bit-rate is so low that all details within the residual signal are lost: in general, a voiced sound residual signal is modeled by a periodic pulse train at the speech fundamental period. For an unvoiced sound, the residual signal is simply replaced by a white noise with the same power. Such a rough modeling of the residual cannot yield a high speech coding quality. Indeed, while a good intelligibility is generally preserved, the coded speech often sounds metallic, with occasional buzzing, as if speaking into a saucepan.

However, for low bit-rate speech coding, LPC-based vocoders remain the best candidates. A US Government standard is based on this principle: the famous LPC-10 at 2,400 bits/s [TRE 82]. For the last 15 years, several studies have been conducted to improve the quality of vocoders and significant progress has been achieved. The main research track consists of using a mixture of pulses and white noise as an excitation signal [MAK 78]. According to recent work, a vocoder operating at 2,400 bits/s can be achieved with a speech quality comparable to that of a CELP coder at 4,800 bits/s [MCC 95].

For years to come, the design and implementation of low bit-rate coders offering a quality close to telephone speech will remain a challenge for engineers and researchers. It would certainly be worthwhile to take advantage of all the existing coding techniques, CELP, MBE, vocoders and others, and to optimize each aspect. Other new techniques with better performance are also expected in the near future.

2.3. Wideband speech coding

Even though the telephone bandwidth offers good conditions for voice communication, it is widely agreed that telephone speech does not sound natural. The bandwidth limitation alters the voice timbre and may cause intelligibility difficulties in the message itself (for example, it is not easy to distinguish [f] and [s] on the phone).

Today, with the incredible development of telecommunications, many applications such as teleconferencing require a level of speech quality far beyond telephone quality. In this context, the use of a wider bandwidth is a must, and it is sometimes referred to as extended bandwidth (50-7,000 Hz) and FM bandwidth (20-15,000 Hz). Bandwidth widening enables a considerable improvement in speech naturalness and a good rendering of the surrounding noise, which increases the communication quality and can create the feeling that the remote speaker is nearby.

In order to transmit wideband speech without requiring too much bit-rate, the use of coding techniques is essential. Even though the basic principles of the techniques used for telephone speech remain globally valid for wideband speech, the development of speech coders for such bandwidths shows a number of specificities.

First of all, the high frequency bands of speech cannot be viewed as a simple "copy" of telephone speech. The energy of speech being essentially concentrated in the telephone band, the higher band of the signal is less energetic. The spectral dynamic is therefore stronger for wideband speech. Moreover, the harmonic structure of voiced sounds, which is usually well marked on the telephone bandwidth, tends to become less obvious in higher bands. In parallel, even though the information contained in the higher bands contribute to the overall improvement of speech quality, it does not play the same role as the information contained in the telephone band: the latter corresponds mostly to the linguistic content of speech (as a result of the formant structure), while the former are mainly linked to the voice characteristics.

As a general rule, it is necessary to double the transmission rate when the bandwidth is doubled, in order to maintain an equivalent quality level. This may *a priori* seem paradoxical, since the high band seems to contain less information and should therefore require less coding bits. However, it must be kept in mind that, when listening to wideband speech, the overall quality is much better and the ear is much more demanding, i.e. the smallest defect is not tolerated very well. This is why a higher coding rate is required. Moreover, let us add that a wideband coder is rarely used solely for voice transmission: it has to accommodate multiple-speaker speech as well as high quality music. The rendering of surrounding noise is also essential for these coders. For all these reasons, it is not obvious to design a satisfying wideband coder, even with a double bit-rate.

Finally, another factor must be taken into account when designing a wideband coder: the masking phenomenon. Given that the bandwidth is wider and the spectral dynamic stronger, this perception phenomenon manifests itself in a more obvious way. If it is well exploited, it can be used to further reduce the coding noise.

Today, only one reference wideband coder exists: the CCITT G.722 standard, established in 1986 [CCI 88b]. This is an ADPCM-based coder, combined with a sub-band coder (SB-ADPCM). It codes almost transparently extended bandwidth speech at 64 kbit/s. It can also code music with a good coding quality. Moreover, this coder can run at 56 or even 48 kbit/s with a slightly lower quality. The key principle of this coder is to divide the full bandwidth into two sub-bands of equal widths, and to code the corresponding signals with the ADPCM technique (originally developed for telephone speech). Note that, because of the energy difference in the two sub-bands, the binary resources are not allocated evenly between them. In the 64 kbit/s version of the coder, 6 bits per sample are allocated to the lower band, but only 2 bits per sample for the higher band [MAI 88, MER 88].

In the past decade, many new coding schemes have been developed for this band. Today, the quality of the 64 kbit/s SB-ADPCM coder can be achieved with

32 kbit/s coders [DIA 93, ORD 91, QUA 91], or even at 24 kbit/s[13]. Moreover, the performance of recent coders at 16 kbit/s is approaching that of SB-ADPCM at 56 kbit/s [FUL 92]. Among the corresponding techniques, the transform coding scheme is one of the most promising. It is presented in more detail in the next section.

2.3.1. *Transform coding*

Transform coding consists of transforming the time-domain signal into another representation space in which redundancy, due to correlations in the signal, is easier and more straightforward to remove. Let us first consider a simple example, with a signal such as a sine curve. The direct quantization of this signal is obviously inefficient because of the strong correlation between the signal's samples. If a frequency-domain representation of the signal is used, a strong concentration of energy is located around the frequency of the sine curve. Therefore, the binary resources can be focused on the quantization of this high amplitude spectral region, thus yielding a good coding accuracy.

We will now explain how a transform-based coder functions. Let us denote as x the vector containing a block of signal samples (typically, a frame of a few tens of milliseconds), and T a transform matrix. First, a vector y is calculated, which contains the transformed values[14]:

$$y = Tx \qquad\qquad [2.13]$$

Then y is quantized and the result is transmitted. In the receiver, the signal is regenerated by applying the inverse transform T^{-1} to the quantized version of y.

For the quantization of y to be efficient, two important properties must be fulfilled:

1) y must be as uncorrelated as possible;

2) y must show an energy concentration, i.e. only a few coefficients in y show high amplitude values, whereas the others remain low.

Here, the first property implies that y contains little redundancy. It can be shown that the second property enables the design of a bit allocation scheme that minimizes

---

13 Recently, the ITU has issued a new recommendation, the G.722.1 standard, for extended band telephone applications (in particular, for hand-free systems). This standard operates at 32 and 24 kbit/s and is based on transform coding.

14 In general, y is the same size as x, so as neither to lose information nor to create additional redundancy. Thus, T is generally a square matrix.

the average power of the coding noise after transformation. Note that the difference in the coding noise power with and without transform constitutes the transform coding gain.

The matrix **T** that satisfies the two constraints stated above is the ortho-normal matrix that diagonalizes the covariance matrix of vector **x**. This is the Karhunen-Loëve transform (KLT). In this case, the coding gain tends towards its maximal value $1/\xi_X$, when the transform size becomes larger. It is worth noting that, in a similar way to predictive coding, this gain is uniquely governed by the redundancy caused by the signal correlation.

However, the use of the KLT matrix is not convenient, as it depends on the signal: it even needs to be variable if the signal is non-stationary! Fortunately, other fixed transforms exist which are close enough to the optimal transform. This is the case for the discrete Fourier transform (DFT) and the discrete cosine transform (DCT). The latter is particularly useful, as it yields a real-valued vector **y** and it reduces side-effects due to block-wise processing. Another advantage of both transforms is that vector **y** has a very strong physical meaning: it can be interpreted as a representation of the signal in the frequency domain. This simplifies the exploiting of masking effects, as this property of the human ear is modeled in the frequency domain.

A speech coder based on transform coding is illustrated in Figure 2.3. The signal first undergoes a transformation. Then the inaudible frequency components are removed, taking the masking effect into account. In a next step, the transformed coefficients are quantized at the same time as the shaping of the coding noise. We present in the next sections the most essential components of a transform-based coder[15].

---

15 Based on this principle, a high quality music coder at 64 kbit/s has been developed [MAH 92] as well as the ASPEC (Adaptive Spectral Perceptual Entropy Coding) coder [BRA 91], which is now integrated in an ISP/MPEG standard for the coding of audio-frequency signals.

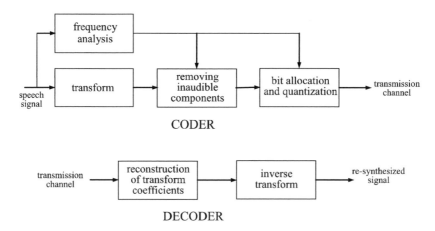

Figure 2.3. *Block diagram of a transform coder*

2.3.1.1. *Transform*

As explained above, the use of a transform such as the DCT is perfectly adapted. However, this type of transform has two drawbacks:

1) it has a poor spectral resolution because of rectangular windowing;

2) it may generate discontinuities at the frame edges when vector **y** is quantized, as it may vary from one frame to the next.

In order to avoid these problems, an appropriate weighting window can be used, with some overlap between neighboring frames: the transform size is then larger than the frame length, which will cause some increase in the bit-rate.

Another technique is particularly efficient to solve this problem: the time domain aliasing cancellation or TDAC [PRI 86]. Under this approach, a particular transform is used, which is twice as long as the speech frame. Nevertheless, only half of the transformed vector **y** is transmitted, i.e. a sub-sampled version of **y**. This causes some distortions in the reconstructed signal, if **y** is considered on its own. However, the strength of the technique lies in the fact that these distortions are completely compensated with the next (transformed) frame.

In a transform coder, the size of the transform is a fundamental parameter. Too small a size does not yield a good spectral resolution, which decreases the coding gain and makes it difficult to exploit frequency masking properties. On the other hand, a large transformation creates a significant algorithmic delay, which must also be avoided. Moreover, the transform size is intrinsically limited by the signal non-stationarity. Finally, for extended bandwidth coding, a good compromise is to use a window length of 512 samples (32 ms).

2.3.1.2. *Frequency masking*

This chapter has already mentioned the masking effect several times. Here is a more accurate description of this phenomenon: a loud sound with a given frequency F can "mask", i.e. make inaudible, weaker sounds in the neighborhood of F, if their intensity is below a given threshold which depends on the intensity of the masking sound. Psychoacoustic studies have revealed that the threshold (also called *signal-to-mask ratio*) is rather constant whatever the value of F and the extension of the frequency area masked around F varies proportionally with F. Moreover, this area is wider for frequencies above F and narrower below F.

In the spectrum of a speech signal, there is some contrast in relation to formant areas and harmonic locations. This suggests that some of the frequency components can be removed if they are made inaudible by the masking effect. However, speech spectrum (or, more generally, music spectrum) is rather complex and very different from a single spectral line. This is why the global masking effect of a complex sound should be considered as the addition of the masking effects of all its individual components. In this way, a *masking curve* can be calculated, below which the frequency components can be assumed to be inaudible.

Experiments have shown that, for speech signals, up to 55% of the frequency components can be masked without causing a noticeable difference [DIA 93]. If these components are not transmitted, a significant proportion of the binary resources can be saved.

2.3.1.3. *Quantization and noise shaping*

The bit allocation process and the various quantization steps form the most complex part of a coder. They will not be described here in detail. However, the general principles are easy to understand: the available bits should be shared out between the transform coefficients so that they can be quantized in such a way that the quantization noise remains comparable across the coefficients. This means that the coding noise will be shaped with a relatively flat spectral distribution (white noise). This is the optimal solution in theory, which guarantees the minimization of the average noise power. However, as soon as the perceptual aspects of the noise are taken into account, the flat spectrum criterion is no longer optimal. Indeed, whereas a "white" coding noise has a minimal power, its shape is not ideal for being masked by the signal itself [JOH 88].

In order to shape the coding noise by accounting for the masking effect, a weighting filter can be used. However, in transform coding, the fact that quantization takes place in the frequency domain offers new possibilities. It is indeed possible to directly modify the spectral behavior of the noise by adjusting the number of bits allocated to the different transform coefficients. This is implemented in such a way that the noise remains completely below the masking curve: the noise would be totally masked by the signal!

It should however be underlined that the calculation of the masking curve is not exact, because of the model's simplifying hypotheses. In fact, the masking curve is only indicative: a noise can remain audible even if its level is below the masking curve. Moreover, for a noise to be completely masked, it is necessary to allocate a sufficient number of bits, otherwise the noise level will locally remain above the masking curve. However, experiments have shown that, even in unfavorable situations, it is beneficial to shape the coding noise according to the masking curve.

2.3.2. Predictive transform coding

Transform coding has been presented in the previous section as one of the most promising techniques for wideband speech coding. However, it is possible to combine transform coding and predictive coding in a single coder. Such a scheme was proposed in the 1990s by several authors [CHE 96, LEF 94] for extended bandwidth signals[16]. The core of this approach is transform coding. However, the transform is not directly applied to the signal but to a filtered version of it, using a perceptual filter $W(z)$ as seen above. The estimation of $W(z)$ is based on linear prediction, hence the "predictive" component of the coder. Under this coding approach, the noise generated by the "transform + quantization" steps shows an approximately flat spectrum but the spectrum of the noise after the complete coding process follows the shape of the transfer function $1/W(z)$. Thus, the perceptual filter plays the role of noise shaping, which facilitates the use of vector quantization on the transform coefficient. It is still possible to use the masking curve for guiding the quantization for some of the parameters (for instance, the phase of the transform coefficients).

In forthcoming years, wideband speech coding is going to become more and more widespread in various applications. New techniques can be expected to emerge and lead to ever more efficient coders.

---

16 A hierarchical coder was proposed for FM band audio signals [MOR 00]. It is essentially based on predictive transform coding.

2.4. Audiovisual speech coding

Speech is not only a sound: it is also a visual process. Seeing one's interlocutor is essential in many situations in everyday life. On the one hand, visual information brings additional intelligibility, especially in a noisy environment: looking at a speaker helps to understand better the content of the spoken message. On the other hand, visual information brings added value to the communication (comfort, pleasure in looking at the other person, better transmission of emotions, gesture information, etc.). The introduction of the speaker's image in telecommunications and man-machine communication systems is a major stake, well illustrated by the development of videophones, videoconferences and multimodal voice interfaces.

In spoken language, sound and face movements are two distinct manifestations of the same source. A real *coherence* therefore exists between audio and visual speech information[17]. For several years, studies have been conducted to migrate the knowledge gathered in the domain of audiovisual speech perception and modeling towards applied techniques in signal processing so as to improve their performance. For instance, a system of speech denoising using filters estimated essentially from geometrical measurements taken from the lip movements has provided promising results [GIR 01]. Moreover, sustained efforts are being invested in the now conventional field of audiovisual speech recognition (see Chapter 11).

Today, digital communications impose a breathless race towards an increasingly efficient compression of signals: in this context, the exploitation of the coherence and the complementary sides of audiovisual speech appears as a relevant challenge, with the goal of reducing the transmission or the resource storage. This section presents the domain of audiovisual coding: we will first illustrate what could be a transmission channel using audiovisual speech compression; then we will show some basic results obtained in joint audio and video coding (more detailed and extended results can be found in [GIR 04]); we will finally conclude on a few perspectives in this domain.

2.4.1. *A transmission channel for audiovisual speech*

To start with, the compression of audio and video speech signal is not a symmetric problem, in the sense that the image signal is generally more resource consuming than the audio signal. For instance, conventional videophone algorithms

---

17 It can even be stated that both information sources are complementary: studies show in particular that the degraded audio information in a noisy environment corresponds to the most visible information on the lips (see, for instance [ROB 98]).

operate at around 100 kbit/s for image coding whereas the audio information can be efficiently coded at rates of approximately 4 kbit/s. In these conditions, the benefit of a joint audio + video coding scheme seems useless, if the latter has a negligible cost compared to the former. A key to bridging this gap lies in the specificities of a talking face: most of the image content varies slowly (the background and most of the face) while only a few informative zones vary rapidly (the mouth and, to a lesser extent, the eyes and eyebrows). An approach therefore consists of isolating these zones, in particular the mouth and the surrounding area, as this source of information is the most correlated with the audio information.

In the present study, we have isolated and extracted the mouth information by measuring a few parameters of the lip contour. These parameters, which are detailed below, can be extracted by various automatic systems which operate on lips with colored makeup, developed at ICP (Institut de la Communication Parlée), a portable real-time version of which is developed by the company Ganymédia, based on a headset camera. This approach can be considered as a first step towards a complete parametric approach for the coding and the synthesis of talking faces, which is an expanding research area (see Chapter 11). For example, the MPEG4 standard comprises a specific part for the parametric description of talking faces: i.e. parameters that describe the geometry and texture of the various parts of the face are extracted from the speaker's picture; once transmitted, they can be used to reconstruct a synthetic face from a generic model.

In this context, the parametric coding of faces is costless and the design of an analysis-by-synthesis audiovisual transmission channel can be envisaged, with a specific intermediate stage that implements a wide range of interactions and optimal formatting between audio and visual data (see Figure 2.4).

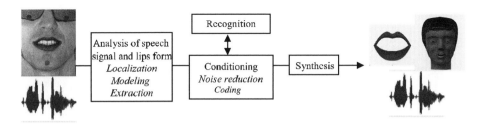

Figure 2.4. *A basic architecture for audiovisual speech processing*

We are now going to illustrate more accurately the principles and the interest residing in audiovisual speech coding, by presenting a study which proposes a strategy for jointly coding audio and video parameters.

2.4.2. *Joint coding of audio and video parameters*

2.4.2.1. *An audiovisual vocoder*

For this preliminary study, we place ourselves in a framework focused on the exploitation of additional intelligibility from visual speech (as opposed to a pure quality goal): the low bit-rate LPC vocoder (at approximately 2 kbit/s). The audiovisual coder derived from this reference keeps the conventional principle of a frame-based analysis-by-synthesis scheme (see Figure 2.5). First of all, an LPC analysis provides the gain coefficients of filter $G/A(z)$, which model the spectrum envelop of the signal. The prediction error, obtained by inverse filtering of the signal through $A(z)$ is modeled by a pulse train for voiced sounds and by a white noise for unvoiced sounds. The algorithms used for this purpose, as well as those used for the quantization of the fundamental frequency F_0 and the gain factor G (transmitted separately) are traditional and they will not be detailed here. On the other hand, the coefficients characterizing the filter $1/A(z)$ are of primary interest as they are coded jointly with the video parameters using an audiovisual VQ (vector quantization) algorithm, which is described in the next section. The video parameters, denoted as A, B, A', B', refer respectively to the internal and external widths and heights of the lip contour. Statistical studies have shown that they contain the largest part of visual intelligibility [ADJ 96, BEN 92]. On the decoder level, the audio and video parameters are obtained by an inverse VQ. The audio signal is reconstructed by filtering the excitation model through the filter derived from the LPC parameters. The video parameters can be used efficiently for animating a 3D lip model, i.e. keeping a large part of visual intelligibility [LEG 96].

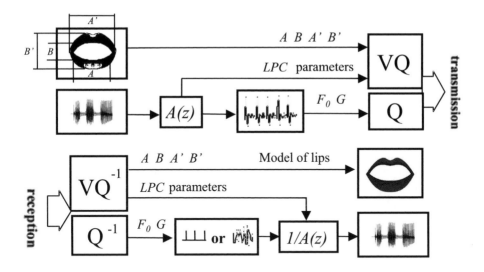

Figure 2.5. *An audiovisual coder with a joint audio + video VQ*

2.4.2.2. *Audiovisual vector quantization*

Vector quantization (VQ) is a very efficient method for coding multidimensional data. The principle is to determine a codebook of prototype vectors which are substituted to the actual data during the coding step: the index of the prototype is stored or transmitted instead of the original data. The prototypes must reflect optimally the distribution of the data according to a given criterion. VQ turns out to be very efficient for reducing redundancy between vector components. This is why we have chosen this method in our study to exploit the natural cross-consistency between audio and video parameters.

The determination of the codebook requires a training data set, a vector metric and an optimization procedure. The latter will not be detailed here. We will simply mention that we use the Linde-Buzo-Gray algorithm, together with an initialization phase based on the *splitting* method [GER 92a]. We will focus instead mainly on the definition of an audiovisual metric, accounting for the difference in nature between the signals from both modalities. We propose the following definition:

$$d_{AV} = \alpha\, d_V + (1-\alpha)\, d_A \qquad\qquad [2.14]$$

d_A is a spectral distance based on the LPC model, namely the Euclidean distance between LSP (*line spectrum pairs*) coefficients, which characterize the $A(z)$

polynomial [SUG 86]. d_V is the Euclidean distance between the lip parameters, i.e., if denoting two sets of lip measurements by A_1, B_1, A'_1, B'_1 and A_2, B_2, A'_2, B'_2:

$$d_V = \sqrt{(A_1 - A_2)^2 + (A'_1 - A'_2)^2 + (B_1 - B_2)^2 + (B'_1 - B'_2)^2}$$

α is a weighing factor which has to be set so as to balance the coding accuracy between audio and video.

Regarding the training data, a set of *phonetically balanced sentences* in French is used to train the codebook of prototypes. It consists of 80 sentences uttered by 8 speakers (4 female and 4 male). The signal is sampled at 8 kHz. Video parameters are extracted at a frame rate of 20 ms, which also corresponds to the window length for the LPC analysis, synchronized and centered on the instant of the video capture. The LPC analysis order is fixed to 10 (LSP coefficients) and, as 4 video coefficients are extracted, the audiovisual vectors are 14-dimensional. The total number of vectors used for training the codebook is approximately equal to 45,000.

2.4.2.3. *Main results*

The performance of the quantization step is measured in terms of transmission rate, average quantization error and subjective quality of the coded speech. The average quantization error (both in the audio and video domains) is defined as the average of the (respectively audio and video) distances over all the training vectors to their corresponding prototypes.

The main result arising from this study is to show that audiovisual VQ turns out to be more efficient than audio and video VQ separately, and this in reference to all measured criteria. To support this point, we first consider the audio and video quantization errors obtained respectively with a reference audio VQ (i.e. applied solely to the audio data, with distance d_A), using 10 bits (i.e. 1,024 prototype LSP vectors) and a reference video VQ (i.e. applied solely to the video data, with distance d_V), using 5 bits (i.e. 32 prototypical lip shapes). The average quantization errors are respectively 0.0494 rad/s for audio VQ and 1.48 mm for video VQ. In parallel, we conduct a VQ experiment on 12 bits, and we vary the weighting factor α between 0 and 1. We observe of course that the audio and video quantization errors vary in the opposite direction from each other, as α governs the coding accuracy of one modality at the expense of the other modality. The important point is that we observe an interval of values taken by α for which both the audio and video quantization average errors achieved with audiovisual VQ are below the reference values: for instance, for $\alpha = 0.1$, values of 0.0490 rad/s for the audio error and 1.35 mm for the video errors are achieved. It is essential to note that this result is obtained with a 12 bit VQ, whereas two separate audio and video VQ would require a total of 10+5 = 15 bits. A gain of 3 bits is obtained, while the quantification error is kept lower for both modalities!

In order to check that these positive results are not only due to the quantization of vectors of larger dimensions, an audiovisual VQ has been carried out after generating artificial sets of audiovisual parameters, by pairing randomly audio and visual parameters from distinct frames, in the training corpus. In this case, the quantization error is much higher, by approximately 20% (for $\alpha = 0.1$). This confirms that the gain achieved by audiovisual VQ is indeed the result of a real correlation between audio and visual parameters, which is efficiently exploited by our method.

Moreover, a first series of informal perceptive tests have confirmed that the objective results presented above also have a subjective impact: no difference can be perceived between signals resulting from an audiovisual VQ and the same signals coded by the two separate reference VQ, whether the comparison is made at the auditory level or the visual level (via a 2D lip model animated by the coded parameters). These first tests have to be confirmed by a series of formal tests.

2.4.2.4. *Multistage audiovisual vector quantization*

An important property of the audiovisual coder structure concerns its complexity, which is essentially governed by the size of the codebook. In fact, the most expensive step in the coding algorithm lies in the search for the closest prototype in the codebook. A classical solution to this problem used in audio coders consists of conducting a multi-stage VQ with several hierarchical codebook layers: a "rough" VQ in a primary codebook guides the coder towards the choice of a second (sub)-codebook in which the result is refined, and so on. The cost reduction relies on the fact that each sub-codebook contains a number of prototypes far smaller than a global codebook would, even though the total number of prototypes is larger.

In our study, the idea is to apply this method assuming that if the audio and video data show a certain degree of cross-consistency, the quantization of a parameter set from one modality must help to limit the prototype search for the other modality. Let us take the example depicted in Figure 2.6: The reference video VQ on 5 bits is used as a first step. During the training phase, the data are assigned to one out of 32 prototypes, on the basis of the video components and for each of the 32 classes, a second stage VQ is carried out on the audio component. The number of bits allocated for this second stage VQ is variable: it fluctuates between 4 and 8 bits, according to the number of data within each video class, with an average of 7.1 bits (corresponding to 156 prototypes) for an average quantization error equal to 0.0541 rad/s. During the coding step, the selection of the audio codebook is governed by a prior selection within the video codebook.

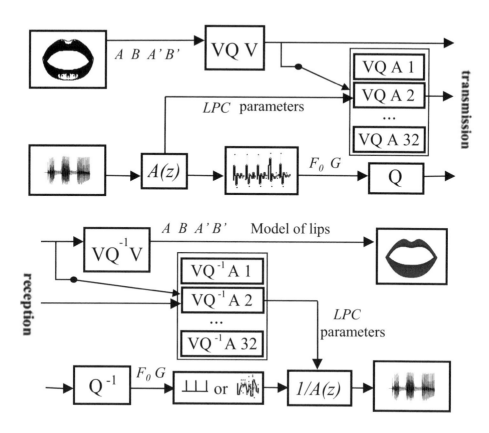

Figure 2.6. *An audiovisual coder with a two stage video-audio VQ (here VQ V means vector quantization for video parameters and VQ A means vector quantization for audio parameters)*

The audio error for this new structure is thus slightly higher than the reference audio VQ with 10 bits (0.0494 rad/s). However, new informal listening tests show that the difference between both coded signals cannot be perceived. A saving of almost 3 bits out of 10 is thus obtained, with a complexity reduction by approximately 6.5 (156 audio prototypes versus 1,024). Compared to the global audiovisual VQ (previous section), audio and video quantization errors are slightly higher, for a quasi-identical number of bits: this is the price to pay for the sub-optimality resulting from the two-stage structure of the coder, with respect to the global VQ. However, the search effort in the codebooks is reduced by a factor of 22 on average (32 video prototypes + 156 audio prototypes versus 4,096 audiovisual prototypes). Moreover, in this particular case, memory saving is also achieved, with a total of 4,064 10-dimensional audio prototypes and 32 4-dimensional video prototypes, as opposed to 4,096 14-dimensional audiovisual prototypes.

2.4.3. *Prospects*

Audiovisual speech coding can be revisited in the light of the constraints and protocols specific to transmission systems and networks for videophones (bit-rate, latency, bandwidth, error probabilities, loss of data). In this context, the techniques presented here can indeed respond to a number of objectives, among which are:

– bit-rate decrease: for instance, by a joint coding of two information sources (audio and video) with the same algorithm, as illustrated by the audiovisual VQ;

– lower complexity: this is illustrated by the two-stage video-audio VQ;

– improved robustness: beyond the possibility of denoising audio speech using lip information, as in the previously mentioned work by [GIR 01], a more general approach consists of exploiting the redundancy between audio and video channels, in order to remove noise, correct errors, and compensate losses and latencies on each channel, owed to the network limitations. The estimation of a lost, degraded or incomplete modality can be envisaged, using the remaining modality. With the view of reducing the bit-rate, this approach can translate into the estimation of expensive pieces of information on the basis of less expensive ones;

– true scalability: adapting the information resolution and quality to the system's capacity and the expectations of the users. This idea is related to the hierarchical coding structure, with several layers of hierarchy according to the quality level. A videophone could become capable of commuting with respect to the available transmission rate and computing resources, between a traditional mode using a real coding of the speaker's image to an intermediate mode using a synthetic face animated by a complete set of face parameters and, furthermore, to a low-resource mode using only a lip-model animated by the lip parameters only. Ultimately, in the case when no visual information is available, the commutation could include a "prediction" mode, for which the lip movements are predicted approximately, from the audio information.

2.5. References

[ADJ 96] ADJOUDANI A., "Reconnaissance automatique de la parole audiovisuelle: stratégie d'intégration et réalisation du LIPTRACK, labiomètre temps-réel", *Thesis INP*, Grenoble, France, 1996.

[ADO 87] ADOUL J.P., MABILLEAU P., DELPRAT M., MORISSETTE S., "Fast CELP coding based on algebraic codes", *Proc. IEEE ICASSP*, p. 1957-1960, 1987.

[ATA 70] ATAL B.S., SCHROEDER M.R., "Adaptive predictive coding of speech signals", *Bell Syst. Tech. J.*, vol. 49, p. 1973-1986, 1970.

[ATA 71] ATAL B.S., HANAUER S.L., "Speech analysis and synthesis by linear prediction of the speech wave", *J. Acoust. Soc. Amer.*, vol. 50, p. 637-655, 1971.

[ATA 79] ATAL B.S., SCHROEDER M.R., "Predictive coding of speech signals and subjective error criteria", *IEEE Trans. on ASSP*, vol. 27, p. 247-254, 1979.

[ATA 82] ATAL B.S., REMDE J.R., "A new model of LPC excitation for producing natural-sounding speech at low bit rates", *Proc. IEEE ICASSP*, p. 614-617, 1982.

[ATA 89] ATAL B.S., COX R.V., KROON P., "Spectral quantization and interpolation for CELP coders", *Proc. IEEE ICASSP*, Glasgow, p. 69-72, 1989.

[BEN 92] BENOÎT C., LALLOUACHE T., MOHAMADI T., ABRY C., "A set of visual French visemes for visual speech synthesis", in Bailly G., Benoît C. and Sawallis T.R. (ed.), *Talking Machines: Theories, Models and Designs*, Elsevier Sc. Publishers, Amsterdam, p. 485-504, 1992.

[BEN 86] BENVENUTO N., BERTOCCI G., DAUMER W.R., SPARRELL D.K., "The 32 kbit/s ADPCM coding standard", *AT&TTech. J.*, vol. 65, p. 12-22, 1986.

[BOU 95] BOURAOUI M., GLASS W., FENG G., "Fast codebook search algorithm based on Hamming ECC for algebraic CELP speech coding", *Proc. Eurospeech'95*, p. 719-722, 1995.

[BRA 91] BRANDENBURG K., HERRE J., JOHNSTON J.D., MAHIEUX Y., SCHROEDER E.F., "ASPEC: Adaptive spectral perceptual entropy coding of high quality music signals", *90th AES-Convention*, Paris, preprint 3011, 1991.

[BRA 95] BRANDSTEIN M.S., MONTA P.A., HARDWICK J.C., LIM J.S., "A real-time implementation of the improved MBE speech coder", *Proc. IEEE ICASSP*, p. 5-8, 1995.

[CAM 91] CAMPBELL J.P., TREMAIN T.E., WELCH JR. V.C., "The Dod 4.8 kbps standard (proposed Federal Standard 1016)", in Atal B.S., Cupperman V. and Gersho A. (ed.), *Advances in Speech Coding*, Kluwer Academic Publishers, 1991.

[CCI 84a] CCITT, Recommendation G.711, "Pulse code modulation (PCM) of voice frequencies", *Red Book*, vol. III.3, p. 85-93, Malaga-Torremolinos, 1984.

[CCI 84b] CCITT, Recommendation G.721, "32 kbit/s adaptive differential pulse code modulation (ADPCM)", *Red Book*, vol. III.3, p. 125-159, Malaga-Torremolinos, 1984.

[CCI 88a] CCITT, Recommendation G.726, 40-, 32-, 24-, and 16-kb/s adaptive differential pulse code modulation, 1988.

[CCI 88b] CCITT, Recommendation G.722, "7 kHz audio coding within 64 kbit/s", *Blue Book*, vol. III.4, p. 269-341, Melbourne, 1988.

[CCI 92] CCITT, Recommendation G.728, Coding of speech at 16 kbit/s using low delay codeexcited linear prediction (LD-CELP), September 1992.

[CHE 92] CHEN J.-H., COX R.V., LIN Y.-C., JAYANT N., MELCHNER M.J., "A low-delay CELP coder for the CCITT 16 kb/s speech coding standard", *IEEE J. on Select. Areas in Comm.*, vol. 10, p. 830-849, 1992.

[CHE 95] CHEN J.-H., GERSHO A., "Adaptive postfiltering for quality enhancement of coded speech", *IEEE Trans. on Speech and Audio Proc.*, vol. 3, p. 59-71, 1995.

[CHE 96] CHEN J.-H., WANG D., "Transform predictive coding of wideband speech signals", *Proc. IEEE ICASSP*, p. 275-278, 1996.

[COX 97] COX R.V., "Three new speech codecs from the ITU cover a range of applications", *IEEE Comm. Mag.*, vol.35, p. 40-47, September 1997.

[DAV 86] DAVIDSON G., GERSHO A., "Complexity reduction methods for vector excitation coding", *Proc. IEEE ICASSP*, p. 3055-3058, 1986.

[DIA 93] DIA H., FENG G., MAHIEUX Y., "A 32 kbit/s wideband speech coder based on transfrom coding", *Proc. Eurospeech'93*, Berlin, p. 1111-1114, 1993.

[DVO 88] DVORAK C.A., ROSENBERGER J.R., "Deriving a subjective testing methodologie for digital circuit multiplication and packetized voice systems", *IEEE J. on Select. Areas in Comm.*, vol. 6, p. 235-241, 1988.

[FUL 92] FULDSETH A., HARBORG E., JOHANSEB F.T., KNUDSEN J.E., "Wideband speech coding at 16 kbit/s for a videophone application", *Speech Comm.*, vol. 11, p. 139-148, 1992.

[GAR 90] GARCIA-MATEO C., CASAJUS-QUIROS F.J., HERNANDEZ-GOMEZ L.A., "Multi-band excitation coding of speech at 4.8 kbps", *Proc. IEEE ICASSP*, p. 13-16, 1990.

[GER 92a] GERSHO G., GRAY R.M., *Vector Quantization and Signal Compression*, Kluwer Academic Publishers, Boston, 1992.

[GER 92b] GERSON I.A., JASIUK M.A., "Techniques for improving the performance of CELP-type speech coders", *IEEE J. on Select. Areas in Comm.*, vol. 10, p. 858-865, 1992.

[GIR 01] GIRIN L.,SCHWARTZ J.-L., FENG G., "Audio-visual enhancement of speech in noise", *J. Acoust. Soc. Am.*, vol. 109, p. 3007-3020, 2001.

[GIR 04] GIRIN L., "Joint matrix quantization of face parameters and LPC coefficients for low bit rate audiovisual speech coding", *IEEE Trans. on Speech and Audio Proc.*, vol. 12, no.3, p. 265-276, 2004.

[GRI 88] GRIFFIN D.W., LIM J.S., "Multiband excitation vocoder", *IEEE Trans. on ASSP*, vol. 36, p. 1223-1235, 1988.

[HEL 89] HELLWIG K., VARY P., MASSALOUX D., PETIT J.P., GALAND C., ROSSO M., "Speech codec for the European mobile radio system", *GLOBECOM Conference*, p. 1065-1069, 1989.

[INM 91] DVSI Inc., Inmarsat-M voice coding system description, Draft 1.3, Inmarsat, 1991.

[ITA 75] ITAKURA F., "Line spectrum representation of linear predictive coefficients of speech signals", *J. Acoust. Soc. Amer.*, vol. 57, p. S35, 1975.

[ITU 96a] ITU, Recommandation P.800, Méthodes d'évaluation subjective de la qualité de transmission, Geneva, 1996.

[ITU 96b] ITU, Recommandation P.830, Evaluation subjective de la qualité des codecs numériques à bande large, Geneva, 1996.

[ITU 96c] ITU-T, Recommendation G.729, Coding of speech at 8 kbit/s using conjugate-structure algebraic code-excited linear prediction (CS-ACELP), 1996.

[JAY 84] JAYANT N.S., NOLL P., *Digital Coding of Waveforms. Principles and Applications to Speech and Video*, Prentice Hall, New Jersey, 1984.

[JAY 92] JAYANT N., "Signal compression: technology targets and research directions", *IEEE J. on Select. Areas in Comm.*, vol. 10, p. 796-818, 1992.

[JOH 88] JOHNSTON J.D., "Transform coding of audio signals using perceptual noise criteria", *IEEE J. on Select. Areas in Comm.*, vol. 6, p. 314-323, 1988.

[KLE 88] KLEIJN W.B., KRASINSKI D.J., KETCHUM R.H., "An efficient stochastically excited linear predictive coding algorithm for high quality low bit rate transmission of speech", *Speech Comm.*, vol. 7, p. 305-316, 1988.

[KLE 90] KLEIJN W.B., KRASINSKI D.J., KETCHUM R.H., "Fast methods for the CELP speech coding algorithm", *IEEE Trans. on ASSP*, vol. 38, p. 1330-1342, 1990.

[KLE 94] KLEIJN W.B., RAMACHANDRAN R.P., KROON P., "Interpolation of the pitch-predictor parameters in analysis-by-synthesis speech coders", *IEEE Trans. on Speech and Audio Proc.*, vol. 2, p. 42-54, 1994.

[KRO 91] KROON P., ATAL B.S., "On the use of pitch predictors with high temporal resolution", *IEEE Trans. on Signal Proc.*, vol. 39, p. 733-735, 1991.

[KRO 86] KROON P., DEPRETTERE E.F., SLUYTER R.J., "Regular-pulse excitation: a novel approach to effective and efficient multi-pulse coding of speech", *IEEE Trans. on ASSP*, vol. 34, p. 1053-1063, 1986.

[LAM 88] LAMBLIN C., Quantification vectorielle algébrique sphérique par le réseau de Bardes-Wall. Application à la parole, PhD thesis, University of Sherbrooke, Canada, 1988.

[LEF 94] LEFEBVRE R., SALAMI R., LAFLAMME C., ADOUL J.-P., "High quality coding of wideband audio signals using transform coded excitation (TCX)", *Proc. IEEE ICASSP*, p. 193-196, 1994.

[LEG 96] LE GOFF B., GUIARD-MARIGNY T., BENOÎT C., "Analysis-synthesis and intelligibility of a talking face", in Van Santen J.P.H., Sproat R.W., Olive J.P. and Hirschberg J. (ed.), *Progress in Speech Synthesis*, Springer Verlag, New York, p. 235-244, 1996.

[MAH 92] MAHIEUX Y., "High quality audio transform coding at 64 kbit/s", *Ann. Télécomm.*, vol. 47, p. 95-106, 1992.

[MAI 88] MAITRE X., "7 kHz audio coding within 64 kbit/s", *IEEE J. on Select. Areas in Comm.*, vol. 6, p. 283-298, 1988.

[MAK 78] MAKHOUL J., VISWANATHAN R., SCHWARTZ R., HUGGINS A.W.F., "A mixed-source model for speech compression and synthesis", *J. Acoust. Soc. Amer.*, vol. 64, p. 1577-1581, 1978.

[MAK 85] MAKHOUL J., ROUCOS S., GISH H., "Vector quantization in speech coding", *Proceedings of the IEEE*, vol. 73, p. 1551-1588, 1985.

[MAR 76] MARKEL J.D., GRAY JR. A.H., *Linear Prediction of Speech*, Springer Verlag, Berlin, 1976.

[MCC 95] MCCREE A.V., PARNWELL III T.P., "A mixed excitation LPC vocoder model for low bit rate speech coding", *IEEE Trans. on Speech and Audio Proc.*, vol. 3, p. 242-250, 1995.

[MER 88] MERMELSTEIN P., "G.722, A new CCITT coding standard for digital transmission of wideband audio signals", *IEEE Comm. Mag.*, January, p. 8-15, 1988.

[MOR 95] MOREAU N., *Techniques de Compression des Signaux*, Masson, Paris, 1995.

[MOR 00] MOREAU N., DYMARSKI P., "Low delay coder (< 25ms) of wideband audio (20 Hz – 15 kHz) scalable from 64 to 32 kbit/s", *Ann. Télécom.*, vol. 55, p. 493-506, 2000.

[NOL 93] NOLL P., "Wideband speech and audio coding", *IEEE Comm. Mag.*, p. 34-44, November 1993.

[ORD 91] ORDENTLICH E., SHOHAM Y., "Low-delay code-excited linear predictive coding of wideband speech at 32 kbps", *Proc. IEEE ICASSP*, paper S1.3, 1991.

[PAL 93] PALIWAL K.K., ATAL B.S., "Efficient vector quantization of LPC parameters at 24 bits/frame", *IEEE Trans. on Speech and Audio Proc.*, vol. 1, p. 3-14, 1993.

[PAN 98] PAN J.-P., FISCHER T.R., "Vector quantization of speech line spectrum pair parameters and reflection coefficients", *IEEE Trans. on Speech and Audio Proc.*, vol. 6, p. 106-115, 1998.

[PRI 86] PRINCEN J.P., BRADLEY A., "Analysis/synthesis filter bank based on time domain aliasing cancellation", *IEEE Trans. on ASSP*, vol. 34, p. 1153-1161, 1986.

[QUA 91] QUACKENBUSH S.R., "A 7 kHz bandwidth, 32 kbps speech coder for ISDN", *Proc. IEEE ICASSP*, paper S1.1, 1991.

[RAM 95] RAMACHANDRAN R.P., SONDHI M.M., SESHADRI N., "A two codebook format for robust quantization of line spectral frequencies", *IEEE Trans. on Speech and Audio Proc.*, vol. 3, p. 157-168.

[ROB 98] ROBERT-RIBES, J., SCHWARTZ, J.L., LALLOUACHE,T., ESCUDIER, P. "Complementarity and synergy in bimodal speech: auditory, visual, and audiovisual identification of French oral vowels in noise", *J. Acoust. Soc. Am.*, 103, p. 3677-3689, 1998.

[SAL 98] SALAMI R., *et al.*, "Design and description of CS-ACELP: a toll quality 8 kb/s speech coder", *IEEE Trans. on Speech and Audio Proc.*, vol. 6, p. 116-130.

[SCH 85] SCHROEDER M.R., ATAL B.S., "Code-excited linear prediction (CELP): high-quality speech at very low bit rates", *Proc. IEEE ICASSP*, p. 937-940, 1985.

[SOO 84] SOONG F.K., JUANG B.H., "Line spectrum pair (LSP) and speech data compression", *Proc. IEEE ICASSP*, San Diego, p. 1.10.1-1.10.4, 1984.

[SOO 90] SOONG F.K., JUANG B.H., "Optimal quantization of LSP parameters using delayed decisions", *Proc. IEEE ICASSP*, p. 185-188, 1990.

[SUG 86] SUGAMURA N., ITAKURA F., "Speech analysis and synthesis methods developed at ECL in NTT – from LPC to LSP", *Speech Comm.*, vol. 5, p. 199-215, 1986.

[TRE 82] TREMAIN T.E., "The government standard linear predictive coding algorithm: LPC10", *Speech Technol.*, p. 40-49, 1982.

[VIS 75] VISWANATHAN R., MAKHOUL J., "Quantification properties of transmission parameters in linear predictive systems", *IEEE Trans. on ASSP*, vol. 23, p. 309-321, 1975.

[XIE 96] XIE M.-J., ADOUL J.P., "Algebraic vector quantization of LSF parameters with low storage and computational complexity", *IEEE Trans. on Speech and Audio Proc.*, vol. 4, p. 234-239, 1996.

[XYD 99] XYDEAS C.S., PAPANASTASIOU C., "Split matrix quantization of LPC parameters", *IEEE Trans. on Speech and Audio Proc.*, vol. 7, p. 113-125, 1999.

[ZWI 81] ZWICKER E., FELDKELLER R., *Psychoacoustique: L'oreille récepteur de l'information* (translated into French from German by C. Sorin), Masson, Paris, 1981.

Chapter 3

Speech Synthesis

3.1. Introduction

Nowadays, the telecommunication sector is experiencing a strong growth in services proposed to clients by large telephone operators. Ultimately, these services rely on the transmission of a voice signal over a network and require ergonomic interfaces that are becoming increasingly complex and more and more similar to human interaction. A man-machine dialog engine coupled to speech recognition and spoken message generation systems constitute the technological backbone of the vast majority of existing systems handling spoken information.

Most voice servers replay pre-recorded messages[1] spoken by a person who has lent his or her voice to the application beforehand. From the service provider's viewpoint, two categories of services can be distinguished: firstly, services for which the entire message has been stored as an acoustic signal, possibly in a compressed form, and which is generally replayed at 32 or 64 kbit/sec; secondly, services delivering messages which result from the concatenation of fixed portions and variable parts. The latter approach is slightly more flexible than the former and offers a wider range of messages for the application. While the quality of the messages are very close to those of natural, spoken utterance, it is difficult to imagine such a solution being tractable for large vocabulary applications (for example, reverse directories), because storage costs and the expenses involved in updating the inventory of recorded names would clearly be prohibitive.

Chapter written by Olivier BOËFFARD and Christophe D'ALESSANDRO.
1 They can consist either of complete sentences or of message portions that are concatenated in order to form sentences.

A text-to-speech (or TTS) synthesis system aims at generating a speech signal that corresponds as best as possible to the pronunciation of a given written message. Such a system can easily overcome the limitations of voice servers, which rely on storage/replay techniques. In fact, updating a textual database is much easier than maintaining its counterpart in an acoustic form. Many logistical problems can also be avoided, for instance those induced by repeated recording operations for a particular speaker. However, the current quality of synthetic speech is far from adequate, and in most cases the resulting output is clearly discernable from a natural speech signal. Therefore, it remains essential to evaluate the quality of the synthesis system by submitting the output to a panel of listeners, to determine the level of acceptability of the service.

The goal of this chapter is to present the fundamental and technological aspects of the various domains required when designing a speech synthesis system, together with the main methodological procedures for evaluation.

Firstly, we will try to answer the following question: how can we generate, from a text, an acoustic speech signal? The answer is based on two observations: firstly, speech is not only acoustic and phonetic material, but also a melodic and rhythmic pattern. Secondly, we will deal with the structural description of a speech synthesis system, and further elaborate on linguistic and acoustic processing. This technical description will be followed by a typology of services that are based on speech synthesis. This presentation will be complemented by an in-depth description of algorithms such as PSOLA ("Pitch Synchronous OverLap and Add") used for prosody modifications.

Finally, we will address the issue of speech synthesis system assessment, which can be either global or analytic. Global approaches can be used both for measuring the intelligibility of the synthetic signal and to evaluate global quality on a multidimensional scale (e.g. naturalness, rate, clearness, pronunciation/accentuation errors, pleasantness), or with respect to semantic contrasts (e.g. fluent/hesitant, smooth/rough). Analytical approaches specifically consider the system components (phoneme-to-grapheme transcription, prosody generation, signal processing). The chapter ends with a list of ongoing national and international projects on speech synthesis evaluation.

3.2. Key goal: speaking for communicating

The concept of a talking machine is completely embedded in a communication framework. A speech synthesis system provides an oral message that conveys the information that is expected by a receiver, most often a human being.. Under the assumption that the listener is co-operative, this section describes the linguistic and acoustic structures required for constructing a spoken message. It is essential for the system to provide an acoustic message that can be understood by the listener in the same way as the speaker intended it. In other words, the various steps in constructing the speech signal must minimize ambiguities in the interpretation of the spoken message between the transmitter and the receiver. The largest factor of complexity in a speech synthesis system is that the underlying mechanisms generating a spoken message are inherent to the cognitive process that governs the whole sequence of operations, from the mental representation to the control of the phonation organs.

A leading paradigm of the field considers the input of a *text-to-speech* synthesis system as a text which is constrained by the linguistic rules of a given language[2].

3.2.1. *What acoustic content?*

The pronunciation of a text utterance starts with a *transcription* step. Indeed, we write by combining elements from a set of signs (the orthographic alphabet) and similarly, we speak by combining elementary acoustic signs. The description of this universe of elementary sounds by abstract signs is the phonemic alphabet. *Phonemes* are descriptive symbols that form, at the linguistic level, the existence of our spoken acoustic universe. The phoneme can be considered as the counterpart of the *grapheme*, which refers to symbolic signs used in writing.

Of course, there is a link between the universe of written messages and its counterpart in the spoken domain; the spoken message must contain the same information as the written or read message. This is the primary goal of a speech synthesis system: the spoken message produced by the machine and perceived by a human being must correspond to an equivalent yet hypothetical human utterance. Given the combinatory complexity of a language, a speech synthesis system must be able to produce messages that have never been uttered before by any human being.

Consider, for example, the last word of the following sentence: "A nice boat". The word "boat" is composed of a single syllable, itself resulting from the combination of three phonemes: the consonant [b], the diphthong [əʊ] and the final stop [t]. The acoustic material created by synthesis will have to be perceived as such, or the meaning of the sentence could be altered. If the acoustic material corresponding to the last word was to be perceived as the sequence composed of the consonant [k], the diphthong [əʊ] and the stop [t], the perceived message would be written "A nice coat". We define as the *phonetic content* of the speech signal, the

---

2 This hypothesis is not too restrictive; concept-to-speech synthesis systems also rely on a textual surface representation at their input.

acoustic properties which correspond to the articulation of phoneme sequences which cannot be perceived as ambiguous: for example, what makes it impossible for a [k] to be perceived as a [t] at the linguistic level.

Besides the linguistic dimension, a second aspect must also be considered when explaining the phonetic content of a sound: its physiological dimension, i.e. what relates to the organ that creates the signal. Within a given linguistic community, different physiological characteristics[3] yield different acoustic realizations for a same phoneme. The term *phone* refers to the actual acoustic observation corresponding to the abstract linguistic entity covered by the concept of *phoneme*.

In summary, the phonetic content depends on its twofold heritage:

– a linguistic heritage which makes it possible to distinguish between messages on the basis of phoneme distinctions;

– a physiological and sociological heritage that enables us to distinguish between speakers with different voice timbres.

A primary and fundamental objective for a speech synthesis system is to convert graphemic sequences into phonemic sequences. This step lies on a linguistic level that depends on the characteristics of the language. Later in this chapter, we will explain how the relationship between the linguistic and acoustic contents is not a one-to-one correspondence, because perception ambiguities may be introduced.

3.2.2. *What melody?*

It is often observed that the phonetic level is not the only way to convey spoken information. Let us consider for instance the two following sentences in French: "Il fait beau." (*The weather is fine.*) and "Il fait beau ?" (*Is the weather fine?*). The sole orthographic difference between these two sentences is a question mark instead of a full stop at the end of the sentence. When uttering these two sentences, prosodic differences can be perceived, which enable each message to be interpreted differently. In the first case, the utterance is produced, in French, with a falling pitch contour on the last syllable, whereas, in the second case, the phonetic content is identical, but the pitch rises so as to express a feeling of surprise or doubt.

Prosodic factors such as this are an integral part of the speech signal, but they are interpreted at a different level than the phonetic content. The duration of sounds, the voice pitch, the acoustic level variations and speaking rate are all forms of prosodic information.

---

3 Just as we show diversity in our physiological features, we also produce sounds which differ in their acoustic properties.

Prosodic factors present in a speech signal depend on a twofold heritage:

– a linguistic heritage related to the message. This typically corresponds to the previous example, when a text provides indications about the way to pronounce a phonetic content. The resulting duration and pitch patterns depend on both the syntactic structure of the sentence and the grammatical nature of the words;

– a linguistic heritage linked to the speaker. Distinct speakers will follow different speaking strategies as a function of their own approach to the language – for instance, depending on their regional origin – or on their wish to connote a message that may look neutral in its written form.

The analogy with a musical instrument illustrates the relationship between the phonetic content and the prosody of a speech signal. Instruments have distinct timbres that in general, produce a fixed and single acoustic content. Production of a musical phrase involves creating a sequence of notes that are characterized by their duration and pitch. Human beings are able to use their phonation apparatus as a musical instrument. Singing practice consists of separating explicitly the acoustic content of a note from its sol-fa content, which can be assimilated to the prosodic factors of speech. In spoken voice, the prosodic and phonetic levels are so intimately overlapping that it is obviously difficult to separate them.

We have underlined in the previous section that a primary objective of a speech synthesis system is to create the acoustic matter related to the phonetic content of speech, i.e. the color of the sounds. In this section, we have insisted on a second objective, as fundamental as the former : that of determining the prosodic content of the message. The examples above have illustrated the link between prosody and linguistic content, which are, to some extent, independent of the physiological characteristics of the speaker. The prosodic content has to be integrated in the acoustic content of the message.

3.2.3. *Beyond the strict minimum*

The phonetic content and the prosodic content of a message are two essential prerequisites for the artificial production of a speech signal. However, the rendering of these two dimensions is not sufficient to guarantee a complete disambiguation of the message between the transmitter and the receiver.

The specification for the values that should be taken by prosodic factors for a given message may strongly depend on the syntactic structure of the sentence and its grammatical labeling. With speech synthesis systems currently available on the market, it is very difficult to rely on an extensive and accurate syntactic description of sentences. Some alternative syntactic interpretations may remain that cannot be resolved at this level alone. For instance, a sentence such as *"Bear left at zoo."* can be understood in two distinct ways ("Turn left when you get to the zoo." or "Someone left a bear at the zoo."), depending on whether "bear" is considered as a verb and "left" as an adverb, or "bear" as a substantive and "left" as a verb. Once the proper syntactic form has been determined, a prosodic factor, such as a short pause between "bear" and "left", makes it possible to select one interpretation or the other. However, the syntactic analysis by itself does not provide all the necessary elements for controlling the prosodic level.

Current systems deal with such difficulties by adopting a neutral attitude, and not favoring a particular interpretation. The ambiguity is propagated up to the listener, who is assumed to have all the information at hand to decide between the various possible interpretations. If speech synthesis systems are to provide a more natural output, it is necessary to introduce all the dimensions involved into a dialog situation. It would indeed be possible to overcome the previous difficulty using a synthesis system that could take into account the semantic level. For instance, if the system knows the semantic class for "bass" (fish or low voice), this will enable it to determine how to render that word phonetically ([bæs] or [beɪs]), which is impossible to decide by any other means – an expression like "he sings like a bass" using the pronunciation [bæs] would certainly raise some interest.

3.3 Synoptic presentation of the elementary modules in speech synthesis systems

The most common model for describing a speech synthesis system, which also offers operational solutions, involves distinguishing between two main tasks: 1) the generation of a linguistic (phonetic and prosodic) description of the message; and 2) the generation of the corresponding synthetic signal (see Figure 3.1).

Such a model is advantageous, as it provides a very clear separation between, on the one hand, the universe of linguistic content and structure and, on the other hand, the synthetic speech signal itself, seen as a raw material which depends on the physiological characteristics of the speaker. This makes it possible, in the case of systems based on the concatenation of elementary acoustic units, to design efficient multilingual systems, because their linguistic aspects are easily interchangeable.

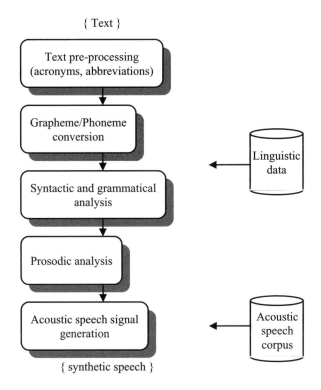

Figure 3.1. *Synoptic description of a text-to-speech synthesis system*

3.3.1. *Linguistic processing*

Linguistic processing aims at providing a pronounceable description of the message initially introduced in a text form to the synthesis system. Four key steps[4] can be distinguished: text pre-processing, grapheme-to-phoneme transcription, morpho-syntactic analysis and prosody generation.

3.3.2. *Acoustic processing*

Acoustic processing has two objectives:

– firstly, to provide a speech signal whose phonetic content is in accordance with the message to be pronounced;

---

4 The complexity of which is subject to variations depending on the system.

– secondly, to generate a speech signal whose prosodic content corresponds to a possible utterance of the message to be pronounced. In most cases, three prosodic parameters are taken into account at the signal level: the duration, the pitch (for voiced sounds) and the energy (i.e. the time-domain envelope of the signal).

Thus, two distinct functions are considered: one generates a speech signal from the sequence of phonemes, whilst the other applies the prosody effects as computed during the linguistic processing step. Most systems handle these two functions separately.

3.3.3. *Training models automatically*

Over the past ten years, mainstream effort in speech synthesis has been focused on building systems capable of processing more and more data. However, if it is feasible for a *human expert* to assimilate a rather large quantity of information, it becomes impossible to handle such databases when they grow beyond a certain point, without the help of data analysis tools. This trend has been reinforced by the ever-growing capacity of machines, both in terms of computational power and storage capabilities.

The direct consequence of the increased amount of data to analyze has been to complicate the analysis grid for the observed phenomena. When this complexity is put together with models which are themselves complex, most people consider the system as a *simple* input/output function.

In this case, the "black box" approach can provide some sort of solution: the system is summarized as its input/output function and its real behavior is mimicked as closely as possible, without trying to understand *how* the system functions. This paradigm can be applied to any type of function in the synthesis system: grapheme-to-phoneme conversion, syntactic analysis, prosody generation, etc.

Starting from an *a priori* constraint on the type of function linking the input to the output, automatic methods aim at learning model parameters from training data, which are representative of the input/output relation.

Such an approach displaces the input/output coupling the entire complexity of the process to be modeled. Given the objective of a synthesis system, which is to *predict* an acoustic output corresponding to the pronunciation of a text a priori unknown to the system; the models involved in the various processing steps are therefore *predictive* models. As a function of the input, the predicted output must be as close as possible to the actual output observed with the corresponding natural phenomenon. It is therefore essential that these models possess *generalization* capabilities.

3.3.4. *Operational constraints*

An operational speech synthesis system is expected to produce a speech signal that can be delivered in the same way as natural speech; for instance, direct to loudspeakers or transmitted over a telecommunication network.

A fundamental constraint that synthesis systems must fulfill to be operational and exploitable as resources for information delivery is to operate in real-time. It is unimaginable that a whole sentence or a paragraph would need to be finished before being processed and listened to. Information processing must take place on-line, following preferably the most adequate data processing structure in such circumstances. In such a pipeline structure, each elementary module handles a minimal waiting line, in which the waiting time depends on the processing being carried out. Such an operational model turns the speech synthesis system structure from a sequence of operations into a set of parallel processes, where each module uses and produces data at its own pace. A priority-handling module takes into account the specificities of speech generation and ensures implementation on sequential machines. In fact, the time granularity of syntactic analysis is at the sentence level, whereas prosodic analysis takes place at the level of breath groups (sequences of phonemes between two pauses) and the signal generation at the phone level, etc.

It must also be noted that, as the processing level becomes closer to the acoustic signal, the information rate becomes higher: consequently the corresponding functions have to handle a larger quantity of data. The underlying information bit rate thus increases from a few dozen of bytes per second to 32 or 64 Kbytes per second for a typical landline telephone network.

In this context, research carried out at FT R&D has yielded a real-time exploitation system called SYC, specific to the characteristics of synthesis systems, whose software modules work on standard hardware and platforms [BIG 93].

3.4. Description of linguistic processing

3.4.1. *Text pre-processing*

Text pre-processing involves turning raw texts (possibly composed of abbreviations, acronyms, numbers, dates and telephone numbers) into a normalized written form, acceptable by the synthesis system. Once a particular structure has been detected, two approaches are possible:

– replace the raw character string by a normalized letter string which corresponds to the pronunciation of the acronym or the number. This is the approach adopted by most text pre-processing systems; or

– carry the detected structure as far as possible through the sequence of linguistic processing. If, for example, a sequence of numbers is analyzed as a telephone number, specific intonation can be applied by the prosody generation module.

Pre-processing operates at the interface between a general-purpose synthesis system and an application that is necessarily targeted to a particular service. Thus, pre-processing names and proper nouns in a reverse directory service[5] is different to pre-processing a fax or e-mail for a reading application.

The most common technical solution to this problem involves scanning exception lexicons containing (for each acronym or abbreviation) a normalized word string which corresponds to a non-ambiguous pronunciation[6]. Detection of structure such as currencies, telephone numbers or dates is generally achieved by rules. It must however be noted that some pre-processing systems rely on probabilistic approaches such as n-grams.

3.4.2. *Grapheme-to-phoneme conversion*

Grapheme-to-phoneme conversion or transcription involves determining the sequence of phonemes corresponding to the pronunciation of a written text.

3.4.2.1. *Problem definition*

The main difficulty in grapheme-to-phoneme conversion lies in the fact that the transformation is not a one-to-one correspondence: for the same sequence of orthographic characters, or graphemes, several distinct pronunciations are possible. For example, the grapheme *x* is pronounced as [ks] in "six" and [gz] in "exam". Conversely, a same phoneme can correspond to several graphemic forms, such as the phoneme [i], which writes *ea* in "leak" and *ee* in "leek".

The aforementioned difficulties reflect problems at the lexical level; a simple dictionary would suffice to provide the phonetic transcription of a word. However, pronouncing a sentence, i.e. a sequence of words, is not only a word-by-word

---

5 The goal is to find the name and address of a person, given his/her telephone number.
6 It is also possible to match directly a phonetic transcription, which of course will be ignored by the grapheme-to-phoneme transcription module.

transcription. Syntactic and phonologic phenomena impact on the phonetic content and alter it with respect to a simple lexical transcription. The main phenomena that operate on the transcription at the sentence level are:

– Co-articulation phenomena that modify the *phonemic* string (sequence of phonemes) into a *phonetic* string (sequence of phones) that may be different from each other. The co-articulation process results from some sort of minimum effort in the articulatory domain. Given the inertia of articulators taking part in phonation, all the phonemes are not actually uttered in their expected form: variants are produced as a result of a particular speaking strategy. For instance, when two consonants with a distinct voicing feature are in contact with each other, the voicing feature is assimilated: the sequence [sg] can become [sk] (progressive assimilation) or [zg] (regressive assimilation).

– The *liaison* phenomenon occurs in French, for example, across words. There are three types:

> - compulsory liaisons: for instance, a liaison consonant [z] is compulsory between [lɛ] (*the*) and [wazo] (*birds*) when pronouncing "les oiseaux" (*the birds*): [lɛzwazo];

> - forbidden liaisons: for example, the production of a liaison consonant [n] between the subject and the verb in "Le garçon aura …" (*The boy will have…*) is forbidden, whereas it is compulsory in "On aura …" (*We will have…*);

> - optional liaisons: in this case, the decision to produce the liaison is left to the speaker. It may also depend on the prosodic structure.

– Homograph heterophones are words that can have distinct pronunciations. Sometimes, the difference depends on the syntactic category: in the sentence "*I resume my resume*", "resume" is either a substantive or a conjugated verb and it is pronounced differently. The difference in pronunciation can also only depend on the tense ("*read*" present or past tense) or on the semantic class ("*to resign*", meaning "*to sign again*" or "*to quit*").

– Finally, it must be mentioned that a schwa (mute *e*) can be introduced across words to avert non-pronounceable sequences of consonants when words are chained in the same breath group.

3.4.2.2. *Implementation strategies*

The previous observations show that grapheme-to-phoneme conversion in a speech synthesis system cannot be addressed in a single module. Three functional levels may contribute to the grapheme-to-phoneme transcription:

– firstly, a grapheme-to-phoneme module, able to handle regular cases;

– secondly, a disambiguation step at the syntactic level and the detection of compulsory liaisons;

– thirdly, a processing step taking place after prosody generation, which provides a phonetic string on the basis of the phonemic string, taking into account the prosodic segmentation (for instance, no liaison at the border of breath groups).

Technically speaking, a distinction can be made between lexicon-based and rule-based grapheme-to-phoneme systems. In transcription lexicons, each entry is described by a graphemic form and its corresponding phonemic form. However, this approach is not very easy to implement: indeed, it requires a storage capacity which is proportional to the size of the lexicon, and the search time for a given word can become prohibitive for real-time or embedded systems. Because of these functional constraints, use of a lexicon is generally reserved for the transcription of morphemes. Rules are however necessary to determine the pronunciation of a word composed of several morphemes. Rule-based transcription systems use a set of rules for the transcription, but also use a lexicon of exceptions.

Several research works have focused on the automatic inference of transcription rules from training examples. The goal is to determine rules and exceptions from a set of transcribed examples. Examples include work based on Hidden Markov Models [COI 91], inference techniques based on analogy in French [YVO 96] and in English [BAG 98], and also stochastic techniques [LUK 96].

To the list of aforementioned techniques, we must add purely functional approaches which do not try to discover a set of transcription rules, but which relate a graphemic input to a phonetic output directly. For example, the NETTALK system, by Rosenberg, models the transcription function by a neural network [SEJ 87].

The current deployment of direct or reverse directories by telephone operators sheds light on the crucial difficulty for synthesis systems to process family names and proper nouns. Technical solutions are essentially based on reference to lexicons, taking into account the etymology of the word for applying specific pronunciation rules [SCH 93].

3.4.3. *Syntactic-prosodic analysis*

The structure of a written text reflects a number of constraints coming from various levels. First of all, the author of a message influences that message according to the particular words and the syntactic structures used to express a given meaning. Secondly, the sequence of words within a sentence in a given language obeys a number of constraints ruled by the syntax of that language: a grammatical structure is a particular way to express a syntactic form.

The spoken form of a message is also constrained by the fact that some phoneme combinations or prosodic patterns are authorized, while others are not, according to certain rules. The goal of the syntactic-prosodic analysis is to determine the most relevant prosodic structure for the message to be uttered, i.e. the most pleasant to the ear and the least ambiguous semantically. The description of the prosodic structure of a sentence is achieved at a symbolic level. This abstract representation of the evolution of prosodic phenomena is then used as the reference for the numerical computation when predicting the numerical control parameters for prosody.

3.4.3.1. *Problem definition*

When studying a text-to-speech synthesis system, the most common approach is to assert the prosodic style should be that of *read* speech. Thus, it can be assumed that the main contribution to the prosodic content is made at the linguistic level and is therefore strongly determined by the syntax.

Three types of information (in increasing order of difficulty to determine), enable specification of the prosody for a spoken message, starting from its written form:

– the structure of its surface textual form, as indicated by its punctuation. For instance, an interrogative sentence will not have the same prosodic pattern as an affirmative sentence;

– the structure of its textual form, highlighted by particular function words: articles, and prepositions. The position of these words will enable the segmentation of a sentence in smaller recurring units in the prosodic organization of the sentence;

– the syntactic structure of the sentences bears information concerning the relation between words and their organization in terms of syntactic components. A hierarchical representation of prosody can be obtained from this information.

As to what concerns the structural description of the prosodic level, the most common model is a 2-level hierarchical model. This comprises:

– a first level, immediately above the lexical level, that corresponds to the accentual phrase. This group of words is characterized by a semantic unity and in French it receives a primary accent on the last syllable;

– a second level, just above the accentual group, that corresponds to the intonational group. An intonational group is composed of accentual groups and it introduces, at its borders, major ruptures of the melody and of the rhythm.

However, the structure of prosodic information does not exclusively depend on the syntax, but also on constraints on the rhythm. These constraints tend to induce balancing processes in the constitution of prosodic words.

3.4.3.2. *Implementation strategies*

The goal of a syntactic-prosodic analysis function is to segment the sentence to synthesize its prosodic components. Depending on the level of complexity of the underlying structural prosodic model, the corresponding approaches can be made more or less sophisticated, as required.

On one level, some systems identify prosodic words on the basis of grammatical words in the sentence and a rough syntactic analysis on a local level [O'S 84], [LAR 89].

Complete syntactic analyses of the sentence yield a hierarchical organization of prosodic components in a more detailed and less ambiguous way than the aforementioned cases. In general, these systems provide more contrasted pitch contours [GEE 83].

Finally, while the previous methods can be considered as part of the rule-based approaches, more recent work has shown the relevance of a prosodic structure model based on a decision tree, the structure of which is determined automatically from training examples [OST 94].

3.4.4. *Prosodic analysis*

Starting from a structural representation of prosody in terms of components, the goal is to determine the time evolution of prosodic phenomena. Here, the focus is the prediction of phoneme duration values and of fundamental frequency contours as a function of time.

3.4.4.1. *Problem definition*

For most synthesis systems, the duration processing step is distinct from that of pitch processing, for a very simple reason: the factors impacting these two phenomena operate at a different time-scale.

As regards pitch prediction, some systems start by *stylizing* the phenomenon prior to modeling it [D'A 95]. The advantage of this approach is to limit the redundancy of the observed phenomenon and to smooth out any measurement errors that may have occurred when analyzing the fundamental frequency.

3.4.4.2. *Implementation strategies*

For duration processing, a class of models termed *add-multiply type* are used. In these, the intrinsic average values of the phoneme durations are modified as a function of factors which are represented in a parametric form and which depend on the linguistic level. The parameters of these models evolve according to *rules*. Statistical analysis techniques are used to determine the main model factors from a corpus of training examples. An abundant literature exists on statistical modeling based on automatic learning, covering for instance neural networks [BAR 95], [CAM 93], [TOU 97] or regression trees [RIL 92].

Concerning pitch contour generation, approaches include non-parametric methods, which involve the storage and restitution of pitch information as a function of the place and the nature of the prosodic word [EME 92]. Other models try to structure a global pitch contour from elementary segmental patterns: Fujisaki's model [FUJ 82] is a parametric model that is governed at the sentence level and at the level of the accentual group. This model has been adapted to French [BAI 86]. Finally, as for duration, many systems are based on automatic learning approaches: they are mostly based on neural networks [TRA 92]. It is also worth noting the original work by [ROS 95] on American English, for which pitch evolution is determined by a Kalman filter, whose parameters follow stochastic laws which are learned from a training data set. Figure 3.2 depicts schematically the linguistic processing steps that take place within a speech synthesis system.

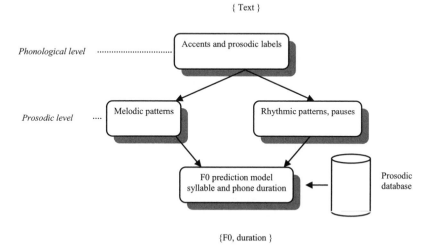

Figure 3.2. *Synoptic representation of the linguistic processing steps within a text-to-speech synthesis system*

3.5. Acoustic processing methodology

This section presents the two main methods that have served as the support for developing new speech synthesis systems. The goal is to relate the acoustic content of the message – which we have so far considered as a phoneme sequence, i.e. from a linguistic viewpoint – to the description of a speech signal that ultimately expresses itself as a variation of acoustic pressure.

As it is not conceivable to consider *a priori* all the possible acoustic configurations of a same phoneme[7], and because the phoneme remains the only link with the text to be pronounced, two methods link the abstract phonetic description and the concrete acoustic reality:

– rule-based synthesis is based on an explicit description of the control parameters over time representing the speech signal. It consists of a set of rules derived from an explicit knowledge of production phenomena, at least at the acoustic level;

– unit-based concatenative synthesis rejects the above hypothesis and proposes a black-box approach. Information which is impossible to model from the speech signal is encapsulated into a speech unit which expresses the link between the linguistic and acoustic levels. The definition of this unit is solely based on external characteristics – for instance, its phonetic description.

3.5.1. *Rule-based synthesis*

Independently from the signal model used in the synthesis system, a rule-based approach for speech synthesis involves modeling the transitions between phonemes in terms of rules dynamically. As opposed to the unit-based concatenative approach, the rule-based approach provides some explicative power to the phonation process [POL 90].

First, a parametric model for the time-domain evolution of the signal must be defined. Then, rules are inferred to describe the parameter behavior, by human expertise applied to a set of training data. For historical reasons and interpretability purposes, the signal model most frequently used in rule-based systems is the formant synthesizer. A set of rules controls a few dozen parameters related to an acoustic model [STE 90].

---

7 Indeed, all types of messages may be synthesized; therefore, they are *a priori* unknown when the speech synthesis system is conceived.

If the inventory of rules is rather complex to elaborate, the set of basic units on which they operate is usually quite limited – typically comparable to the number of phonemes, plus a few allophonic variants. The number of rules can be higher or lower, depending on the sophistication of the targeted model. Liberman proposes a set of minimal rules [LIB 59] that rely on a sub-phonemic modeling of language-independent units, defined as the combination of acoustic cues corresponding to the place of articulation and the mode of production.

The MITalk rule-based system [ALL 87] for American English and the multilingual InfoVox system [CAR 82] are both based on the above methodology.

3.5.2. *Unit-based concatenative synthesis*

Since the acoustic content of the speech signal can be described with reference to a finite set of linguistic units, namely the phonemes, a basic approach involves juxtaposing (or concatenating) an acoustic representative for each phoneme to synthesize. The simplest approach would be to store a single such representative (i.e. a phone), which is used for all occurrences of that phone in the synthetic message. In French, approximately 40 units would be sufficient, each of them having an average duration in the range of 100 ms. This can seem to be a very attractive solution, as 4 or 5 seconds of stored speech would suffice to synthesize any written text in French.

However, as early as in the 1950s, this solution was discarded by an experiment proposed by Harris [HAR 53]: syllables were artificially created by juxtaposing phones which were extracted from different contexts, and these syllables were not perceived as they should have been. For instance, an initial consonant [k] cut from syllable [kik] and replacing the initial consonant [k] in [kæk] is perceived as [tæk] instead of the initial [kæk].

This fundamental experiment impacts on the strategic orientation of current systems, as it demonstrated that speech is above all a dynamic material. Our linguistic filter gives us the erroneous impression of a static speech resulting from the simple concatenation of sounds. This simple view hides the dynamic nature of speech resulting from co-articulation effects.

From the physiological viewpoint, uttering a message calls for the movement of the articulators which are the central actors of the phonation process. These physiological organs see their mass involved in an inertia process. This aspect can become predominant for particular sound combinations and timing constraints for producing the phoneme stream. The acoustic realization of a phone will therefore depend on those that have just been produced before, and those that are coming up next. *Co-articulation* refers to the influence over the realization of a sound, of the sound or the group of sounds that precede and follow that realization.

Initially, the principle of co-articulation has been accounted for in a minimal way, by limiting the time horizon to a sequence of two half-phones. Based on the previous representation, speech corresponds to a quasi-stationary process where transitory areas are inserted in between acoustically stable portions. If a way of accounting for co-articulation involves rendering transitions between phones, the static component of sounds – i.e. what makes an [a] an [a], whatever its context – is assumed to be localized somewhere between two successive transitions, i.e., schematically, in the middle of a phone. This leads to the definition of the *diphone*: an acoustic unit starting in the middle of a phone and ending in the middle of the next phone [DIX 68] – sometimes, the term "dyad" is also used [FUJ 76].

If the use of diphones noticeably improves the synthetic speech quality, intelligibility tests show that some confusion remains between particular sounds or groups. For example, in French, some consonant groups are regularly misperceived by listeners [POL 87]. The defects seem to be due to the large variability of liquids and glides that appear in these groups. In order to minimize these effects, one can add more units to the inventory, including some combinations of 3, 4 or more phonemes. The generalization of this principle leads to the concept of variable-length units [SAG 88].

Speech synthesis systems using the diphone or the polyphone as an acoustic unit emerged at the beginning of the 1970s [LIE 70], [GEN 76] and have proved reliable [BIG 93], [AUB 91], [POR 90]. Figure 3.3 presents the implementation principles for a diphone based system.

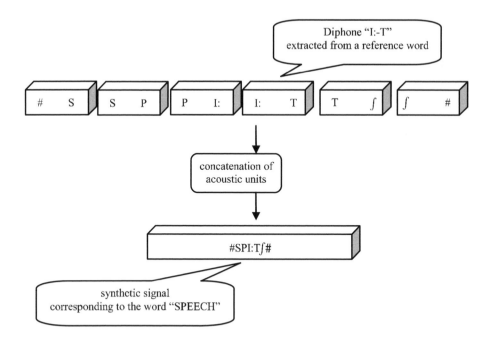

Figure 3.3. *Diphone-based synthesis of word "speech", [spiːtʃ], uttered in isolation. The # sign denotes a pause*

3.6. Speech signal modeling

Whether a rule-based (and explicative) or a concatenative (and descriptive) approach is used, the time signal evolution can either be represented by a model or not. The advantages of using a model are its capacity to reduce the acoustic signal's redundancy and to define parameters that are better fitted to acoustic processing.

A rule-based system cannot avoid such a model, as it is essential to reduce the quantity of information to design algorithmic solutions based on a finite set of controllable rules. Conversely, concatenative approaches based on pre-recorded units usually preserve the signal in it's raw time-domain form, i.e. without modelling.

The advantage of a parametric model is twofold. By reducing redundancy in the signal, it acts as a compression scheme in cases where storage capacities for the synthesis system are limited (for instance, in embedded systems). It also offers a parametric representation that is easier to relate to supra-segmental information.

Many models describing the time-domain evolution of speech signals have been proposed over the past 50 years. Even if the signal to be modeled is considered as the simple variation of a physical quantity with time, most of them are based on an assumption which stems from the speech production mechanisms: the source-filter decomposition assumption.

Models of speech signal representation can be grouped into two categories depending on the corresponding methodological hypothesis:

– models relying on a *production* hypothesis describe the generation of the speech signal on the basis of physiological parameters. The goal is to reproduce the behavior of the human phonation apparatus;

– models relying on *perception* phenomena describe schemes to generate a speech signal that will be perceived *like* a human-produced signal.

The latter type of model is the most popular, in the actual implementation of real speech synthesis systems, as it is easier to formalize. However, the former hypothesis is certainly more explicative in understanding speech production mechanisms.

After a quick review of the source-filter hypothesis, this section will present articulatory models, formant-based methods, auto-regressive approaches and finally, the harmonic plus noise model of the speech signal. Only the articulatory approach places itself in the category of production-based model, whereas the three others resort to the perception hypothesis.

3.6.1. *The source-filter assumption*

The assumption behind source-filter decomposition is that the speech signal results from the combination of some acoustic energy (interaction of the lungs and the larynx) coupled with a transfer function that is determined by the shape of the supra-glottic cavities. This description is founded on J. Müller's work (1848) dealing with phonatory mechanics. Revisited in the context of signal processing by G. Fant [FAN 60], the source-filter model explains the speech signal $s(t)$ as the convolution of an excitation signal $e(t)$ by a time-varying filter $h_t(t)$, which yields: $s(t) = e(t) * h_t(t)$. The excitation signal characterizes the variation of acoustic pressure in the larynx; the filter represents the time-frequency behavior of the vocal tract transfer function.

In practice, it is impossible to access by measurement, simultaneously and non-intrusively, the exact values of $s(t)$, $e(t)$ and $h_t(t)$. Only measurement of $s(t)$ is easy, simply requiring a microphone to capture the speech signal at the output of the lips. In order to constrain the solution of a single equation with two unknown variables, the various models that we will review will make different hypotheses on one of the variables, so as to enable the estimation of the other variable from the known signal $s(t)$.

3.6.2. *Articulatory model*

An articulatory model provides a speech signal production process based on a physiological description. Historically, the initial studies concerning the dynamic behavior of articulators have mostly focused on their external description (lips, jaws, cheeks). The development of X-ray photography has offered possibilities of having access to more accurate information regarding the behavior of internal articulators, such as the vocal tract section, the position of the tongue and of the velum. Nowadays, less dangerous devices, based for instance on magnetic resonance, provide even more efficient measurement tools.

The methodology behind the articulatory approach involves determining a transformation between the articulatory level and the acoustic level, under the constraint of linguistic hypotheses.

Once the configuration of the vocal tract has been determined, several algorithms are possible to compute the speech signal:

– the less computationally expensive method consists of turning back to the acoustic level using linear acoustic techniques [MAE 79];

– some systems compute the variation of the acoustic pressure at the lips by solving the propagation equation for a pressure wave in a dynamic medium [RIC 95].

3.6.3. *Formant-based modeling*

In formant-based approaches, the synthesis process is related to an interpretation hypothesis at both the phonetic and acoustic levels. A formant characterizes a resonance of the vocal tract. The analysis of a natural acoustic signal involves characterizing its resonances, on the basis of its spectrum, by their central frequencies and bandwidths. On the contrary, speech synthesis starts from a finite set of formant-based representations – location, amplitude and bandwidth – and generates a speech signal from an artificial spectrum. The artificial transfer function is obtained by combining second-order resonating cells in cascade or in parallel.

This approach may seem attractive, as it provides a connection between the phonetic interpretation and the acoustic reality; but data acquisition and analysis in this framework remains rather complex. Formant detection from a speech signal is a delicate task, especially when they are close to each other or crossing one another.

3.6.4. *Auto-regressive modeling*

In 1960, Fant introduced a linear model of the time-domain waveform of the speech signal. To the source-filter hypothesis, he added the hypothesis of independence between the glottal waveform and the vocal tract. The vocal tract is modeled as an all-pole filter, also called "auto-regressive". The glottal waveform is modeled roughly as a pulse train with a fundamental period equal to the pitch (for voiced sounds) and as a white noise with zero mean and unit variance (for unvoiced sounds).

This model is widely used in speech processing, especially in speech coding. The rough approximation of the glottal waveform results in a rather mediocre synthetic speech quality. However, the number of model parameters is limited, and prosody control for speech synthesis is relatively easy. Indeed, a simple lengthening or shortening of the excitation signal suffices to modify the duration of a phone, and changing the period of the pulse train results in a pitch modification.

This approach was very popular until the end of the 1980s, after which it gave way to more complex techniques which offered a better signal quality (for example, the harmonic plus noise model).

3.6.5. *Harmonic plus noise model*

The original principle of the harmonic plus noise model involves describing the speech signal as the superposition of a harmonic component and a noise component. This model embraces under a same formalism both the description of voiced signals like vowels and unvoiced signals like fricatives or other unvoiced consonants.

With regard to prosody modifications, the TD-PSOLA approach used in this context (see section 3.7.3.1) does not make any distinction between a voiced and an unvoiced signal. In some situations though, a defect corresponding to a tonal noise can appear when lengthening a noise signal[8]. When prosodic modifications required by the synthesis system invoke too drastic distortions (for instance, if the target fundamental frequency is too far away from the original), the perceived signal is altered, and sounds like an abnormal speech signal (for instance, like a falsetto voice, when the target pitch is too high). The suppression/duplication process of

---

8 It is possible to reduce the effects of this problem by using *ad hoc* solutions such as time-axis reversal.

short-term signals can also cause the total suppression of a segmental unit, if this unit is occurring in a prosodic context that is too different from the target (all short-term signals are then suppressed).

For all these reasons, a large amount of work was dedicated, in the middle of the 1990s, to the design of a new approach aimed at controlling the prosodic parameters in speech. The goal is to describe rather finely the spectral content of the speech signal, so as to be able to modify flexibly both voiced and unvoiced sounds and to access easily the fundamental period, the duration and the spectral envelope (especially for joining acoustic units). These observations have led to a number of research works on harmonic models [STY 95b], [VIO 98], [OUD 98].

The origin of sinusoidal models is probably to be found in the *phase vocoder* [POR 76], [DOL 86]. The principle of decomposing a signal by a sinusoidal model is twofold: to obtain a fine-grain analysis of the signal in both the frequency domain (represented as a distribution of harmonics) and time domain (phase information). Using a Fourier analysis, the signal is decomposed as a sum of 256 to 1,024 complex cosine functions. Then, taking into account the explicit phase information in the model yields better results than the traditional *channel vocoder* does. Several variants of this model [LAR 99] have brought corrections to the main defects that occur when modifying the phase. The key problem with the *phase vocoder* lies in the compromise between time and frequency: large windows provide an interesting frequency resolution but they smooth over time the transitory events. Moreover, the presence of noise added to the periodic component of the signal deteriorates the quality of the phase estimate.

The sinusoidal model represents a speech signal as a sum of sinoidal functions, evolving over time [MCA 86]. This model has the capacity for frequency resolution (sinus distribution) and time resolution (evolution of each sinusoid over time). At a given instant, a sinus function can appear, evolve or disappear.

During the analysis step, the amplitude, phase and frequency coefficients are estimated through a short-term spectral analysis (for instance, at a 10 ms rate):

$$s(t) = \sum_{k=1}^{K(t)} a_k(t) \sin(2\pi f_k n + \phi_k) \qquad [3.1]$$

In the synthesis step, the parameters $a_k(t)$, f_k and ϕ_k are interpolated at the signal's sampling rate. Two approaches can be distinguished: the "OverLap and Add" (OLA) approach and the direct parameter interpolation.

The harmonic model is a sinusoidal model for which a harmonic relation is imposed on the frequencies f_k in equation (3.1). This means that $f_k = 2\pi k F_0(t)$, where $F_0(t)$ is the actual value of the fundamental frequency as estimated at time t. The harmonic plus noise model (referred to as the HNM, H/S, H + N or hybrid model), is a model of the signal which introduces two additive components: the first one, $s_h(n)$, is described by a harmonic model and reflects the periodic component within the signal, while the second one, $s_n(n)$, is described by a noise model and represents the random component: $s(n) = s_h(n) + s_n(n)$.

To make a compromise between time and frequency, some segments in the speech signal such as voiced/unvoiced transitions or some transient events are poorly rendered by a harmonic model. Solutions requiring a larger number of degrees of freedom in the model can be envisaged. For example, [LI 98] propose a mixture of pulses shifted in time. Alternatively, [PEE 99] present a model that gathers the properties of a sinusoidal representation and an OLA processing step, which is acknowledged as being efficient for the rendering of time-localized events.

It is worth underlining that parametric models such as those described above offer the possibility of controlling the spectral characteristics of the signal, which is an important feature for correcting these aspects when joining successive acoustic units. Harmonic models have been quite naturally involved in the development of voice transformation systems [STY 95a], [STY 98], [OUD 98].

3.7. Control of prosodic parameters: the PSOLA technique

The PSOLA technique ("Pitch Synchronous OverLap and Add") is not, strictly speaking, a speech signal modeling technique. It is rather a speech signal processing technique, which can be applied to natural and synthetic signals and whose objective is to enable the control and modification of prosodic parameters. The originality and efficiency of such a technique consists of modifying independently the fundamental frequency and the duration of the signal, at a local level and without altering the timbre of the original voice. These functionalities are indeed of primary importance for a speech synthesis system.

This paragraph presents both from an historical angle and in a comparative way, the foundations and algorithms for modifying prosodic parameters in a speech signal. However, this presentation cannot be exhaustive, and we will focus mostly on the algorithms that are currently used in existing speech synthesis systems[9].

---

9 Note that most of the theoretical background and some of the applications are common to other domains, such as speech coding.

A text-to-speech synthesis system must have the capacity to control three essential prosodic parameters: the duration of sounds, the pitch contour and the energy over time. The role of the linguistic processing step (see section 3.4) is to predict, from the written text, sequences of numerical values for each of the three prosodic parameters. It is therefore necessary to design operators that can be applied to the speech signal so as to reach particular prosodic targets. The fundamental frequency F_0 is a parameter specific to the glottal excitation. We have already discussed the difficulty of extracting this information from the speech signal. Concerning duration, it is also difficult to slow down or to speed up locally the speech signal without altering the fundamental frequency and the voice timbre. A simple dilation of the time axis is not adequate: the frequency scale of the spectral content becomes compressed in an inverse proportion[10].

Historically speaking, parametric analysis-synthesis models of the speech signal based on the source-filter decomposition paradigm have advantageously offered the possibility of yielding explicit access to duration and pitch parameters. For instance, the modification of the lag between successive pulses in the excitation of a Linear Prediction (LP) model make it simpler to lower or heighten the fundamental frequency [OLI 77], [STE 83]. Duration control is a function of the length of the excitation signal. Whereas these models allow for an explicit control of prosody parameters, they do not offer a fully satisfying speech quality (buzzing timbre of the LP model). The goal to obtain a high-quality speech synthesis has motivated a large part of the research during the 1980s. At that time, the effort was rightly put on the intrinsic quality of the synthetic signal. By setting aside the prediction of prosodic parameters, two major factors indeed impact the quality of a speech synthesis system: the choice of the acoustic units (see section 3.5.2) and the modifications of the pitch and of the duration.

Following [MOU 95], modifications of duration and pitch can be formalized as follows:

− concerning pitch modification, the goal is to modify locally the value of F_0 without altering either the timbre (the spectral envelope) or the time-domain evolution of the signal (the apparent speech rate must not change). If $F_0(t)$ denotes the original fundamental frequency value and $\tilde{F}_0(t)$ the modified value, and by letting $\alpha(t)$ denote the time-distortion factor, we have the following relationship: $\tilde{F}_0(t) = \alpha(t)F_0(t)$;

− concerning duration modification, the goal is to locally apply a speed up or slow down factor to the apparent articulation rate of the uttered speech. The spectral

---

10 An easy way to verify this property is to play an audiotape on a tape recorder at a speed which is different from that used when the tape was recorded. The result is a distorted voice, both in terms of pitch (lower or higher pitch) and timbre (e.g. the voice sounds stern when the tape is slowed down).

envelope of the transformed signal must remain as close as possible to the original one. Slowing down the speech signal by a factor of 2 is equivalent, with respect to the fundamental frequency, to resampling the speech contour so as to preserve its shape over a doubled length of time. A simple dilation of the time-scale by a factor of 2 (by a resampling operation on the signal) would lower the fundamental frequency by the same factor. Here, the transformation that we want to formalize is slightly different: we want to define a compound operation rather than a multiplicative form, as is the case for F_0. We denote as t, the time index for the original signal and $\tilde{t}$, the modified time-axis. We denote as D the dilation function that transforms t into $\tilde{t} = D(t)$. In practice, it is useful to introduce a local dilation factor $\beta(\tau)$, where τ is a time variable. We then have the following relationship:

$$\tilde{t} = \int_0^t \beta(\tau)d\tau \qquad [3.2]$$

To understand this equation, we can consider its behavior when the modification factor is constant, i.e. $\beta(\tau) = \beta$. In this case, we obtain $\tilde{t} = \beta t$, i.e. a linear dilation of the time axis. In practice, a speech synthesis system has to be able to modify the duration of some sounds more than others, hence the need to formulate time scale modifications as in equation [3.2], where the gradient function $\beta(\tau)$ can have one form or another. This correspondence between the two time axes means that the properties of a given event at an instant t in the original signal (for instance, its short-term spectrum and it local F_0 value) are projected at the instant $\tilde{t}$ in the modified signal.

3.7.1. *Methodology background*

At the beginning of the 1990s, the expansion of techniques belonging to the PSOLA family in many speech synthesis systems based on acoustic units concatenation is due to a combination of a very high quality of the synthetic signal and a remarkable algorithmic efficiency, with respect to the state of the art at that time. However, it is important to keep in mind that prosodic modification techniques such as FBS (Filter Bank Summation) [POR 81], [SEN 82], [CHA 88] also yield very good results. The filter bank approach is based on two elementary processing steps: a resampling operation by a factor α, and a multiplication of the instantaneous frequency by a factor β. For instance, in order to modify only the duration of the signal by a factor α, the signal is first resampled by that factor. Then, the implicit modification of the frequency axis must be compensated by a factor $\beta = 1/\alpha$. At that stage, the spectral envelope is processed separately in order to preserve the timbre property of the original signal.

Before going into more detail on the various methods, here are a few additional notations:

– $s(n)$ corresponds to the analysis signal in the time domain, $\tilde{s}(n)$ is the synthetic signal, where n is the time index. $X_k(n)$ denotes the complex amplitude of the short-term Fourier transform of $s(n)$, the analysis window being centered on n ;

– t_m denotes an instant on the time axis of the analysis signal. This instant refers to the center of an analysis window;

– similarly, $\tilde{t}_m$ denotes an instant on the time axis for the synthetic signal (i.e., the signal reconstructed by one of the methods). This instant can correspond to the center of a synthesis window;

– we denote as $s_m(n)$ a sample from a short-term signal, the analysis window being centered on instant t_m, i.e. $s_m(n) = h(t_m - n)s(n)$;

– we denote as h_m a weighting window whose center is located at instant t_m .

3.7.2. The ancestors of the method

[CHA 86] proposes the PSOLA technique in order to modify the rhythm of a speech signal, but also so as to be able to alter the spectrum, in particular, the fundamental frequency. The principles behind the approach, which can be explained in the short-term Fourier analysis framework, involves synchronizing the position of analysis and synthesis time markers on the fundamental periods of the signal. PSOLA can be understood as the result of a double heritage: the heritage of simple and efficient methods for modifying the speech rate and the heritage of block-wise short-term Fourier analysis/synthesis methods. We present here a brief overview concerning these two families of approaches and we will detail the PSOLA techniques and its subsequent variants in a second step.

[CHA 88] reports that block-wise methods have been initially applied for the modification of speech rate and rhythm. In particular, [LEE 77] proposes to slice the signal into fixed-length or variable-length blocks. Some segments are then duplicated or removed so as to apply the desired time-scale modification. [NEU 78] corrects some discontinuity problems by proposing a processing rate that is synchronized with the fundamental period. A short-term signal block, extracted over a time interval related to the fundamental period, can then be connected to other such short-term signals with some time domain smoothing, without introducing obvious discontinuities. The advantage of this method resides in the simplicity of the corresponding algorithms. For compressing the signal bandwidth, [MAL 79] proposes an algorithm for duration modification, which can be interpreted in the framework of the short-term Fourier transform. Starting from markers placed on the analysis signal, short-term signals are obtained by a windowing operation. The interval between two marks is chosen to be proportional to the local period and to the dilation factor on the time axis. The synthesis signal is reconstructed using an "overlap and add" operation applied to the short-term signals. The relationship between the fundamental period and the dilation factor make this algorithm poorly adapted to modification factors close to one. [ROU 90] proposes the SOLA method (Synchronized OverLap and Add), dealing with rhythm modification in the speech signal. The analysis markers are asynchronous but the synthesis markers are synchronized on local periodicity information (correlation between successive short-terms signals).

From the interpretation of the block-wise Fourier analysis/synthesis method stem the "simple" OLA [ALL 77] and the WOLA [CRO 80] algorithms.

Concerning the OLA method, the synthetic signal is reconstructed using an overlap and add operation on the short-term analysis signals:

$$\tilde{s}(n) = \sum_{m=-\infty}^{m=+\infty} s_m(n) = \sum_{m=-\infty}^{m=+\infty} h(n-t_m)s(n)$$

The previous equality leads to the "simple" OLA synthesis equation:

$$\tilde{s}(n) = \frac{\sum_{m=-\infty}^{m=-\infty} s_m(n)}{\sum_{m=-\infty}^{m=-\infty} h(n-t_m)}$$ [3.3]

This approach sums directly the contribution of each block s_m; a weighting factor is however necessary to neutralize the effect of the analysis windows h_m. In this formulation, there is no particular constraint on the analysis instants t_m. However, a minimal overlap of the windows is necessary for the reconstructed signal $\tilde{s}(n)$ to be reasonably similar to the original signal $s(n)$.

The WOLA (Weighted Overlap and Add) algorithm [CRO 80] originates from the interpretation of the Fourier analysis/synthesis method. The synthetic signal is reconstructed in the frequency domain, as opposed to the previous methods that operate in the time domain. The approach is to sum the various contributions of the transforms $X_k(n)$ after they have been re-sampled by a constant factor S. It is indeed possible to retrieve the values of $X_k(n)$ from a sub-sampling of $X_k(mS)$, as long as $S < N/4$. From the sequence of $X_k(n)$, only one vector out of S is preserved. The spectrum is then reconstructed using a re-sampling filter:

$$X_k(n) = \sum_{m=-\infty}^{m=+\infty} f(n - mS) X_k(mS) \tag{3.4}$$

It is easy to show the following equivalence in the time domain:

$$\tilde{s}(n) = \sum_{m=-\infty}^{m=+\infty} f(n - mS) s_m(n) \tag{3.5}$$

The previous equation clearly shows that the WOLA method reconstructs the synthetic signal by weighting the short-term signals with a smoothing window f_m. Compared to OLA, these signals are windowed during the synthesis step. There is no weighting factor taking part in the process. For an exact reconstruction of the signal, i.e. $\tilde{s}(n) = s(n)$, it is however necessary to impose some constraint on the relationship between the analysis and the synthesis window. Moreover, the filter bank method suffers from a constraint of uniformity on the synthesis markers, as $t_m = mS$.

The theoretical framework for the OLA and WOLA methods provides a rather simple paradigm for applying linear modifications on the short-term signal s_m before synthesis. For non-linear modifications, i.e. the most common situation in speech synthesis, a more general framework is necessary. [GRI 84] proposes a solution by introducing a least-square minimization criterion. By letting $Y_k(t_m)$ denote the spectrum obtained after transformation of the spectrum $X_k(t_m)$, corresponding to the signal $s(n)$ and calculated at time t_m, the proposed approach establishes an error measure between the modified spectrum $Y_k(t_m)$ and that corresponding to $\tilde{s}(n)$, constructed from the sequence of $Y_k(t_m)$. The minimization of this criterion leads to the so-called "least-square OLA" formulation:

$$\tilde{s}(n) = \frac{\sum_{m=-\infty}^{m=-\infty} f(t_m - n) \tilde{s}_m(n)}{\sum_{m=-\infty}^{m=-\infty} f^2(t_m - n)} \tag{3.6}$$

It is worth noting that this approach applies the OLA principle to the *synthesis signals* ($\tilde{s}(n)$ resulting from a windowing of the reconstructed signal). The LSOLA method benefits from the same advantage as the basic OLA, namely that the markers t_m are unconstrained, while guaranteeing a reconstructed spectrum close to the original.

In this context, the PSOLA technique is to be considered as a block-wise process that extends the principles of the SOLA method to the analysis axis. We have seen above that the SOLA approach is asynchronous at the analysis step and synchronous at the synthesis step. As regards PSOLA, the approach is synchronous both during analysis *and* synthesis. The analysis instants t_m are synchronous with the fundamental period. Duration modifications take place in the time domain and pitch and envelope modifications are done in the frequency domain. The synthetic signal is reconstructed by applying a least-square OLA.

We have just presented what, according to [CHA 86], distinguishes the PSOLA approach from the other methods. We have intentionally skipped details concerning pitch modification in the frequency domain. In that respect, several solutions are possible in the PSOLA framework (see [CHA 88]).

Let us note a few points concerning the notion of *pitch-synchrony*. Pitch-synchronous processing (or, more accurately, synchronous with the fundamental period) is not new in speech processing. For instance, for some approaches related to linear prediction models, the analysis windows are centered on time markers chosen to correspond to glottal closure (as this is an optimal instant for decoupling the larynx and the vocal tract). This can be viewed as *absolute* pitch synchrony. However, locating accurately these glottal closure instants is a difficult task, as it requires blind deconvolution, i.e. to estimate the source signal from the speech signal without any knowledge of the filter. Conversely, pitch synchrony such as the one used in the context of the OLA methods is a *relative* pitch-synchrony. What matters is not the exact location of the glottal closure instant, but the value of the period; some time shift can occur, corresponding to a slight phase shift. However, it is important that the local interval between two time markers reflect the local period of the signal. The markers can then be *synchronized* on the waveform by aligning them with events that are easier to locate on the signal (for instance, a maximum of energy or amplitude). It could be preferable in these conditions to qualify this approach as a *quasi-synchronous* process, in order to distinguish it from the former one.

3.7.3. *Descendants of the method*

Keeping the main feature of the PSOLA technique, i.e. the synchronous analysis and synthesis scheme, several variants have been proposed. Let us start with the technique we will refer to as FD-PSOLA (FD for Frequency Domain), i.e. the full solution as proposed in [CHA 86][11]. After FD-PSOLA was presented, a subsequent variant was to displace the pitch-processing step from the frequency to the time

---

11 It must however be kept in mind that, in the FD-PSOLA approach, unvoiced signals are entirely processed in the time domain.

domain. This has yielded the lineage of TD-PSOLA (TD for Time Domain) algorithms [HAM 89][12], [CHA 89], [MOU 90b], [MOU 95].

3.7.3.1. *TD-PSOLA*

TD-PSOLA, or the application of the PSOLA algorithm in the time domain[13], was designed in the 1990s and can be viewed as the most efficient compromise between quality versus complexity [CHA 88], [HAM 89], [MOU 95]. Indeed, this algorithm enables modifications of the prosodic components of a speech signal without altering too perceptibly the voice timbre [CHA 88]. It also shows an extremely favorable computational complexity with respect to the original implementation in the frequency domain (FD-PSOLA): only 2 multiplications and one addition per sample are required for a complexity in $O(n)$ (whereas frequency domain transforms are in $O(n \ln n)$). The principle is to synchronize the sequence of analysis markers (itself pitch-synchronous) on the sequence of synthesis markers, derived from the rhythm and pitch transformations applied to the time scale of the analysis signal. In other words, the goal is to find a mapping function between the short-term analysis and synthesis signals (either by suppression/duplication or by interpolation).

Once the synthesis markers and the mapping function have been determined, the synthesis signal is reconstructed by LS-OLA[14]. A synthesis window is necessary, since a re-sampling operation takes place[15]. The whole scenario is almost identical to FD-PSOLA, except for the fact that the step for going into the frequency domain for modifying F_0 is omitted. It is therefore necessary to perform a wide-band analysis so that the time-resolution is too rough to resolve the fundamental period from the short-term signal: this constraint imposes the analysis windows to be rather short. Otherwise, harmonics at frequencies multiple of the fundamental frequency of the analysis signal would be introduced in the synthetic signal (this phenomenon sounds as the presence of a second voice or of some form of reverberation).

What is gained in terms of complexity by the TD-PSOLA technique is lost with respect to the impossibility of applying explicit transformations to the spectrum. [MOU 90a] proposes a model for interpreting the errors caused by TD-PSOLA, when voiced and noise signals are processed. It is worth noting that the process is sensitive to the window size. This width depends on the voice tessitura and on the modifications applied. As mentioned above, in a narrow band scenario, i.e. when the

---

12 Hamon's algorithm is initially referred to as KDG.

13 TD-PSOLA has been patented by France-Télécom R&D.

14 Alternatively, the less complex WOLA approach can also be used, if the analysis and synthesis window are chosen judiciously.

15 For instance, when the duration is shortened by a factor of 2, in the suppression/duplication mode, one short-term signal out of two is lost.

analysis windows contain several fundamental periods, the resulting signal shows reverberation because of the presence of harmonic components of the original fundamental frequency. In wideband analysis, the windows are sufficiently short and the harmonics of the analysis signal cannot be resolved. The signal processed with TD-PSOLA shows some spectral spreading, which enlarges the formant's bandwidth. A solution to this problem, proposed by [HAM 89], consists of applying some pre-emphasis to the signal before it is processed.

3.7.3.2. *PIOLA, WSOLA and MBROLA*

PIOLA, for pitch inflected OLA [VOT 91], is a method for modifying prosodic parameters that is very similar to TD-PSOLA. The authors of the patent do not establish any explicit phase relationship between the analysis windows and the signal to modify.

WSOLA, for Weighted Synchronized OverLap and Add, is a method used to modify the duration of a speech signal. It is directly inspired from the SOLA method apart from the Overlap and Add operation, which is synchronized on the short-term analysis signals (and not on the synthesis signals, as in the case of the SOLA method). The synchronization is done on the basis of a similarity measure derived from the correlation between short-term analysis signals. During the analysis step, the short-term signals are defined by applying a fixed-length window yielding narrow band analysis conditions. During synthesis, the interval between short-term signals and the window length are constant, so that the normalization coefficient of the OLA method can be set to a constant value; as a result, the complexity of the algorithms is lower [VER 93]. However, this technique is only able to deal with duration and it cannot be used in the context of speech synthesis.

The MBROLA approach, initially called MBR-PSOLA [DUT 93a] [DUT 94], draws its originality from the context in which the PSOLA-like solutions are used in speech synthesis. The force of the PSOLA approach, and also its weakness, resides in the pitch-synchrony during analysis and synthesis. This is a rather delicate operation, especially in the voiced/unvoiced transitions. In text-to-speech synthesis, the PSOLA modified signals are speech segments corresponding to the acoustic units. These units are known in advance and they can be submitted to a number of prior processing steps, outside of real-time constraints. The MBROLA technique addresses the synchrony constraint on the analysis axis by pre-processing the inventory of acoustic units with a MBE (Multi-Band Excitation) type vocoder [GRI 88]. As to what concerns the quality of the reconstructed signal, [DUT 93b] reports slightly higher intelligibility rates for TD-PSOLA but more fluidity for MBR-PSOLA. The gain is significant in terms of time complexity.

3.7.4. *Evaluation*

Many evaluations of the PSOLA and PSOLA derived techniques have been taking place in the past years. In most cases, the evaluation concerns a complete system. The PSOLA component is assessed within a full text-to-speech synthesis system [BIG 93], [DUT 94], [HUA 96] or in the context of natural speech modifications [CHA 89], [BAI 00].

To our knowledge, very few studies are strictly focused on the behavior of TD-PSOLA and its proper operating conditions with respect to perceptual sensitivity (length of the analysis window, maximal prosodic range, etc). On this particular point, [KOR 97] and [KOR 99] report experiments in psycho-acoustics concerning the perception of test signals which have been modified by TD-PSOLA. The authors submit stimuli, which have been generated by Klatt's model, to prosodic manipulations (pitch and duration). The authors do not describe TD-PSOLA in detail, or it implementation. [KOR 97] presents an experiment on a stimulus defined by a single formant, for which very small pitch modifications are perceived (in the order of 2%). These modifications are no longer perceived anymore when the values of the fundamental in the analysis and synthesis signals are within such a ratio that the distribution of harmonics is in phase (which validates the theoretical analysis of spectrum re-sampling – see above). Moreover, the modifications in the position of the analysis markers have much less influence in the perception of differences: variations within 20% are not perceived. The authors propose an interesting interpretation on the basis of the JND (just noticeable difference) notion, concerning the level of the harmonics. We believe that it is absolutely necessary to conduct such studies so as to distinguish between the variants of such or such algorithm and, above all, to get some experimental feedback for proposing more efficient models. In this respect, perceptually based notions such as JND at the spectral level have not been sufficiently integrated in solutions such as PSOLA.

3.8. Towards variable-size acoustic units

We briefly presented in section 3.3, the principles of a text-to-speech synthesis system. Today, major technological advances in speech synthesis systems involve the concatenation of pre-recorded acoustic units, extracted from a corpus of continuous speech. This is known as *corpus-based synthesis*.

Applications with a limited vocabulary and with constrained syntactic sentence structures can be based on speech synthesis systems that concatenate words, breath groups or sentence fragments. For these systems the *units* directly stem from the specification of the application, and the consequence is a rigidity of the system [MOB 01]. Conversely, speech quality is usually very close to that of natural speech: co-articulation effects are relatively limited as they occur at the border of the units [STÖ 99], [LEN 02], [BLA 01]. In this section, we will focus exclusively on studies concerning speech synthesis with an unlimited vocabulary.

In the framework of unlimited vocabulary synthesis, the initial systems were based on phones (which are the acoustic counterparts of the phonemes, defined in the phonological domain). It was therefore natural to consider the phone as the *ideal* acoustic unit to fabricate sound material: linguistic efficiency and very small inventory of units. However, [HAR 53] showed by a number of experiments on allophones, that assembling phones yields a disastrous result in terms of intelligibility, because the continuity between phones is a key factor in speech perception (co-articulation effects).

We know that *diphones*, already mentioned in section 3.5.2, have been introduced in order to overcome these co-articulation defects. The diphone is an acoustic unit which contains the transition between two phones: a diphone corresponds to two successive half-phones. Diphones offer a relatively intelligible speech but they occasionally provide annoying defects and a lack of naturalness of the synthetic speech. At the end of the 1980s, digital processing systems became powerful enough to enable solutions with larger unit than diphones. This leads to systems [OLI 90] adding larger units, or *polyphones*, to a set of diphones. The next decade experienced an intense activity towards the solution of finding the optimal set of acoustic units [SAG 88], [BLA 95], [BIG 93], [BEU 99], always adding more, longer units in more and more diverse contexts, etc.

This permanent headlong rush has to be tempered. Systems using variable-length and/or multi-reference units (i.e., the same unit in various contexts) tend to generate a speech signal of unstable quality. The more heterogenous the concatenation points are spread, the more this effect is audible. Diphone-based synthesis provides an average yet homogenous quality: once the phoneme sequence is known, there is no ambiguity to form the sequence of units (the number of diphones is equal to the number of phonemes plus one). At the other extreme, corpus-based synthesis systems search for the longest sound elements in a corpus of continuous speech that allow for reconstructing the utterance while minimizing the number of concatenation points. For some portions of the message, the quality is excellent (which is logical, as this part of the message indeed corresponds to real human speech) but for other portions, the quality can be mediocre. The contrast within a given message is perceptually annoying.

Today, it is technically possible to design a continuous speech database of more than 10 hours; several gigabytes of data are indeed easy to handle or a standard computer. However, a first difference between an acoustic database of diphones and of continuous speech concerns the vocal quality over time. It is indeed possible to control the pronunciation of a list of logatomes[16] by a reference speaker. This is much more difficult for a continuous speech database. When the text is semantically predictable, it is not always certain that the segmental or supra-segmental acoustic target will be reached by the speaker. Indeed, with the intention of delivering a relevant and meaningful message, the speaker produces some acoustic parameters with a relative instability, affecting for instance the phonetic accuracy of some phones, including particular effects such as accentuation, affectation, etc. During synthesis, when two pieces of sound are concatenated, acoustic differences can be relatively significant at the border and can cause audible artifacts.

With diphone sets or even enriched diphone sets, the search problem for finding the optimal concatenation sequence is relatively trivial. For synthesis systems that use continuous speech databases, the algorithmic search complexity grows exponentially. Quite often, the units are not even defined. To simplify, let us consider a continuous speech database as a very long sequence of (say one million) phonemes, which can be viewed as a reference string. The concatenation process can be summarized as the following goal: given a phoneme string to synthesize, how can it be generated as the concatenation of substrings extracted from the reference string? This is an NP-hard problem. It is our opinion that the corpus-based synthesis approach is not a miracle solution: the problem is shifted but the difficulties remain, even though they have changed nature.

For diphone-based synthesis, which can be understood as an *a priori* defined corpus-based approach, the expertise lies in the corpus. Conversely, for continuous speech based synthesis, the database is commonplace, and is composed in large quantity but with little expertise. During the synthesis step, the difficulty is to find the adequate sequencing. In a laboratory experimental context, [FRA 01] shows that a sequence of 15 phonemes or so can generate a lattice of concatenation sequences comprising over a billion of possibilities, even though the speech database is rather modest in quantity. It is therefore impossible in practice, for spatial and temporal complexity reasons, to explore exhaustively all the possible concatenations and to choose the best one, for instance, according to acoustic criteria (continuity between segments, proximity to a natural target, etc.).

For corpus-based approaches, the core of the problem is not the database, but the algorithm for selecting the optimal concatenation sequence. Finding the solution is computationally intractable. Most systems overcome the difficulty by choosing *a priori* a heuristic approach to solve the sequencing difficulty. The processing step is usually carried out from the beginning to the end of the sequence, by searching the longest sub-sequences, adjoining some hypothesis on the optimality of the

---

16 A logatome is a meaningless word that contains a particular diphone or polyphone.

solution[17]. This hypothesis of optimality leads to dynamic programming algorithms such as the Viterbi algorithm.

Two major problems can therefore be distinguished: 1) how do we constitute a speech database which contains relevant acoustic units?; 2) once the database of acoustic units is defined, what is the best strategy to obtain a synthesis sequence?

3.8.1. *Constitution of the acoustic database*

A database for speech synthesis must fulfill the following criteria:

– It must contain exhaustively all possible couples of successive sounds in a given language. We have seen that co-articulation effects prevent a simple categorization of sound elements. Phone sets must therefore be discarded and the baseline unit for a minimal rendering of co-articulation is the diphone. Besides the phonemic description of the unit, it is possible to associate a prosodic description (segmental units recorded in different prosodic contexts).

– It must be accompanied with *explicit* descriptive information at the phonological level. The linguistic steps in the synthesis system indeed provide a phonemic string together with prosodic information. A difficulty resides in the correspondence between a sequence in the phonemic string and its corresponding acoustic image in the database. If the sound segments in the database cannot be described with information available at the phonological levels of the synthesis system, this correspondence cannot take place efficiently.

– It must have a constant level of vocal quality over the whole database. The vocal quality can be influenced by a variation of the recording procedure[18] or by the modification of some extra-linguistic factors that are specific to the speaker[19].

For a concatenative synthesis system, the first point is essential, but the control of prosody at the unit level is optional. Most speech synthesis systems indeed rely on a prediction model for prosody. However, the current evolution tends to address the segmental and the supra-segmental issues together, at the initial step of the acoustic database creation.

In this section, we start by presenting databases designed on segmental considerations only. We will deal with *inventories of phonological units*, putting the stress on the need for phonetic expertise. We will then consider *inventories of multi-*

---

17 In this respect, any optimal solution achieved on a sub-sequence is part of the optimal solution for the entire sequence.
18 A simple change of microphone can cause enough heterogeneity to alter a database, especially by introducing phase distortions.
19 A cold, caught between two recording sessions, for instance.

reference units, which try to account for syntactic and prosodic constraints and variants; these inventories can indeed contain several acoustic representations for a single unit. Finally, the ultimate stage is reached when segmental and supra-segmental conditions can be varied in similar proportions. In this case, we will refer to *corpora of continuous speech*. Speech synthesis systems based on continuous speech are rather recent. However, most of the work dedicated to them does not deal with a fine description of their content. They generally consist of recordings of texts in sufficiently large quantity (a few hours of speech). It is possible to make the relationship between the construction of the corpus of continuous speech and the quality of synthetic speech more systematic, via phonological and prosodic *coverage* criteria [FRA 01].

3.8.1.1. *Inventory of phonological units*

The notion of *inventory of phonological units* covers corpora of acoustic units that have been designed with a particular attention. A distinction must be made between inventories of diphones, di-syllables, half-syllables, polyphones and mixed units (mixture of diphones, half-syllables) [POR 94].

The concept of diphone emerges in the 1950s, when [KÜP 56], [PET 58] proposed creating sound matter by juxtaposing a particular type of speech unit that contains the transition between two successive phones. [EST 64] implemented this technique for the English language. [DIX 68] took up the idea and improved the quality of the synthetic signal by adding some prosodic modifications: a natural prosodic contour is applied onto the raw synthesis output. For the French language, [LEI 68] was the first to achieve a full synthesis system based on diphone concatenation. The acoustic signal is created by the excitation of a bank of oscillators. [LIE 77] describe a voice response unit that generates synthetic signals from a text; the system is reported to use an inventory of 627 diphone units. [JOS 74] describe initial experiments carried out at ICP (*Institut de la Communication Parlée*) in Grenoble. Simultaneously, [LAR 76] and [EME 77] proposed a diphone-based synthesis system in French. The quality of the speech signal is improved by using a channel vocoder. Until the middle of the 1990s, the diphone remained the dominant type of synthesis unit used in most systems. The system developed by CNET (*Centre National d'Etudes en Télécommunications*) [BIG 93], extended to a multilingual framework, uses an inventory of 1,300 diphones for French, 2,500 for Spanish, 2,400 for English, 3,000 for Russian, 2,800 for German and 1,100 for Italian.

For the synthesis of Japanese, [SAI 68] proposed using longer units than diphones, constructed on a similar basis, but at a higher phonological level. The *disyllable* is an acoustic unit which ranges from the vocalic kernel of a syllable to the vocalic kernel of the next syllable. The main difficulty raised by this approach comes from the large number of such units that must be defined and recorded. Whereas Japanese has only approximately 150 distinct syllables [SAG 86], English has over 10,000 [KLA 87]. The advantage of the disyllable approach is that the concatenations only take place on vocalic kernels, which are intrinsically more stable spectrally than consonantic kernels. More recently, [CHE 00] applied this approach to the synthesis of Arabic. [FUJ 76] proposes to use half-syllables. The goal is to reduce the overlarge number of di-syllables by cutting them into two pieces and by grouping particular consonantic clusters or vocalic kernels. This procedure makes it possible to bring back the inventory of units to approximately 1,000 for English. More recently, other studies aiming at increasing the inventory of diphones to longer units, lead to the definition of mixed units [POR 96]. The resulting inventories contain diphones, longer syllable-based units, half-syllables and di-syllables [JÜR 95], [POR 90], [CHO 97], [CHO 98].

The concept of *polyphone* is an intermediate alternative, somewhere between the diphone and the syllabic unit. [OLI 77] and [OLI 90] propose to increase an inventory of diphones with longer units so as to protect certain vowels and some functional words in English. This idea is similar to that of the *polysound* ("polyson", in French) proposed by [LAF 85], [BIM 88]. Like the polyphone, the polysound is composed of a half-phone at each extremity and can contain one or several phones inside the unit. Tests have shown some improvement in terms of intelligibility and synthetic voice quality [BOË 93], but these gains are not systematic across voices and languages. Moreover, other evaluations on French with an inventory of approximately 1,300 diphones and 1,000 triphones do not lead to a definite advantage of the polyphone. Two reasons may explain this disappointing result: 1) radically different logatomes which carry the units and 2) an inadequate prosody model with respect to the raw synthesis material. Polyphone units can be recorded in a neutral context (logatomes) or within natural words. In any case, the longer the unit, the more distinct the original prosody of the unit with respect to the target prosody (i.e. the prosody determined by the system), because the speaker who recorded the units and the speaker whom the prosodic model was based on are different. The prosodic pattern applied to these units can generate significant segmental distortions (for instance, the suppression of a very short phone within the unit). An acceptable solution would be to define a relatively conservative prosodic model for these units. [LEM 94] proposes to address the problem by modifying the PSOLA algorithm, to make it more elastic and deal better with prosodic modifications when too many short-term signals are suppressed with the standard algorithm. The major drawback with this solution comes from the fact that the exact prosodic pattern in not applied to the segment: this can be a problem, for experimental prosodic studies, for instance. The optimal solution consists of defining a new prosodic model for each new synthetic voice. A reference speaker will have to donate not only a segmental material but also a supra-segmental content. However, as of now, this approach has not been experimented with yet, as there is no known automatic system to capture the prosody of a speaker.

Finally, solutions based on sub-phonemic units must be mentioned. The goal is to create an inventory of acoustic elements which can be found in the realization of various phones: for instance, the acoustic signal corresponding to the explosion of a [t] can be used in several distinct allophonic contexts. [CON 99] proposes the use of semi-allophones so as to favor non-discontinuous concatenations. In the framework of continuous speech corpus-based synthesis, [BAL 99] sets the semi-phone as the minimal association element between a phonological description and an acoustic segment. As shown by all these studies, a sufficient number of semi-phone or phone variants are necessary as a function of the phonemic context, for the approach to be effective. This is the reason why this type of unit is essentially used in the context of continuous speech databases or multi-reference units. It is worth noting that some research is using a formal statistical description of this type of acoustic events by associating a sub-phonemic element with the state of hidden Markov chain [DON 96]. This sort of modeling approach is rather close to the notion of *senone*, which is met in the field of speech recognition [HWA 96].

3.8.1.2. *Inventory of multi-reference units*

Diphones, polyphones and other strongly constrained units such as those mentioned above are usually recorded in a single context. The concept of multi-reference units comes from studies on allophonic or sub-phonemic units. Indeed, we have seen that it is necessary for these units to be realized in several acoustic contexts. [NAK 88], [HIR 90] propose an allophone-based synthesis system for Japanese. The authors describe an automatic procedure for selecting the acoustic units, called COC (for context-oriented-clustering). [CHU 01a], [CHU 01b] propose a similar approach for the synthesis in Mandarin.

A corpus of multi-reference units is interesting because it offers several pronunciation variants for a same phonetic unit. The corpus therefore provides a strong potential that is not so easy to exploit fully. Indeed, the very concept of unit concatenation relies on a functional description of phonetic constituents in speech. It is therefore illusory to try to finely characterize the acoustic evolution, except for building a rule-based system. Most of the systems thus keep the phonemic description as the central representation.

In order to select more accurately a candidate unit, two aspects must be taken in consideration (under the condition to satisfy the phonemic constraints):

– above the phonemic level, the corpus must be labeled with phonological and linguistic information (position of the unit in the syllable, of the syllable in the word, prosodic characteristics, etc.) [TAY 99], [BLA 95], [CON 99]. If this information is determined according to the same rules as those applied for the constitution of the corpus, the system is able to use them for discriminating an acoustic unit as a function of its linguistic context;

– below the phonemic level, acoustic information can be taken into account – for instance spectral information, energy, pitch, voicing, formants, etc. – so as to minimize the acoustic discrepancies between acoustic units [WOU 98].

3.8.1.3. *Continuous speech corpus*

A continuous speech corpus can be considered as the ultimate stage of evolution reached by acoustic databases for synthesis. To build such corpora, a large number of natural sentences are recorded; for the speaker, there is no more *artificial* constraints compared to a corpus of logatomes. However, a natural speech corpus cannot contain everything and the distribution of the events follows an exponential law [TUB 85]. For these heavy-tailed random phenomena, it can be stated that a few events are extremely frequent and many are very rare. However, the concept of *Out-Of-Vocabulary* is irrelevant in speech synthesis. The system must be able to synthesize any sentence, without any acoustic *gap*.

To form a continuous speech corpus, two strategies are possible:

–   to collect a sufficient number of sentences *chosen at random* and to rely on the law of large numbers, in the hope the selected sample will reflect the statistical properties of the language;

–   to select *intelligently* a corpus of sentences from a large linguistic source (this source being only in text form, it is thus possible to process several dozen of million sentences). *Intelligently* may cover two approaches: either to take into account linguistic, phonologic and phonetic knowledge on the language in general, or to use the information used in practice by the synthesis system. The second approach is certainly less general, but it usually provides a more efficient adaptation of the corpus to the potential of the speech synthesis system.

[BUC 96], [SAN 97a] and [SAN 97b] have focused on the linguistic content of continuous speech databases, by applying greedy algorithms.

3.8.2. *Selection of sequences of units*

The principle behind a selection algorithm is to search the reference speech corpus (acoustic unit inventory or continuous speech corpus) for a sequence of acoustic segments and can be considered as a specific production of the phonological target [PRU 01].

The objective is twofold [IWA 92a]. On the one hand, the goal is to find a match between a sub-sequence of the phonemic string to synthesize and a plausible instance in the reference corpus. This is referred to as discrimination based on *target criteria* [HUN 96]. A match with the target is not sufficient as the decision is made at the unit level. An additional mechanism is necessary to guarantee that the proposed sequencing fulfills acoustic continuity criteria (at the segmental and/or at the supra-segmental levels). This corresponds to the *concatenation criteria*. The difficulty lies in the fact that both criteria must be combined. The selection of a sub-sequence which matches a unit in the corpus depends on its past (left context) and future (right context). Once again, the problem is combinatory, which can be formulated as the search for an optimal path in a graph.

Almost all operational solutions call for a *dynamic programming* approach [BAL 99]. The goal is to quantify the cost of a sequence as the accumulation [IWA 92b], [BLA 95] or the multiplication [HOL 98] of local costs. However, from a perceptual point of view, [GRO 01] shows that the characterization of the global quality of a synthetic sentence is more complex than the accumulation of local costs. An important stake involves establishing a relation between an objective characterization of the quality of a synthetic sentence and its perception from the psycho-acoustic viewpoint. Little work has concentrated on the subject: let us mention results from [TOD 02] which show that the best correlation between a MOS perceptual scale and an objective measure is obtained by a multiplicative accumulation.

3.8.2.1. *Conformity to the target*

Several solutions have been proposed as a response to the target conformity criterion. It is essential, on the one hand, to have an acoustic description of the corpus and on the other hand, to establish a global mapping function, at the sequential level, between the phonological level (known to the system) and the acoustic description. Two sub-problems must be distinguished:

– the phonological and acoustic description of the corpus; and

– a local mapping function between sub-sequences.

Conventional acoustic descriptions take into account the segmental nature of speech via the use of phonetic labels [SAG 86], [NAK 88], [OLI 90], [BOË 92]. This operation of phonetically labeling and marking segments in the speech signal is referred to as the *segmentation* task. This segmental description can be accompanied by spectral parameters: formants, MFCC coefficients, LPC envelope, etc. Supra-segmental information can also be added to the segmental description, such as voicing features, energy and pitch values, etc. [WAN 93], [CAM 94]. Once the description is determined in terms of acoustic attributes, it is necessary to establish a mapping function between this level of representation and the phonological level as used by the synthesis system. This type of function is based on a metric; the goal is indeed to search for the unit that is *closest* to the target synthesis unit. Two strategies are possible: one is to define the metric on the basis of expertise and confront the result with subjective tests [BRE 98], [PRU 01], [LEE 01], [PEN 90], the other is to implement a learning procedure from examples[20] [BLA 95], [LEM 96], [HUN 96].

[HIR 89] defines a contextual mapping function based on segmental (Japanese allophones) and supra-segmental (mean F_0, F_0 contour and allophone duration) criteria. [TAK 90], [TAK 92] present a similar solution relying on symbolic contextual information that takes into account prosodic information. They propose an original decision algorithm called "top-down hypothesis", which works in three steps. First, discontinuities between units are minimized. The number of candidate units is then reduced by applying a selection function using contextual criteria. Finally, a second selection function based on a similarity measure between F_0 contours is used to decide on the optimal unit.

[BLA 95] defines a match function as a linear combination of distances between acoustic attributes. This distance is then used to make a decision on the basis of a nearest neighbor rule. A learning technique is called for to estimate an optimal set of weights. A similar learning principle based on a similarity measure between synthetic and natural sentences was also proposed by [LEM 96] for French. [BLA 97], [NAK 95], [HUA 96] propose to design the match function using a classification tree [BRE 84]. The tree leaves define homogenous sets of acoustic segments. They are reached by following a path in the branches of the binary tree. The selection of the branches in the path is guided by phonetic criteria. The difficulty of such approaches is to define a tree that possesses a generalization capability, which is nothing else but a classical learning problem. If the depth of the tree is too high, the risk is to simply describe a contextual specificity of the training database, which is unlikely to reflect on new sentences to be synthesized. If the tree is not deep enough, the contextual description of the acoustic segments is too poor and leads to a synthesis result that is impossible to exploit (consider, in the extreme case, a one-node tree for which all acoustic segments in the database are equivalent candidates for any target unit etc.). [DON 03] specifies however that this type of representation is not adapted to model non-linear phenomena between the phonological and the acoustic descriptions.

---

20 In this case, an objective distance must be defined between the synthetic sentence and a reference sentence stemming from natural speech.

3.8.2.2. *Selection of an optimal sequence*

We have just discussed the first objective that is to describe the relationship between a candidate unit and a target unit in the reference corpus. We are now addressing the second objective, namely to find a sequence of candidate units which minimizes a global acoustic quality criterion. Taking into account continuity properties, when concatenating acoustic units, relies both on acoustic and symbolic criteria. Costs derived from symbolic characteristics were historically the first to be implemented [SAG 88], [BIG 93], [OLI 98]. The advantage of this approach is its simplicity, but a phonological description is not able to completely account for the acoustic variability (see section 3.4). Acoustically based costs try to overcome the modeling difficulties specific to the symbolic approach. The idea is to measure the continuity of synthesis units on the basis of metrics defined at the acoustic level. Most of the time, the metric is computed on acoustic features (LPC parameters, LSF or MFCC coefficients, etc.) as a Euclidean or non-Euclidean distance [NOM 90], [BOË 93], [HAN 98], [KLA 96], [DON 01]. [CON 96] proposes the notion of *optimal coupling*. The problem met with this type of minimization is related to the objective itself. In fact, speech spectral parameters evolve constantly over time. It can therefore be dangerous to search for absolutely stationary time intervals; it would be preferable to take into account the dynamics of acoustic parameters.

The search for a sequence of units relies generally on a graph of potential sequences together with a metric that quantifies the global cost of any sequence. The first generation of systems of that kind tried to build graphs for corpus of pre-recorded acoustic units [BIG 93]. In this case, the nodes of the graph correspond to the description of a speech unit that is *already* known. In that case, it is an atomic element that cannot be decomposed.

For continuous speech corpora, the core representation is the phone, as there is no notion of predefined acoustic units. In that case, the problem is much more difficult to deal with, as it is necessary both to sequence the linguistic representation (i.e. to find the relevant units which are still unknown) and to form optimal sequences (i.e. to minimize the acoustic cost). Once defined the target match function and the acoustic metric, the task consists of finding a path in a valued graph. In most cases, the costs are greater or equal to zero, and shortest path algorithms such as Dijkstra [DIJ 59] or Floyd [FLO 62] can be used [HUS 96]. [BIG 93] implements a variant, known as A*, which operates according to a "width first" strategy and, as opposed to Dijkstra's algorithm, which ranks the hypotheses according to some heuristics. It can be shown that, if the heuristic (which estimates the cost at the exit node) never overestimates the true cost, the solution is admissible and this algorithm is the most efficient process for finding it [MAR 77].

3.9. Applications and standardization

Enumerating exhaustively all industrial products using speech synthesis would be extremely fastidious. However, it is interesting to sketch the taxonomy of the application domains of this technology.

First, it is essential to recall that a speech synthesis system produces information that is distributed via an acoustic channel. A major feature of this channel is to be rather inadequate for communication multiplexing. In a conversation, the acoustic channel is used almost exclusively by one speaker at a time. The speech synthesis system will turn out to be a good technological choice if the application is guaranteed to be the only one to use the acoustic space available around the information delivery system.

Today, the quality and the intelligibility of synthetic speech is less effective than that of natural speech. Users of information delivery by voice will only accept synthetic speech if they have no other choice.

In summary, the ideal prototype for a service using speech synthesis is a service that must deliver a large quantity of unpredictable information in an ergonomic context where the acoustic channel is as user-friendly as possible.

In the telecommunication sector, several operators propose two leading services based on speech synthesis: reading e-mails on the phone (notably for mobile phones, for which the reading surface is limited) and information delivery concerning telephone numbers (direct or reverse directories).

In the sector of embedded devices, the miniaturization of components makes it possible to design navigation tools for vehicles or more specific systems such as reading fault report in the context of system control.

Finally, it is important to mention some historical applicative frameworks for speech synthesis, concerning reading aids for the blind and the visually-impaired [MOU 03].

Some research also deals with the problem of how to adapt speech synthesis system to a specific application [BLA 00].

Since the 1990s, a multitude of text-to-speech synthesis systems have emerged. The developments were mainly software based, even if, occasionally, dedicated processors were called for when the acoustic processing steps required a relatively important computation load (electronic devices for signal processing, for instance). From a technological viewpoint, it is necessary to base the developments on common *interfaces*, so as to functionally structure the synthesis system into consistent processing units. Here, the objective of standardization is essential. It can be a *de facto* standard (by sharing a knowledge corpus or as the result of the will of a consortium) or a predefined standard (designed by international standardization bodies).

In practice, a simultaneous standardization of data resources (structuring of the linguistic or prosodic information, for instance) and of processes (a software being accompanied with a set of programming methods dedicated to speech synthesis) took place. SSML, i.e. speech synthesis markup language [TAY 97] is an example of information structuring in the syntactic and prosodic domains, using a set of labels located in the text stream to synthesize. This first attempt then evolved towards SABLE, which served as a base to the recent development of VoiceXML in W3C.

Many standardization attempts have been taking place, as a result of initiatives stemming from the industrial sector: MS-SAPI (Microsoft Speech API) and JSAPI (Java Speech API) are the most common implementations of Speech API (application programming interface). Speech API (or SAPI) describes procedures for speech processing (synthesis, recognition, coding, etc). It is composed of three interfaces. The first interface, *Voice Text*, describes software methods for starting, stopping, suspending, etc., the vocal synthesis of a text. The second interface is a configuration interface, *Attribute Interface*, for specifying some behaviors of the synthesis, i.e. the characteristics of the audio stream at the output of the system or the type of synthetic voice to use. Finally, the third interface is the *Dialog Interface*, for parametrizing the behavior of some functions of the synthesis system, for instance the modification of a lexicon of exceptions for grapheme-to-phoneme conversion. The SAPI model also proposes a set of markers that can be integrated in a text. These markers enable the modification of various parameters of the synthesis systems: emphasis on a word, insertion of a pause, modification of the average laryngeal frequency, speech rate measured in terms of number of words per minute, pronunciation of a word, etc.

An ever-growing part of the texts to synthesize are in the HTML format (hyper-text markup language). HTML is part of the wider language family SGML (standard generalized markup-language) and relies on the use of markers that are inserted in the text. An HTML analyzer, thanks to specific escape sequences, can locate these markers. For instance, the code <p> associated with </p> specifies the opening and closing of a paragraph. SSML has been a first response to the objective of a marker-oriented language for speech synthesis. The *phrase* marker, for instance, enables the definition of an entity that will be syntactically interpreted as a phrase by the synthesis system. The *define* marker is used to define the grapheme-to-phoneme transcription of the word within a sentence. As a last example, let us mention the *language* marker, for switching dynamically from one language to the other. Bell Laboratories took up these ideas and proposed a richer version: STML for spoken text markup language. Today, the VoiceXML description language has been normalized by W3C: it integrates speech synthesis and recognition. The wide use of XML has been very advantageous, as XML parsers are widely used and XML markers have a strict semantic.

As regards ISO, MPEG-4 provides a very interesting standardization of speech-oriented components. In particular, MPEG-4 normalizes the data stream dedicated to speech synthesis. This coding scheme accounts at the same time for the linguistic, segmental and acoustic levels of the signal. MPEG-4 was designed as a scene integration tool. In that respect, it provides parameter sets for synchronizing the speech synthesis stream with a visual stream (position of the lips in the sentence).

More recently, in the sector of network protocols, the IETF (Internet Engineering Task Force) has proposed a draft standard for the access to speech synthesis and recognition resources. The MRCP protocol (for Multimedia Resource Control Protocol) provides an elegant solution to the problem of integrating software resources stemming from distinct sources. Even if they are SAPI compliant, the various hardware and software environments can vary and pose integration problems. MRCP is a protocol that responds efficiently to a client/server integration strategy [L'H 04].

3.10. Evaluation of speech synthesis

3.10.1. *Introduction*

3.10.1.1. *Synthesis systems*

The evaluation of speech synthesis is a research and application domain in direct connection with the current development of synthesis. Indeed, if a method existed to precisely evaluate and diagnose the quality defects of a speech synthesizer, it would be easier to overcome them or at least to find solutions. Therefore, progress made in evaluating and analyzing synthetic voice quality conditions intimately that of the synthesis algorithm themselves.

Let us recall that the term *speech synthesis* can cover very different types of systems:

– restitution of a pre-recorded message;

– concatenation of words or syntactic groups, with a fully determined vocabulary and phraseology;

– text-to-speech synthesis, i.e. the pronunciation of any text with an unlimited vocabulary; this is a much more complex task and its result is much less pleasant to the ear.

In the first two cases, the evaluation of speech synthesis is necessarily global. Conversely, for text-to-speech synthesis, the evaluation can be both global and specific to certain aspects of the system. Referring to the description of sections 3.4 and 3.5, three functional components can be identified in a text-to-speech synthesis system:

– text analysis, for transforming the written text into a sequence of phonetic symbols (which represent the pronunciation of the text) and prosodic symbols (for marking accentuation and intonation). This first component is primarily related to the automatic language-processing field;

– the symbolic description of the text must be converted into numerical parameters (pitch contour, pauses, voice strength, etc.) using phonetic-acoustic procedures. This second component uses several types of algorithms, such as rule-based systems, stochastic automata or neural networks;

– the last component is the acoustic synthesizer, which computes the speech signal from the numerical parameters. This component is merely related to signal processing.

Text-to-speech synthesis is, so to speak, a chain of processes from the text to the acoustic signal. The weakest link in this chain is the limiting factor for the quality. It is therefore important to evaluate each link, at the same time as the global system. This is why we will address in this section both the analytical evaluation, i.e. the evaluation of the separate modules according to a *glass box* approach and the global evaluation of the systems, i.e. the *black box* scheme [BEN 96].

3.10.1.2. *Why evaluate?*

Evaluating speech synthesis is a necessity and it will remain so, as long as the quality of synthetic speech will be inferior to that of human speech (which is bound to be the case for a long time, as speech calls for all the physical, psychical and spiritual faculties of the human being). When designing a system, it is important to be able to measure the progress made at different phases, which can only be done by objective means that do not involve the opinion of those who are designing the system. It is also important to evaluate systems in order to choose and decide what system to use and how. System designers and users thus have the same needs, for diagnosis, comparison and standardization:

– diagnostic tests of full systems and system components, so as to assess their quality level;

– comparative tests to evaluate the merits and to rank the systems or their components;

– standardization tests to choose a system for a given application by measuring its acceptability, to obtain its homologation and to promote standards.

In general, the reference for evaluating synthesis systems is natural speech, either in clean form or intentionally degraded. In order to compare the results of a same test with different systems, it is recommended to proceed to test calibration by using common references. Therefore, the production of resources for evaluation is an important aspect (list of sentences, words, phonetized texts, test software, etc).

The vast majority of tests are based on the judgment of a group of human subjects. This requires methodological precautions (for instance, taking into account familiarization and learning effects, linguistic and auditory abilities of the subjects, the type of tasks, the motivation of the subjects, etc.). This implies that only very few evaluation tasks can indeed be fully automatic, as opposed to speech recognition, for instance. Therefore, speech synthesis evaluation depends on human factors and it is therefore a lengthy and costly process, as soon as objectivity is sought out rather than impressions [CAL 89], [SOR 96].

3.10.2. *Global evaluation*

Global evaluation refers to the evaluation of a speech synthesis system output, without taking in consideration its internal functioning and without trying to understand the source of its defects. This type of evaluation will thus be useful to the designer of the system and to their users. Two families of tests for the global evaluation of speech synthesis systems are commonly used in the literature. The first family corresponds to the quantitative and qualitative intelligibility of synthesis, whereas the second one aims at a global appreciation in terms of a mark on the subjective quality of synthetic speech, according to multi-scale or semantic analyses.

3.10.2.1. *Intelligibility*

Above all, a speech synthesis system must provide intelligible speech. Therefore intelligibility tests are the primary type of tests that must be implemented to evaluate an application before any other more subtle quality issues are considered. Several test protocols have been proposed for the evaluation of intelligibility (or clearness).

Clearness tests using logatomes. This test involves using logatomes (meaningless pseudo-words following the phonotactic rules of the language), with a CVC structure (where V is a vowel and C a consonant or a group of consonants). 4 to 8 lists of 50 logatomes, synthesized by several systems, are listened to by a set of subjects who transcribe phonetically what they have heard. Occasionally, some noise is added to a natural reference, for comparison purposes. The subjects have to be trained for phonetic listening and transcription, since the logatomes do not correspond to any known words.

Clearness tests using syllables. The test by Fletcher and Steinberg [FLE 29] uses CVC syllables, by groups of 66 syllables. Similarly, the listeners must be trained.

Clearness tests using words. The original article [EGA 48] used 20 lists of 50 common monosyllabic words. The test requires some training and it seems that the use of either logatomes or words does not make any significant difference.

Rhyme tests (MRT, DRT). These tests aim at focusing the attention of the subjects on a unique sound by using a forced choice paradigm. Proposed by Fairbanks in 1958, the FRT (*Fairbank Rhyme Test*) is intended to test initial consonants by using monosyllabic stimuli (and responses). The goal of the listener is to complete the initial consonant of words presented in the form VC (open answer). The *Modified Rhyme Test* (MRT) uses CVC stimuli and the response is chosen among 6 possibilities. Only one consonant differs between the options (either the initial one or the final one) and the stimulus is included in a carrying sentence. Some noise may be added to the speech signal. This test is not very difficult and it is frequently used. The *Diagnostic Rhyme Test* (DRT) allows for the study of phonetic confusions. The subject can choose between two words among which only one consonant is different by a single distinctive feature. Rhyme tests are targeted towards naive subjects who do not have any particular phonetic knowledge, while providing accurate diagnostic possibilities on the type of confusions and on the overall system intelligibility.

Semantically unpredictable sentences (SUS). This test is, so to speak, analogous to logatomes at the sentence level. The stimuli are composed of meaningless sentences, i.e. sentences which are unpredictable from a semantic viewpoint. They are formed with words and syntactic structure compatible with the language [BEN 90]. The five most frequent syntactic structures have been retained. The sentences are formed with a maximum of 7 words (which is the maximum number of words which can be processed globally by a human brain). The stimuli are composed of three types of words (CV, CV and CVC) across which the consonants vary. A complete modular test system has been integrated in the SOAP system [HOW 92].

Intelligibility and cognitive load. For all the aforementioned tests, the listeners must focus exclusively on the evaluation task, which is not a very realistic situation. In real life, speech messages can be listened to while doing something else at the same time, without being too strained. Therefore, measurements have been proposed to monitor the cognitive load created by a listening task [SIL 90] so as to evaluate the synthetic speech intelligibility. The subject must simultaneously answer questions related to a synthesized text and carry out some visual and motor task (to follow with a mouse a mobile object on a computer screen, for example). This type of speech synthesis quality measure evaluates jointly the intelligibility of the message and the effort that the listener must dedicate to comprehend it.

Intelligibility and telephone. The previous evaluation schemes are generally independent from a particular application. The results may not be valid for a degraded transmission channel. In [SPI 90], a study was carried out to measure the influence of telephone transmission on the intelligibility of synthetic speech.

In summary, five types of intelligibility tests are generally used: logatomes, words in phonetically balanced lists, MRT, DRT and SUS. If it is not possible to carry out several tests, if the subjects have awareness in phonetics and if a numerical figure is expected from the evaluation, syllable or logatomes are to be preferred to a rhyme test. If, conversely, a global evaluation has demonstrated that no overall comprehension problem has taken place, the MRT will help to detect specific problems and the DRT will provide diagnostic elements (in particular, for rule-base speech synthesis), in situations where a particular phenomenon is under study [LAR 90]. To proceed to fast and cheap performance estimation, the MRT is the best solution.

3.10.2.2. *Global quality*

Even if its intelligibility is acceptable, synthetic speech can be of mediocre quality. It is therefore necessary to test the global subjective quality. Several procedures have been set up in this respect:

Task-oriented conversation. This test consists of simulating a conversation between the subject and the synthesis system. The task of the subject is to resolve a particular problem. At the end of the conversation, the subject fills out a questionnaire concerning the quality of the synthesis.

Pair-wise preference judgment. This method is generally used to choose one system over others, whose general level of quality is known. In this case, an A-B test or an A-B-X test can be implemented.

Degradation Category Rating (DCR). The DCR method enables a direct comparison of several systems. Stimuli are composed of A-B-A-B sequences, where A is a sentence stemming from a reference system and B the same sentence synthesized by the system under test. For each test, the subject provides a judgment on the degradation of B with respect to A, on a 5-level scale: degradation is inaudible = 5, degradation is audible but not annoying = 4, degradation is slightly annoying = 3, degradation is annoying = 2, degradation is very annoying = 1.

Generally, the corpus is composed of 2 sentences chosen among a list of 10 phonetically balanced sentences and 16 subjects with no particular competence are selected. Several sets of comparison can be presented, by varying the level of noise superimposed on the reference signal between each stimulus.

Absolute Category Rating (ACR). After having listened to synthetic sentences or texts, the subjects fill out a questionnaire in which they express their opinion. There are several types of such tests. The ACR method is used to evaluate and compare the quality of systems with respect to a reference, by listening to the system output separately. The comparison between systems is therefore indirect. The speech material is produced by a single system; it lasts between 6 and 10 seconds and it contains 2 or 3 sentences separated by a pause of at least 1 second. A number of 8 such test samples is produced for each system, and the acoustic level is equalized. At least 12 subjects participate in the test. A noisy natural speech reference can be used. For each test sample, the listeners choose one rating on the following scale:

a)   listening-quality test: excellent = 5, good = 4, fair = 3, poor = 2, bad = 1;
b)   listening-effort scale: complete relaxation possible; no effort required = 5, attention necessary; no appreciable effort required = 4, moderate effort required = 3, considerable effort required = 2, no meaning understood with any feasible effort = 1;
c)   loudness-preference scale: much louder than preferred = 5, louder than preferred = 4, preferred = 3, quieter than preferred = 2, much quieter than preferred = 1.

Global quality (UIT-T P.85). A multi-dimensional evaluation procedure of the global quality was normalized in 1994 by the ITU (International Telecommunication Union, formerly CCITT). The speech test samples last between 10 and 30 seconds and a human reference (possibly degraded) is recommended. Subjects listen twice to each test, with a total duration of approximately one hour (for instance, 4 stimuli for 4 systems and 3 references), including instructions and learning. The 8 analysis dimensions are:

1. Overall impression: how do you rate the quality of the sound of what you have just heard? Excellent – good – fair – poor – bad.

2. Listening effort: how would you describe the effort you were required to make in order to understand message? Complete relaxation possible; no effort required – attention necessary; no appreciable effort required – moderate effort required – effort required – no meaning understood with any feasible effort.

3. Comprehension problems: did you find certain words hard to understand? Never – rarely – occasionally – often – all of the time.

4. Articulation: were the sounds distinguishable? Yes, very clear − yes, clear enough − fairly clear − no, not very clear − no, not at all.

5. Pronunciation: did you notice any anomalies in pronunciation? No − yes, but not annoying − yes, slightly annoying − yes, annoying − yes, very annoying.

6. Speaking rate: the average speed of delivery was: much faster than preferred − faster than preferred − preferred − slower than preferred − much slower than preferred.

7. Voice pleasantness: how would you describe the voice? Very pleasant − pleasant − fair − unpleasant − very unpleasant.

8. Acceptance: do you think this voice could be used for such an information service by telephone? Yes − no.

Global quality (Verbmobil). A multi-dimensional assessment procedure for the evaluation of the global quality of synthetic speech was proposed in 1995 for German in the context of the Verbmobil project [KRA 95]. The 8 analysis dimensions and the corresponding scales are:

1.   Naturalness: 1 − very natural, 2 − natural, 3 − rather natural, 4 − rather unnatural, 5 − unnatural, 6 − very unnatural.

2.   Intelligibility: 1 − very easy, 2 − easy, 3 − rather easy, 4 − rather hard, 5 − hard, 6 − very hard.

3.   Comprehensibility: 1 − very easy, 2 − easy, 3 − rather easy, 4 − rather hard, 5 − hard, 6 − very hard.

4.   Pleasantness: 1 − very pleasant, 2 − pleasant, 3 − rather pleasant, 4 − rather unpleasant, 5 − unpleasant, 6 − very unpleasant.

5.   Distinctness: 1 − very clear, 2 − clear, 3 − rather clear, 4 − rather unclear, 5 − unclear, 6 − very unclear.

6.   Speed: 1 − much too slow, 2 − too slow, 3 − somewhat too slow, 4 − somewhat too fast, 5 − too fast, 6 − much too fast.

7.   Pronunciation: 1 − not annoying, 2 − slightly annoying, 3 − rather annoying, 4 − annoying, 5 − very annoying.

8.   Stress: 1 − not annoying, 2 − slightly annoying, 3 − rather annoying, 4 − annoying, 5 − very annoying.

Global quality (JEIDA). For Japanese, a 6-dimensional grid has been designed in 1995, in coordination by the JEIDA (Japan Electronic Industry Development Association) [OFF 95]. The grid is composed of bipolar semantic scales using pairs of opposite words, which is a rather different approach from other multidimensional approaches:

1. *Descriptive words for the intelligibility:* easy/hard to understand, easily misread/hardly misread;

2. *Descriptive words for the sound quality:* beautiful/dirty, smooth/rough, glossy/lifeless, sharp/dull, full of life/nasal, articulate/muffled, thick/thin, powerful/weak, rich/poor, grave/light, sweet/metallic sound, soft and full/harsh, bright/somber, soft/hard, clear/turbid;

3. *Descriptive words for the temporal factors:* natural/unnatural rhythm, fast/slow, continuous/choppy;

4. *Descriptive words for the intonation:* natural/unnatural intonation, natural/unnatural accent, fluent/halting;

5. *Descriptive words for the overall goodness:* human-like/artificial, preferable/not preferable, excellent/poor;

6. *Descriptive words for the suitability:* easy/hard to hear, comfortable/frustrating, pleasant/annoying, Japanese/foreign, male/female, high voice/low voice, young/old, suitable/unsuitable for the purpose.

3.10.3. *Analytical evaluation*

Global evaluation provides information as of the level of quality reached by a system in reference to more or less degraded (or coded) natural speech. For the system designer, or for specific applications, it is also worthwhile to test each system component separately [SAN 93].

3.10.3.1. *Grapheme-to-phoneme conversion*

Even though grapheme-to-phoneme conversion seems rather simple to evaluate, as compared to other system components (such as prosody generation), many questions have to be raised. Which phonetic alphabet should be used? How do we handle pronunciation variants? Should the reference be a text or a lexicon? Should evaluation text be found or fabricated by a linguist in order to concentrate difficulties? What entry format should be used? What about spelling mistakes? Should morpho-syntactic labels also be provided? What output format should be used? How do we represent liaisons? What metric do we use to score the results? Practical answers have been proposed for French by [YVO 98], which describes the evaluation of grapheme-to-phoneme transcription systems in French, carried out in the context of the francophone AUPELF program. A remarkable feature of this type of evaluation is its automatic nature: there is no need to call for human subjects to evaluate the transcription, as soon as a reference transcription (with pronunciation variants) has been established.

3.10.3.2. *Prosody*

Research on the evaluation of speech synthesis has shown that tested subjects are generally more confident in their own judgment when the comparison deals with global acoustic differences rather than differences focused on supra-segmental or prosodic levels. Moreover, it is important to take into account the interaction between the phonemic and prosodic levels: indeed, scores obtained for tests dealing with prosody intimately depend on the segmental system quality, and vice-versa. It was shown that, when comparing two prosody generation systems A and B by using two distinct synthesis systems at two segmental levels 1 and 2, A may be judged better than B when using 1 whereas B will be judged better than A when using 2 [BEN 91]. Prosody evaluation *per se* is therefore delicate and no real standard test exists today. Some tests use a "delexication" step, involving the replacement of the original text by sequences of "mamama". Prosodic tests can focus either on the prosodic form or on the prosodic function. Concerning the form, a judgment of the naturalness by naïve listeners is sought out for the prosody of sentences [GRI 91]. Concerning the function, the objective is to evaluate whether the synthetic prosody yields a proper perception of the linguistic content (for instance, assertion, question, doubt, etc.).

3.10.3.3. *Acoustic synthesizer*

The acoustic synthesizer plays a part in the synthesis system for analyzing, modifying and synthesizing the speech signals used in concatenative synthesis, or for the computation of the signal through a formant-based synthesizer. In the evaluation context, the difficulty is in measuring the ability of the analysis/synthesis or synthesis systems to implement diverse natural signal transformations, or to produce realistic synthesis signals, for prosodic tasks in the wide sense (pitch, duration, intensity and voice quality modifications) [BAI 00].

In the potential list of synthesizers, we can find systems which operate directly on the time-domain signal, those which carry out some source/tract separation, those which split the signal into a harmonic component and a noise component and finally, those based on formant synthesis. Prosodic modifications consist of transforming a synthetic reference signal so as to approach at best the characteristics extracted from another signal (the prosodic target). The reference signal must be either a natural signal or a signal composed of segments (diphones, polysounds, etc.) extracted from a database.

Many comparisons have already been made as to what concerns the quality assessment of such or such concatenation method, but a more ambitious approach is considered in [BAI 00]. Pairs of sentences are designed with various speaking rates, intonations, articulation strengths, styles, so as to evaluate the capacity of the system to transfer prosody from one to the other. An advantage is to define a task that can be calibrated with an absolute reference. This benchmark can include several levels of evaluation (pitch or duration transfer, but also types of voice (breathy, creaky, stressed), vowel reduction, etc.).

3.10.4. *Summary for speech synthesis evaluation*

Speech synthesis evaluation is a difficult and costly procedure. It is not well suited to automation, because the factors to evaluate resort to comprehension, opinion and pleasantness as judged by human beings. Human perception of synthetic speech is discussed in details in [PIS 96] which recommends that cognitive and behavioral aspects related to the targeted application are taken into account.

As speech synthesis has an important application sector in the telecommunication domain, standardization work and international evaluation campaigns have been taking place. Regarding telecommunications, recommendations have been proposed by the CCITT [CCI 85b] [CCI 85a], and then by the ITU-T [UIT 94]. The Esprit project SAM 2589 (Speech Assessment Methodologies) of the European Community ended in 1992 [SAM 92] [POL 92] [CAR 90]. This project's final report proposed several test protocols for the different levels (segmental level, supra-segmental level, prosodic level and global evaluation). Test software has been released for several European languages. Another European project called EAGLES (European Advisory Group on Language Engineering Standards, LRE-61-100 project) produced a detailed report [GIB 97] [BEZ 95] and recommendations of evaluation, yet without proposing any original procedures. Similar work has been published on the Japanese side, by the JEIDA, more specifically on the Japanese language [OFF 95]. For French, a francophone project conducted within the AUPELF took place between 1996 and 1999 [D'A 98] [D'A 00] [YVO 98]. An analytical approach was favored and results have been obtained in this framework for the evaluation of grapheme-to-phoneme transcription, prosody generation and acoustic synthesis systems. For German, an evaluation effort was carried out by the Verbmobil project [KRA 95] [KLA 967], concerning intelligibility, comprehension and multi-scale evaluation. Finally, the international COCOSDA project (International Coordinating Committee on Speech Databases and Speech Input/Output Systems Assessment) and its European branch

(EuroCOCOSDA, project LRE 62-057) has a "speech synthesis" group which proposed a questionnaire to research laboratories working in the field of speech synthesis, concerning their common practice in terms of evaluation tests, in particular which tests they use, envisage or would like to have [POL 96]. This is indeed a vast program.

3.11. Conclusions

Because it materializes a human communication action, the finest details and subtleties of speech will never be completely unveiled. A simple tape-recorder is sufficient to capture a speech signal, but its study calls for both physical and linguistic concepts.

Current text-to-speech synthesis systems are able to generate speech with an acceptable quality, if the message is neither too long nor too poor in terms of information content. However, it is worthwhile noting that synthetic speech quality has been constantly improving over the years. The technological developments responsible for this progress have mainly concerned acoustic signal processing and computational power speed-ups, for processing ever-larger quantities of linguistic data.

However, will we ever be able to have a machine read Verlaine's poems convincingly? For sure, the current research fields will have to be complemented by more complex formulations and models of our cognitive system. Meanwhile, it is already possible today, with our current level of knowledge, to have speech synthesis systems produce rather surprising utterances.

3.12. References

[ALL 77] J.B. ALLEN. "Short term spectral analysis, synthesis, and modifications by discrete Fourier transform", *IEEE Transaction on Acoustics, Speech and Signal Processing (ASSP)*, 25(3): pp 235-238, 1977.

[ALL 87] S. ALLEN, S. HUNNICUT and D. KLATT. *From Text To Speech, The MITTALK System*, Cambridge University Press, 1987.

[AUB 91] V. AUBERGE. La synthèse de la parole, des règles aux lexiques, PhD thesis, University of Grenoble, 1991.

[BAG 98] P. C. BAGSHAW. "Phonemic transcription by analogy in text-to-speech synthesis: novel word pronunciation and lexicon compression", *Computer Speech and Language*, 12(2): pp119-142, 1998.

[BAI 86] G. BAILLY. "Multiparametric generation of French prosody from unrestricted text", *Proceedings of the IEEE International Conference on Acoustics, Speech and Signal Processing*, pp 2419-2422, 1986.

[BAI 00] G. BAILLY. "Evaluation des systèmes d'analyse-modification-synthèse de parole", in *Actes des Journées d'Étude sur la Parole (JEP'2000)*, pp 109-112, 2000.

[BAL 99] M. BALESTRI, A. PACCHIOTTI, S. QUAZZA, P.-L. SALZA and S. SANDRI. "Choose the best to modify the least: A new generation concatenative synthesis system", in *Proceedings of the European Conference on Speech Communication and Technology (EUROSPEECH'99)*, pp 2291-2294, 1999.

[BAR 95] P. A. BARBOSA and G. BAILLY. "Generation of pauses with the z-score model", in *Progress in Speech Synthesis*. Springer Verlag, New York, 1995.

[BEN 90] C. BENOÎT. "An intelligibility test using semantically unpredictable sentences: towards the quantification of linguistic complexity", *Speech Communication*, 9(4):293-304, 1990.

[BEN 91] C. BENOÎT, F. EMERARD, B. SCHNABEL and A. TSEVA. "Quality comparison of prosodic and of acoustic components of various synthesisers", *Proceedings of Eurospeech* 91, 2:875-pp 878, 1991.

[BEN 96] C. BENOÎT. "Evaluation inside or assessment outside?"*Progress in Speech Synthesis*, in R. VAN SANTEN, R. SPROAT, J. HIRSCHBERG and J. OLIVE (Eds.), Springer Verlag, 1996.

[BEU 99] M. BEUTNAGEL, A. CONKIE, J. SCHROETER, Y. STYLIANOU and A. SYRDAL. "The AT&T Next-Gen TTS system", in *Proceedings of the Joint Meeting of ASA, EAA and DEGA*, 1999.

[BEZ 95] R. VAN BEZOOIJEN and V.J. VAN HEUVEN. "Assessment of speech output system", *Eagles* (LRE-61-100) report, 1995.

[BIG 93] D. BIGORGNE, O. BOËFFARD, B. CHERBONNEL, F. EMERARD, D. LARREUR, J.L. LE SAINT-MILON, I. MÉTAYER, C. SORIN and S. WHITE. "Multilingual PSOLA text-to-speech system", in *Proceedings of the IEEE International Conference on Acoustics, Speech and Signal Processing (ICASSP'93)*, pp 187-190, 1993.

[BIM 88] F. BIMBOT. "Synthèse de la parole: des segments aux règles, avec utilisation de la décomposition temporelle ", PhD thesis, Ecole Nationale Supérieure des Télécommunications (ENST), Paris, 1988.

[BLA 95] A.W. BLACK and N. CAMPBELL. "Optimising selection of units from speech databases for concatenative synthesis", in *Proceedings of the European Conference on Speech Communication and Technology (EUROSPEECH'95)*, pp 581-582, 1995.

[BLA 97] A. BLACK and P. TAYLOR. "Automatically clustering similar units for unit selection in speech synthesis", in *Proceedings of the European Conference on Speech Communication and Technology (EUROSPEECH'97)*, pp 601-604, 1997.

[BLA 00] A.W. BLACK and K. LENZO. "Limited domain synthesis", in *Proceedings of the International Conference on Spoken Language Processing (ICSLP'00)*, 2000.

[BLA 01] A.W. BLACK. "Perfect synthesis for all of the people all of the time", in *Proceedings of the IEEE/ESCA TTS Workshop, 2001.*

[BOË 92] O. BOËFFARD, L. MICLET and S. WHITE. "Automatic generation of optimal unit dictionaries for text-to-speech synthesis", in *Proceedings of the International Conference on Spoken Language Processing (ICSLP'92)*, pp 1211-1214, 1992.

[BOË 93] O. BOËFFARD, B. CHERBONNEL, F. EMERARD and S. WHITE. "Automatic segmentation and quality evaluation of speech unit inventories for concatenation based multilingual PSOLA text-to-speech systems", in *Proceedings of the European Conference on Speech Communication and Technology (EUROSPEECH'93)*, pp 1449-1452, 1993.

[BRE 84] L. BREIMAN, J. FREIDMAN, R. OLSHEN and C. STONE. *Classification and Regression Trees*, Wadsworth and Brooks, 1984.

[BRE 98] A.P. BREEN and P. JACKSON. "Non-uniform unit selection and the similarity metric within BT's laureate TTS system", in *Proceedings of the ESCA Workshop on Speech Synthesis*, pp 373-376, 1998.

[BUC 96] A.L. BUCHSBAUM and J.P.H. VAN SANTEN. "Selecting training inputs via greedy rank covering", in *Proceedings of the Annual Symposium on Discrete Algorithms (SODA)*, pp 288-295, 1996.

[CAL 89] CALLIOPE. *La parole et son traitement automatique*, Masson, Paris, 1989.

[CAM 93] N. CAMPBELL. "Detecting prosodic boundaries in a speech signal", *ATR Research Activities of the Speech Processing Department*, 1993.

[CAM 94] N. CAMPBELL. "Prosody and the selection of units for concatenation synthesis", in *Proceedings of the ESCA-IEEE Workshop on Speech Synthesis*, pp 61-64, 1994.

[CAR 82] R. CARLSON, B. GRANSTRÖM and S. HUNNICUT. "A multi-language text-to-speech module", *Proceedings of the IEEE International Conference on Acoustics, Speech and Signal Processing*, 3:1604-1607, 1982.

[CAR 90] R. CARLSON, B. GRANSTRÖM and L. NORD. "Segmental evaluation using the Esprit/SAM test procedures and mono-syllabic words", in *Talking Machines*. North Holland, G. Bailly and C. Benoît ed., 1990.

[CCI 85a] CCITT. "Evaluation subjective de la qualité de codeurs numériques par la procédure de jugements par catégories de dégradations (DCR)", Annexe B au supplément No 14, livre rouge, vol. V:279-284, 1985.

[CCI 85b] CCITT. "Méthode de jugement par catégories absolues (ACR) pour les essais subjectifs des dispositifs numériques ", Annexe A au supplément 14, livre rouge, vol V:274-279, 1985.

[CHA 86] F. CHARPENTIER and M. STELLA. "Diphone synthesis using an overlap and add technique for speech waveforms concatenation", in *Proceedings of the IEEE International Conference on Acoustics, Speech and Signal Processing (ICASSP'86)*, pp 2015-2018, 1986.

[CHA 88] F. CHARPENTIER. Traitement de la parole par analyse-synthèse de Fourier, application à la synthèse par diphones, PhD thesis, Ecole Nationale Supérieure des Télécommunications (ENST), 1988.

[CHA 89] F. CHARPENTIER and E. MOULINES. "Pitch-synchronous waveform processing techniques for text-to-speech synthesis using diphones", in *Proceedings of the IEEE International Conference on Acoustics, Speech and Signal Processing (ICASSP'89)*, pp 13-19, 1989.

[CHE 00] N. CHENFOUR, A. BENABBOU and A. MOURADI. "Étude et évaluation de la disyllabe comme unité acoustique pour le système de synthèse arabe Paradis ", in *Proceedings of the International Conference on Language Resources and Evaluation (LREC'00)*, 2000.

[CHO 97] F.C. CHOU, C.Y. TSENG, K. CHEN and L. LEE. "A Chinese text-to-speech system based on part-of-speech analysis, prosodic modelling and nonuniform units", *Proceedings of the IEEE International Conference on Acoustics, Speech and Signal Processing (ICASSP'97)*, pp 923-926, 1997.

[CHO 98] F. C. CHOU and C.Y. TSENG. "Corpus-based mandarin speech synthesis with contextual syllabic units based on phonetic properties", *Proceedings of the IEEE International Conference on Acoustics, Speech and Signal Processing (ICASSP'98)*, pp 893-896, 1998.

[CHU 01a] M. CHU, H. PENG and E. CHANG. "A concatenative mandarin TTS system without prosody model and prosody modification", in *Proceedings of the ESCA Tutorial and Research Workshop on Speech Synthesis (SSW)*, 2001.

[CHU 01b] M. CHU, H. PENG, H. YANG and E. CHANG. "Selecting nonuniform units from a very large corpus for concatenative speech synthesizer", in *Proceedings of the IEEE International Conference on Acoustics, Speech and Signal Processing (ICASSP'01)*, 2001.

[COI 91] B. VAN COILE. "Inductive learning of pronunciation rules with the DEPES system", in *Proceedings of the IEEE International Conference on Acoustics, Speech and Signal Processing*, 2:745-748, 1991.

[CON 96] S. CONKIE, A. and S. ISARD. "Progress in Speech Synthesis", Chapter "Optimal coupling of Diphones", pp 293-304. Springer Verlag, 1996.

[CON 99] A. CONKIE. "Robust unit selection system for speech synthesis", in *Proceedings of the Joint Meeting of ASA, EAA, and DAGA (AED)*, pp 52-55, 1999.

[CRO 80] R.E. CROCHIERE. "A weighted overlap-add method of short-time Fourier analysis/synthesis", *IEEE Transactions on Acoustics, Speech and Signal Processing (ASSP)*, 28(1):99-102, 1980.

[D'A 95] C. D'ALESSANDRO and P. MERTENS. "Automatic pitch contour stylization using a model of tonal perception", *Computer Speech and Language*, 9:257-288, 1995.

[D'A 98] C. D'ALESSANDRO and B3 PARTNERS. "Joint evaluation of text-to-speech synthesis in French within the Aupelf Arc-B3 project", *Proceedings of the 3rd International Workshop on Speech Synthesis,* pp 11-16, 1998.

[D'A 00] C. D'ALESSANDRO, V. AUBERGE, G. BAILLY, F. BECHET, P. BOULA DE MAREÜIL, S. FOUKIA, J.P. GOLDMAN, E. KELLER, P. MERTENS, V. PAGEL,D. O'SHAUGHNESSY, G. RICHARD and F. YVON. "Evaluation de la synthèse de parole à partir du texte dans la francophonie: premiers résultats ", in *Ressources et évaluation en ingénierie des langues*, K. CHIBOUT, J. MARIANI, N. MASSON and F. NÉEL (eds), Editions Duculot, AUPELF-UREF, 2000.

[DIJ 59] E.W. DIJKSTRA. "A note on two problems in connection with graphs", *Numerishe Matematik,* 1:269-271, 1959.

[DIX 68] N.R. DIXON and H.D. MAXEY. "Terminal analog synthesis of continuous speech using the diphone method of segment assembly", *IEEE Transactions on Acoustics, Speech and Signal Processing (ASSP)*, 16(1):40-50, 1968.

[DOL 86] M. DOLSON. "The phase vocoder: a tutorial", *Computer Music Journal,* 10(4):14-27, 1986.

[DON 96] R.E. DONOVAN. Trainable Speech Synthesis, PhD thesis, Cambridge University, 1996.

[DON 01] R.E. DONOVAN. "A new distance measure for costing spectral discontinuities in concatenative speech synthesisers", in *Proceedings of the ESCA Tutorial and Research Workshop on Speech Synthesis (SSW4)*, 2001.

[DON 03] R.E. DONOVAN. "Topics in decision tree based speech synthesis", *Computer Speech and Language*, 2003.

[DUT 93a] T. DUTOIT. MBR-PSOLA. "Text-to-speech synthesis based on an MBE re-synthesis of the segments database", *Speech Communication*, 13(3/4):pp435-440, 1993.

[DUT 93b] T. DUTOIT and H. LEICH. "High quality text-to-speech synthesis: A comparison of four candidate algorithms", *Speech Communication,* 13(3/4):435-440, 1993.

[DUT 94] T. DUTOIT. "A comparison of four candidate algorithms", in *Proceedings of the IEEE International Conference on Acoustics, Speech and Signal Processing (ICASSP'94)*, pp 565-568, 1994.

[EGA 48] J.P. EGAN. "Articulatory testing methods", *Laryngoscope,* 58:955-991, 1948.

[EME 77] F. EMERARD. "Synthèse par diphones et traitement de la prosodie ", PhD thesis, University of Grenoble III, 1977.

[EME 92] F. EMERARD, L. MORTAMET AND A. COZANNET. "Prosodic processing in a text-to-speech synthesis system using a database and learning procedures", in *Talking Machines: Theories, Models and Designs* in G. BALLY, C. BENOÎT and T. R. SAWALLIS (eds.), Amsterdam: North Holland, 1992.

[EST 64] S.R. ESTES, H.R. KERBY, H.D. MAXEY and A.M. WALKER. "Speech synthesis from stored data", *IBM Journal of Research and Development,* 1964.

[FAN 60] G. FANT. *Acoustic Theory of Speech Production*, Mouton, The Hague, 1960.

[FLE 29] H. FLETCHER and J.C. STEINBERG. "Articulatory testing methods", *Bell Systems Technical Journal*, 8:806-853, 1929.

[FLO 62] R.W. FLOYD. "Algorithm 97: Shortest path", *Communications of the ACM*, 5(6):345, 1962.

[FRA 01] H. FRANÇOIS and O. BOËFFARD. "Design of an optimal continuous speech database for text-to-speech synthesis considered as a set covering problem", in

Proceedings of the European Conference on Speech Communication and Technology (EUROSPEECH'01). 2001.

[FUJ 76] O. FUJIMURA. "Syllables as concatenated demisyllables and axes", *The Journal of the Acoustical Society of America (JASA)*, 59(S1(A)), 1976.

[FUJ 82] H. FUJISAKI and K. HIROSE. "Analysis and synthesis of voice fundamental frequency contours of spoken sentences", *Proceedings of the IEEE International Conference on Acoustics, Speech and Signal Processing*, pp 950-93, 1982.

[GEE 83] P. GEE and F. GROSJEAN. "Performance structures: a psycholinguistic and linguistic appraisal", *Cognitive Psychology*, 15:411-458, 1983.

[GEN 76] J. GENIN. "Les études de synthèse de la parole au CNET", *L'écho des recherches*, 85:40-49, 1976.

[GIB 97] D. GIBBON, R. MOORE and R. WINSKI (Editors). *Handbook of Standards and Resources for Spoken Language Systems*, Mouton de Gruyter, Berlin, 1997.

[GRI 84] D.W. GRIFFIN and J.S. LIM. "Signal estimation from modified short-time Fourier transform", *IEEE Transactions on Acoustics, Speech and Signal Processing (ASSP)*, 32(2):236-243, 1984.

[GRI 88] D.W. GRIFFIN and J.S. LIM. "Multi-band excitation vocoder", *IEEE Transactions on Acoustics, Speech and Signal Processing (ASSP)*, 36(8):1223-1235, 1988.

[GRI 91] M. GRICE, K. VAGGES and D. HIRST. "Assessment of intonation in text-to-speech synthesis systems – a pilot test in English and Italian", *Proceedings of Eurospeech-91*, 2:879-882, 1991.

[GRO 01] L. GROS and N. CHATEAU. "Instantaneous and overall judgements for time-varying speech quality", *Acta Acoustica,* 87:367-377, 2001.

[HAM 89] C. HAMON, E. MOULINES and F. CHARPENTIER. "A diphone synthesis system based on time-domain prosodic modifications of speech", in *Proceedings of the IEEE International Conference on Acoustics, Speech and Signal Processing (ICASSP'89)*, pp 238-341, 1989.

[HAN 98] J.H.L. HANSEN and D.T. CHAPPELL. "An auditory-based distortion measure with application to concatenative speech synthesis", *IEEE Transactions on Speech and Audio Processing (SAP)*, 6(5):489-495, 1998.

[HAR 53] C. M. HARRIS. "A study of the building blocks in speech", *Journal of the Acoustical Society of America*, 25(5):962-969, 1953.

[HIR 89] T. HIROKAWA. "Speech synthesis using a waveform dictionary", in *Proceedings of the European Conference on Speech Communication and Technology (EUROSPEECH'89)*, pp 140-143, 1989.

[HIR 90] T. HIROKAWA and K. HAKODA. "Segment selection and pitch modification for high quality speech synthesis using waveform segments", in *Proceedings of the International Conference on Spoken Language Processing (ICSLP'90)*, pp 337-340, 1990.

[HOL 98] M. HOLZAPFEL and N. CAMPBELL. "A nonlinear unit selection strategy for concatenative speech synthesis based on syllable level features", in *Proceedings of the International Conference on Spoken Language Processing (ICSLP'98)*, 1998.

[HOW 92] P. HOWARD-JONES. "Soap, speech output assessment package", *ESPRIT SAM-UCL-042 Project*, 1992.

[HUA 96] X. HUANG, A. ACERO, J. ADCOCK, H.-W. HON, J. GOLDSMITH, J. LIU and M. PLUMPE. "WHISTLER: a trainable text-to-speech system", in *Proceedings of the International Conference on Spoken Language Processing (ICSLP'96)*, pp 2397-2390, 1996.

[HUN 96] A. HUNT and A.W. BLACK. "Unit selection in a concatenative speech synthesis system using a large speech database", in *Proceedings of the IEEE International Conference on Acoustics, Speech and Signal Processing (ICASSP'96)*, pp 373-376, 1996.

[HUS 96] J.-L. HUSSON and Y. LAPRIE. "A new search algorithm in segmentation lattices of speech signals", in *Proceedings of the International Conference on Spoken Language Processing (ICSLP'96)*, pp 2099-2102, 1996.

[HWA 96] M.-Y. HWANG, U. XUEDONG and F.A. LLEVA. "Predicting unseen triphones with senones", *IEEE Transactions on Speech and Audio Processing (SAP)*, 4(6):412-419, 1996.

[IWA 92a] N. IWAHASHI, N. KAIKI and Y. SAGISAKA. "Concatenative speech synthesis by minimum distortion criteria", in *Proceedings of the IEEE International Conference on Acoustics, Speech and Signal Processing (ICASSP'92)*, pp 65-68, 1992.

[IWA 92b] N. IWAHASHI and Y. SAGISAKA. "Speech segment network approach for an optimal synthesis unit set", in *Proceedings of the International Conference on Spoken Language Processing (ICSLP'92)*, pp 479-482, 1992.

[JOS 74] M. JOST-SAINT-BONNET, M. MRAYATI and L.-J. BOË. "La synthèse par diphones, premier rapport", *Bulletin de l'Institut de Phonétique de Grenoble, III, 1974*.

[JÜR 95] C. JÜRGENS and M. WUNDERLICH. "A comparison of different speech units for the german TTS-system Tubsy", in *Proceedings of the European Conference on Speech Communication and Technology (EUROSPEECH'95)*, pp 1105-1108, 1995.

[KLA 87] D.H. KLATT. "Review of text-to-speech conversion for English", *The Journal of the Acoustical Society of America (JASA)*, 82(3):737-793, 1987.

[KLA 96] E. KLABBERS and R. VELDHUIS. "On the reduction of concatenation artifacts in diphone synthesis", in *Proceedings of the International Conference on Spoken Language Processing (ICSLP'98)*, pp 1983-1986, 1996.

[KLA 97] H. KLAUS, K. FELLBAUM and J. SOTSCHECK. "Auditive bestimmung und vergleich der sprachqualität von sprachsynthesesystemen für die deutsche sprache", *Acustica-Acta Acustica*, 83:124-136, 1997.

[KOR 97] R.W. KORTEKAAS and A. KOHLRAUSCH. "Psychoacoustical evaluation of the pitch-synchronous overlap-and-add speech-waveform manipulation technique using single-formant stimuli", *Journal of the Acoustical Society of America*, 101(4):2202-2213, 1997.

[KOR 99] R.W. KORTEKAAS and A. KOHLRAUSCH. "Psychoacoustical evaluation of Psola. ii. double-formant stimuli and the role of vocal perturbation", *Journal of the Acoustical Society of America (JASA)*, 105(1): 522-535, 1999.

[KÜP 56] K. KÜPFMÜLLER and O. WARNS. *Sprachsynthese aus lauten. Nachrichtentechnische Fachberichte*, 3: 28-31, 1956.

[KRA 95] V. KRAFT and T. PORTELE. "Quality evaluation of five speech synthesis Systems", *Acta Acoustica*, 3: pp351-365, 1995.

[LAF 85] F. LAFERRIERE, G. CHOLLET, L. MICLET and J.-P. TUBACH. "Segmentation d'une base de données de Polysons, application à la synthèse de la parole ", in *Actes des Journées d'Étude sur la Parole (JEP'85)*, pp 107-110, 1985.

[LAR 76] D. LARREUR and F. EMERARD. "Speech synthesis by dyads and automatic intonation processing", in *Proceedings of the IEEE International Conference on Acoustics, Speech and Signal Processing (ICASSP'76)*, 1976.

[LAR 89] D. LARREUR, F. EMERARD and F. MARTY. "Linguistic and prosodic processing for a text-to-speech synthesis system", *Proceedings of the Eurospeech Conference*, pp 510-513, 1989.

[LAR 90] D. LARREUR and C. SORIN. "Quality evaluation of French text-to-speech synthesis within a task, the importance of the mute 'e'", *Proceedings of the 1st ESCA Workshop on Speech Synthesis*, pp 91-96, 1990.

[LAR 99] J. LAROCHE and M. DOLSON. "New phase vocoder technique for pitch shifting, harmonizing and other exotic effects" in *IEEE Workshop on Applications of Signal Processing to Audio and Acoustics,* 1999.

[LEE 77] F.F. LEE. "Time compression and expansion of speech", *Journal of the Audio Engineering Society*, 20(9):738-742, 1977.

[LEE 01] M. LEE, D.P. LOPRESTI and J.P. OLIVE. "A text-to-speech platform for variable length optimal unit searching using perceptual cost functions" in *Proceedings of the ESCA Tutorial and Research Workshop on Speech Synthesis* (SSW4), 2001.

[LEI 68] E. LEIPP, M. CASTELLENGO and J.S. LIÉNARD. "La synthèse de la parole à partir de digrammes phonétiques" in *Proceedings of the International Congress on Acoustics (ICA)*, pp C-5-6, 1968.

[LEM 94] P.Y. LE MEUR. "Protection de segments sub-phonétiques en synthèse Psola" in *Actes des Journées d'Étude sur la Parole (JEP'94)*, pp 137-142, 1994.

[LEM 96] P.-Y. LE MEUR. Synthèse de la parole par unités de taille variable, PhD thesis, Ecole Nationale Supérieure des Télécommunications (ENST), Paris, 1996.

[LEN 02] K. LENZO and A.W. BLACK. "Customized synthesis: blending and tiering", in *Proceedings of the Applied Voice Input/Output Society Conference (AVIOS'02)*, 2002.

[L'H 04] J. L'HOUR, O. BOËFFARD, J. SIROUX, L. MICLET, F. CHARPENTIER and T. MOUDENC. "Doris, a multiagent/IP platform for multimodal dialogue applications", in *Proceedings of the International Conference on Spoken Language Processing (ICSLP'04)*, 2004.

[LI 98] CHUNYAN LI and V. CUPERMAN. "Enhanced harmonic coding of speech with frequency domain transition", in *Proceedings of the IEEE International Conference on Acoustics, Speech and Signal Processing (ICASSP'98)*, pp 581-584, 1998.

[LIB 59] A. M. LIBERMAN, F. INGEMANN, L. LISKER, P. DELATTRE and F. S. COOPER. "Minimal rules for synthesizing speech", *Journal of the Acoustical Society of America,* 31(11): pp1490-1499, 1959.

[LIE 70] J. S. LIENARD. "La synthèse de la parole, historique et réalisations actuelles", *Revue d'Acoustique,* 11: pp 204-213, 1970.

[LIE 77] J.-S. LIENARD, D. TEIL, C. CHOPPY, G. RENARD and J. SAPALY. "Diphone synthesis of French: vocal response unit and automatic prosody from the text", in *Proceedings of the IEEE International Conference on Acoustics, Speech and Signal Processing (ICASSP'77)*, pp 560-563, 1977.

[LUK 96] R. W. P. LUK and R. I. DAMPER. "Stochastic phonographic transduction for English", *Computer Speech and Language,* 10:133-153, 1996.

[MAE 79] S. MAEDA. "An articulary model of the tongue based on a statistical analysis", *The Journal of the Acoustical Society of America,* 65(S22), 1979.

[MAL 79] D. MALAH. "Time-domain algorithms for harmonic bandwidth reduction", *IEEE Transaction on Acoustics, Speech and Signal Processing* (ASSP), 27(2):121-133, 1979.

[MAR 77] A. MARTELLI. "On the complexity of admissible search algorithms", *Artificial Intelligence,* 8:1-13, 1977.

[MCA 86] R.J. MCAULAY and T.F. QUATIERI. "Speech transformation based on a sinusoidal representation", *IEEE Transactions on Acoustics, Speech and Signal Processing (ASSP),* 34(6):1449-1464, 1986.

[MOB 01] B. MOBIUS. "Rare events and closed domains: Two delicate concepts in speech synthesis", in *Proceedings of the ESCA Tutorial and Research Workshop on Speech Synthesis (SSW4),* 2001.

[MOU 90A] E. MOULINES. Algorithmes de codage et de modification des paramètres prosodiques pour la synthèse de la parole à partir du texte, PhD thesis, Ecole Nationale Supérieure des Télécommunications (ENST), Paris, 1990.

[MOU 90B] E. MOULINES and F. CHARPENTIER. "Pitch-synchronous waveform processing techniques for text-to-speech synthesis using diphones", *Speech Communication,* 9(5/6): pp453-467. 1990.

[MOU 95] E. MOULINES and J. LAROCHE. "Non-parametric techniques for pitchscale and time-scale modification of speech", *Speech Communication,* 16(2):175-205, 1995.

[MOU 03] T. MOUDENC and F. EMERARD. "Synthèse vocale et handicap ", *Annales des Télécommunications,* 58(5-6):928-934, 2003.

[NAK 88] S.Y. NAKAJIMA and H. HAMADA. "Automatic generation of synthesis units based on context oriented clustering" in *Proceedings of the IEEE International Conference on Acoustics, Speech and Signal Processing (ICASSP'88)*, pp 659-662, 1988.

[NAK 95] S. NAKAJIMA. "Automatic synthesis unit generation for English speech synthesis based on multi-layered context oriented clustering", *Speech Communication*, 14(4):313-324, 1995.

[NEU 78] E.P. NEUBURG. "Simple pitch-dependant algorithm for high-quality speech rate coding", *The Journal of the Acoustical Society of America (JASA)*, 63(2):624-625, 1978.

[NOM 90] T. NOMURA, H. MIZUNO and H. SAITO. "Speech synthesis by optimum concatenation of phoneme segments", *Proceedings of the First ESCA Workshop on Speech Synthesis*, pp 39-42, 1990.

[OFF 95] JEIDA "Speech Input/Output systems Expert Committee on Standardization of Office Automation Equipment. Jeida guidelines for speech synthesizer evaluation", *The Japan Electronic Industry Development Association*, 1995.

[OLI 77] J.P. OLIVE. "Rule synthesis of speech from dyadic units", in *Proceedings of the IEEE International Conference on Acoustics, Speech and Signal Processing (ICASSP'77)*, pp 568-570, 1977.

[OLI 90] J.P. OLIVE. "A new algorithm for a concatenative speech synthesis system using an augmented acoustic inventory of speech sounds", in *Proceedings of the ESCA Workshop on Speech Synthesis*, pp 25-29, 1990.

[OLI 98] J. OLIVE, J. VAN SANTEN, B. MOBIUS and C. SHIH. "Multilingual text-to-speech synthesis: the Bell Labs approach" in *Synthesis*, pp 191-228. Kluwer Academic Publishers, 1998.

[O'S 84] D. O'SHAUGHNESSY. "Design of a real-time French text-to-speech system", *Speech Communication*, 3:233-243, 1984.

[OST 94] M. OSTENDORF and N. VEILLEUX. "A hierarchical stochastic model for automatic prediction of prosodic boundary location", *Computational Linguistics*, 20(1), 1994.

[OUD 98] M. OUDOT. Application du modèle sinusoïdes et bruit au décodage, au débruitage et à la modification des sons de parole, PhD thesis, Ecole Nationale Supérieure des Télécommunications, Paris, 1998.

[PEE 99] G. PEETERS and X. RODET. SINOLA: "A new analysis/synthesis method using spectrum peak shape distorsion, phase and reassigned spectrum", in *Proceedings of the International Computer Music Conference (ICMC'99)*, 1999.

[PEN 90] H. PENG, Y. ZHAO and M. CHU. "Perceptually optimizing the cost function for unit selection in a TTS system", in *Proceedings of the International Conference on Spoken Language Processing (ICSLP'90)*, pp 1097-2000, 1990.

[PET 58] G.E. PETERSON, W.S.Y. WANG and E. SIVERTSEN. "Segmentation techniques in speech synthesis", *The Journal of the Acoustical Society of America (JASA)*, 30(8):739-742, 1958.

[PIS 96] D.B. PISONI. "Perception of synthetic speech", in R. VAN SANTEN, R. SPROAT, J. HIRSCHBERG and J. OLIVE (eds.), *Progress in Speech Synthesis,* Springer Verlag, 1996.

[POL 87] L. C. W. POLS, J. P. LEFEVRE, G. BOXELAAR and N. VAN SON. "Word intelligibility of a rule synthesis system for French", *Proceedings of Eurospeech Conference,* 1:179-182, 1987.

[POL 90] L. POLS. "Does improved performance of a rule synthesizer also contributes to more phonetic knowledge?", *Proceedings of the ESCA Tutorial Day on Speech Synthesis,* pp 50-54, 1990.

[POL 92] L.C.W. POLS and SAM-PARTNERS. "Multilingual synthesis evaluation methods", *Proceedings of Int. Conf. on Speech and Lang. Proc.*, pp 181-184, 1992.

[POL 96] L.C.W. POLS and U. JEKOSCH. "A structured way of looking at the performance of text-to-speech systems", in R. VAN SANTEN, R. SPROAT, J. HIRSCHBERG and J. OLIVE (eds.), *Progress in Speech Synthesis*. Springer Verlag, 1996.

[POR 76] M.R. PORTNOFF. "Implementation of the digital phase vocoder using the Fast Fourier Transform", *IEEE Transactions on Acoustics, Speech and Signal Processing (ASSP)*, 24(3):243-248, 1976.

[POR 81] M.R. PORTNOFF. "Time-scale modification of speech based on short-time Fourier analysis", *IEEE Transactions on Acoustics, Speech and Signal Processing (ASSP)*, 29(3):374-390, 1981.

[POR 90] T. PORTELE, W. SENDLMEIER and W. HESS. "Hadifix: a system for german speech synthesis based on demisyllables, diphones, and suffixes", *Proceedings of the first ESCA workshop on speech synthesis*, pp 161-164, 1990.

[POR 94] T. PORTELE, F. HÖFER and W. HESS. "A mixed inventory structure for German concatenative synthesis", in *Proceedings of the IEEE International Conference on Acoustics, Speech and Signal Processing (ICASSP'94)*, pp 115-118, 1994.

[POR 96] T. PORTELE, F. HÖFER and W. HESS. *Progress in Speech Synthesis*, in "A mixed inventory structure for German concatenative synthesis", pp 263-277, Springer Verlag, 1996.

[PRU 01] R. PRUDON and C. D'ALESSANDRO. "A selection/concatenation text-to-speech synthesis system: databases development, system design, comparative evaluation", in *Proceedings of the ESCA Tutorial and Research Workshop on Speech Synthesis (SSW4)*, 2001.

[QUA 95] S. QUAZZA. "Predicting durations by means of automatic learning algorithms", *Proceedings of the Fourth Workshop of the Experimental Phonetics Group*, 1995.

[RIC 95] G. RICHARD, M. LIU, D. SINDER, H. DUNCAN, Q. LIN, J. FLANAGAN, S. LEVINSON and S. SLIMON. "Numerical simulations of fluid flow in the vocal tract", *Proceedings of the Eurospeech Conference*, pp 1297-1300, 1995.

[RIL 92] M. D. RILEY. "Tree-bases modelling of segmental duration", in *Talking Machines: Theories, Models and Designs*, in G. BAILLY, C. BENOÎT and T. R. SAWALLIS (eds.), Elsevier Science, 1992.

[ROS 95] K. N. ROSS. "Modelling of intonation for speech synthesis", PhD thesis, Boston University, Boston, 1995.

[ROU 90] S. ROUCOS and A. WILGUS. "High quality time scale modification for speech using fast synchronized overlap-add algorithms" in *Proceedings of the IEEE International Conference on Acoustics, Speech and Signal Processing (ICASSP'90)*, pp 409-412, 1990.

[SAG 86] Y. SAGISAKA and H. SATO. "Composite phoneme units for the speech synthesis of Japanese", *Speech Communication*, 5:217-223, 1986.

[SAG 88] Y. SAGISAKA." Speech synthesis by rules using an optimal selection of non-uniform synthesis units" in *Proceedings of the IEEE International Conference on Acoustics, Speech and Signal Processing (ICASSP'88)*, pp 679-682, 1988.

[SAI 68] S. SAITO and S. HASHIMOTO. "Speech synthesis system based on interphone transition units" in *Proceedings of the International Congress on Acoustics (ICA)*, pp B-5-12, 1968.

[SAM 92] SAM. "Multi-lingual speech input/output assessment, methodology and standardization", *Final Report, Esprit Project 2589*, SAM, 1992.

[SAN 93] J.P.H. VAN SANTEN. " Perceptual experiments for diagnostic testing of text-to-speech systems", *Computer Speech and Language*, 7:49-100, 1993.

[SAN 97a] J.P.H. VAN SANTEN. "Combinatorial issues in text-to-speech synthesis", in *Proceedings of the European Conference on Speech Communication and Technology (EUROSPEECH'97)*, pp 2511-2514, 1997.

[SAN 97B] J.P.H. VAN SANTEN and A.L. BUCHSBAUM. "Methods for optimal text selection", in *Proceedings of the European Conference on Speech Communication and Technology (EUROSPEECH'97)*, pp 553-556, 1997.

[SCH 93] M. SCHMIDT, S. FITT, C. SCOTT, and M. JACK. "Phonetic transcription standards for European names, Onomastica", *Proceedings of the Eurospeech Conference*, pp 279-282, 1993.

[SEJ 87] T. SEJNOWSKY and C.R. ROSENBERG. "Parallel networks that learn to pronounce English text", *Complex Systems*, 1:145-168, 1987.

[SEN 82] S. SENEFF. "System to independently modify excitation and/or spectrum of speech waveform without explicit pitch extraction", *IEEE Transactions on Acoustics, Speech and Signal Processing* (ASSP), 30(4):566-578, 1982.

[SIL 90] K. SILVERMAN, S. BASSON and S. LEVAS S. "Evaluating synthesizer performance: is segmental intelligibility enough?", *Proceedings of the Int. Conf. on Speech and Lang. Proc.*, ICSLP-90, 2:981-984, 1990.

[SOR 96] C. SORIN and F. EMERARD. "Domaines d'application et évaluation de la synthèse de parole à partir du texte " in *Fondements et perspectives en traitement automatique de la parole*. AUPELF-UREF, H. Méloni (ed.), 1996.

[SPI 90] M.F. SPIEGEL, M.J. ALTOM, M.J. MACCHI and K.L. WALLACE. "Comprehensive assessment of the telephone intelligibility of synthesized and natural speech", *Speech Communication*, 9(4):279-291, 1990.

[STE 83] M.G. STELLA, "Speech Synthesis" in F. FALLSIDE and W. WOODS (eds.), *Computer Speech Processing,* Prentice Hall, 1983.

[STE 90] K. N. STEVEN. "Control parameters for synthesis by rule", *Proceedings of the ESCA tutorial day on speech synthesis*, pp 27-37, 1990.

[STÖ 99] K. STÖBER, T. PORTELE, P. WAGNER and W. HESS. "Synthesis by word concatenation" in *Proceedings of the European Conference on Speech Communication and Technology (EUROSPEECH'99)*, pp 619-622, 1999.

[STY 95A] Y. STYLIANOU, O. CAPPÉ and E. MOULINES. "Statistical methods for voice quality transformation" in *Proceedings of the European Conference on Speech Communication and Technology (EUROSPEECH'95)*, 1995.

[STY 95B] Y. STYLIANOU, J. LAROCHE and E. MOULINES. "High quality speech modification based on a harmonic + noise model" in *Proceedings of the European Conference on Speech Communication and Technology (EUROSPEECH'95)*, pp 451-454, 1995.

[STY 98] Y. STYLIANOU, O. CAPPE and E. MOULINES. "Continuous probabilistic transform for voice conversion", *IEEE Transactions on Acoustics, Speech and Signal Processing (ASSP)*, 6(2):131-142, 1998.

[TAK 90] K. TAKEDA, K. ABE and Y. SAGISAKA. "On the unit search criteria and algorithms for speech synthesis using non-uniform units" in *Proceedings of the International Conference on Spoken Language Processing (ICSLP'90)*, pp 341-344, 1990.

[TAK 92] K. TAKEDA, K. ABE and Y. SAGISAKA. "On the basic scheme and algorithms in non-uniform unit speech synthesis" in G. BAILLY, C. BENOIT and T SAWALLIS (eds.), *Talking Machines: Theories, Models, and Designs*, pp 93-105. Elsevier B.V., 1992.

[TAY 97] P. TAYLOR and S. ISARD. SSML: "A speech synthesis markup language", *Speech Communication*, 21:123-133, 1997.

[TAY 99] P. TAYLOR and A.W. BLACK. "Speech synthesis by phonological structure matching" *Proceedings of the European Conference on Speech Communication and Technology (EUROSPEECH'99)*, pp 875-878, 1999.

[TOD 02] T. TODA and M. KAWAI, AND H. TSUZAKI. "Perceptual evaluation of cost for segment selection in concatenative speech synthesis", in *Proceedings of the IEEE/ESCA Workshop on Speech Synthesis*, 2002.

[TOU 97] S. DE TOURNEMIRE. "Automatic identification and generation of prosodic contours for text-to-speech synthesis of French", *Proceedings of the Eurospeech Conference*, pp 191-194, 1997.

[TRA 92] C. TRABER. "F0 generation with a data base of natural F0 patterns and with a neural network" in *Talking Machines: Theories, Models and Designs,* G. BAILLY and C. BENOÎT (eds.), North Holland, pp. 287-304, 1992

[TUB 85] J.-P. TUBACH and L.-J. BOË. "Un corpus de transcriptions phonétiques: constitution et exploitation statistique ", Technical report, Ecole Nationale Supérieure des Télécommunications (ENST), Paris, 1985.

[UIT 94] UNION INTERNATIONALE DES TELECOMMUNICATIONS UIT-T. Méthode d'évaluation subjective de la qualité de parole des serveurs vocaux. Recommendation P.85, 1994.

[VER 93] W. VERHELST and M. ROELANDS. "An overlap-add technique based on waveform similarity (Wsola) for high quality time-scale modification of speech" in *Proceedings of the IEEE International Conference on Acoustics, Speech and Signal Processing (ICASSP'93)*, pp 554-557, 1993.

[VIO 98] F. VIOLARO and O. BOËFFARD. "An iterative algorithm for decomposition of speech signals into periodic and aperiodic components", *IEEE Transactions on Speech and Audio Processing (SAP)*, 6(5):426-434, 1998.

[VOT 91] L.L.M. VOTGEN, C. MA, W.D.E VERHELST and J.H. EGGEN. "Pitch inflected overlap and add speech manipulation", *European Patent 91202044.3. 1991.*

[WAN 93] W.J. WANG, N. CAMPBELL, N. IWAHASHI and Y. SAGISAKA." Tree-based unit selection for English speech synthesis" in *Proceedings of the IEEE International Conference on Acoustics, Speech and Signal Processing (ICASSP'93)*, pp 191-194, 1993.

[WOU 98] J. WOUTERS and M.W. MACON. "A perceptual evaluation of distances measures for concatenative speech synthesis" in *Proceedings of the International Conference on Spoken Language Processing (ICSLP'98)*, pp 2747-2750, 1998.

[YVO 96] F. YVON. "Grapheme-to-phoneme conversion of multiple unbounded overlapping chunks", *Computational and Language E-print*, 1996.

[YVO 98] F. YVON, P. BOULA DE MAREÜIL, C. D'ALESSANDRO, V. AUBERGE, M. BAGEIN, G. BAILLY, F. BECHET, S. FOUKIA, J.P. GOLDMAN, E. KELLER, V. PAGEL, F. SANNIER, J. VERONIS, D. O'SHAUGHNESSY and B. ZELLNER. "Objective evaluation of grapheme to phoneme conversion for text-to-speech synthesis in French", *Computer Speech and Language*, 12(4):393-410, 1998.

Chapter 4

Facial Animation for Visual Speech

4.1. Introduction

Facial animation for visual speech is a topic situated at the confluence of two rapidly emerging technology domains in the area of multimedia: speech technologies and 3D animation. On the one hand, speech is a visible process. In fact, even though the acoustic signal is generally sufficient to guarantee the full comprehension of a spoken message, the visual information carried by the face of an interlocutor enables some compensation of the acoustic degradations that may occur when listening conditions become adverse. This possibility of using lip reading is well known to hearing impaired people, but it is also used by regular listeners. As a consequence, some research areas interested in speech perception, recognition and synthesis have naturally focused on this bimodality. Thus, the design of anthropomorphic visual models of speech articulators enables the testing of scientific hypothesis concerning the integration of acoustic and visual information by human beings and the improvement of speech intelligibility for the current speech synthesis systems. On the other hand, given the level of quality reached today in the rendering and the anthropomorphic character of some 3D face models, the animators and the general public are likely to become less and less tolerant of inconsistencies that may subsist between facial gestures and the speech produced by these virtual actors. It therefore becomes necessary to develop animation techniques and models which account for the complexity and the specificity of speech production.

Chapter written by Thierry GUIARD-MARIGNY.

After a brief description of the current applications in facial animation for visual speech, this chapter presents a few results related to the intrinsic bimodality of speech. Stress is put on the attention which is required to obtain good lip synchrony in virtual animation. A few methods for the animation of synthetic faces are described and the chapter concludes on the future of virtual actors.

4.2. Applications of facial animation for visual speech

4.2.1. *Animation movies*

From the "live" virtual presenter to the video game heroine, the virtual guide for educational CD-ROMs, the avatar in network games, the fake president or simply the good old cartoon character, the use of animated synthetic faces is more and more widespread. Is this a real person or a virtual character? It is becoming more and more difficult to answer this question, given the growing realism of animated characters. In movies, on television, in cartoons, video games and CD-ROMs, they appear in all sorts of contexts. The public is captivated and fascinated by these new technologies but at the same time, customers are becoming more and more demanding.

4.2.2. *Telecommunications*

Another area in which research on facial animation is becoming more and more central is low-rate telecommunications, for applications such as videophones or teleconferencing. In fact, the quantity of information needed to transmit the image of a speaker's face with a size and rate large enough to enable lip reading remains too large with regard to the available bandwidth of today's communication channels. One solution is to use coding algorithms (such as MPEG), so as to reduce the information rate in the video sequence. The principle consists of coding and transmitting the differences between successive images. Another option is to use a face model. Here, the idea is to first transmit the shape and texture of the speaker's face to the communication device of its interlocutor, which can reconstruct the face using a synthesis model. Then, a video analysis system extracts features which characterize the face movement of the speaker and sends them to the interlocutor's device. The movements of the face can then be synthesized synchronously with the speech signal. The reduction of the information quantity is thus very significant.

4.2.3. *Human-machine interfaces*

The face is the main and the most user-friendly communication means between human beings. Research on interfaces for a better dialog between humans and machines has naturally turned itself towards face modeling and animation. Thus, the image of the face can be added to the speech signal in applications where the messages produced by the computers are pre-stored or are generated from a text. This is why a significant research activity on facial animation from a simple text has been taking place. For such audiovisual synthesis systems, the speech signal and the face movements are predicted from a text and synchronized for the animation [BEN 92, BEN 95, COH 90, HIL 88, LEG 96a, LEG 96b, MOH 93, MOR 90, PEA 86, PEL 91, PEL 96, SAI 90].

4.2.4. *A tool for speech research*

The need for designing and using models in research does not need to be proven. This is especially true in the complex domain of speech communication [PER 91]. In some cases, the acquisition of experimental data is difficult, invasive or even dangerous (for instance, measurements on the geometry of the vocal tract). The use of a model designed from a reduced quantity of data allows more extensive simulations and investigations, which lead to new hypotheses. These hypotheses can be subsequently confronted with new data. Models are essential tools for testing hypotheses in speech production and perception. This is how considerable progress has been made in psycholinguistics, thanks to the development of acoustic speech synthesis. A synthesizer makes it possible to produce in a strictly controlled manner various stimuli and to test independently a variety of simple hypotheses. Following this idea, synthetic faces [BRO 83, COH 94, MAS 90] are used to generate controlled stimuli and validate hypotheses in audiovisual speech production and perception through investigations and tests based on the methodology of experimental psychology.

4.3. Speech as a bimodal process

As Cathiard recalls in her PhD dissertation [CAT 94], "to see the face of a person – in the course of a face-to-face interaction, which is usually bimodal – enables us at the same time: 1) to recognize the person, i.e. to determine his/her personal identity; 2) to determine his/her emotional state and more generally to 'read' his/her facial expressions and his/her intentions; 3) to focalize on the speech signal from this person, and separate it from others ('cocktail party' effect) to ensure the origin of the signal; 4) and finally, to understand what the person is saying, by taking advantage of the visible movements of spoken language, i.e. the audiovisual perception of speech". All of these four functions make up a face-to-face bimodal interaction and can be found in the various applications of face synthesis.

Of course, the first function appears in all applications of face synthesis, in which the image of the face is the way to recognize a character. The results which we present in this chapter correspond essentially to the fourth function mentioned above, i.e. to provide an improved intelligibility to the speech signal. Therefore, we are now going to present a few results stemming from the field of experimental psychology applied to audiovisual speech. Our intention is not to present an extensive review on this subject, but simply to illustrate with a few meaningful results the knowledge and precautions to be taken to achieve efficient facial animation with the primary purpose of improving the acoustic intelligibility of speech. The main experiments concerning intelligibility of visible speech will first be reported. Then, three key points for facial animation for visual speech will be addressed: the phonetic configuration of the visible articulators, lip synchronization issues and the consistency between acoustic and visual sources.

4.3.1. *The intelligibility of visible speech*

In most situations related to speech communication, the acoustic speech signal is self-sufficient – the existence of radio and telephone proves this. However, the results of the experiments which we are going to describe show that the image of the speaker's face provides information on the transmitted message. This information may be collected consciously or not by the interlocutor. In the case of a perfectly audible signal, it is irrelevant to evaluate the contribution of the image of the speaker's face to the overall intelligibility, as the understanding level is almost perfect. Nevertheless, as soon as the speech signal is missing (hearing impaired listener) or distorted (by some noise), the advantage becomes significant. Everybody has observed this fact during a conversation in a disco or a lively bar for instance and it has been proven by many experiments for the English language [BIN 74, ERB 69, ERB 75, NEE 56, SUM 54, SUM 79, SUM 89] and the French language [BEN 94b]. Thus, Sumby and Pollack [SUM 54], Binnie *et al.* [BIN 74], Erber [ERB 69] and Benoît *et al.* [BEN 94b] have subjected panels of auditors with neither hearing nor visual deficiencies to series of tests which consist of presenting spoken words added with several noise levels in two distinct conditions: 1) audio-only and 2) audiovisual. In some experiments, a vision only condition has also been tested but the results correspond to those obtained in the audiovisual condition with maximum acoustic degradation. The tested subjects must indicate which words they have perceived. The results of these experiments (as summarized in Figure 4.1) all show that, for high noise levels, the image of the speaker's face brings a significant increase in intelligibility as compared to the audio-only condition. Sumby and Pollack [SUM 54] have used corpora of 8, 16, 32, 64, 128 and 256 words. The curves represented in the figure correspond to the mean of the performance on the various corpora. The corpus tested by Binnie *et al.* [BIN 74] contained 16 syllables, the one used by [ERB 69] being 250 words and the one by Benoît *et al.* [BEN 94b] being 18 words.

Of course, the gain in terms of intelligibility depends on the size and the phonetic content of the corpus. Sumby and Pollack [SUM 54] have thus shown that the gain is inversely proportional to the corpus size. Summerfield *et al.* [SUM 89] have used in their experiments sentences for which lip-reading has a variable degree of difficulty: in this case, the identification scores decrease with the difficulty of the sentence, as could be expected. According to these authors, the reasons for which a word is easy to lip-read depends on three properties: 1) the variation of lip movements must limit the number of possible lexical interpretations; 2) the beginning of the word must contrast with the following syllable; 3) the occurrence frequency of the word in the language must be high. Thus, the word *boy* is easy to recognize visually, whereas *scissors* is a difficult one.

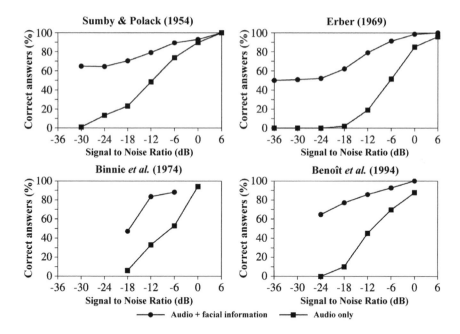

Figure 4.1. *Results from four intelligibility tests on noisy speech. The results have been extracted from the following articles: [BEN 94b, BIN 74, ERB 69, SUM 54]*

A second interesting result showed by these experiments is the complementarity between vision and audition. Thus, Benoit *et al.* [BEN 94b] have shown that, for French, in high noise level conditions, the recognition rate for the vowel [a] in the audio-only case is much higher than for the vowel [y]. Conversely, for the video-only condition, the vowel [y] is easier to identify.

All the experiments that we have mentioned have been conducted by adding noise to the original signal. A study by Reisberg *et al.* [REI 87] shows that, even when the speech signal is perfectly audible, the image of the speaker's face can significantly help out. Their experiment consists of having subjects repeat a message (as and when they hear it), either in the audio-only condition or in the audiovisual condition. The results show some advantage brought about by the visual modality, when the message to be repeated is in a foreign language or pronounced with a strong foreign accent. This is also the case when the linguistic complexity of the message is high. Extracts from Kant's work *Critique of Pure Reason* have been used for this test.

The image of the speaker's face brings increased intelligibility to the spoken message. It would be interesting to observe a similar advantage with the use of a synthetic face. However, speech production is a complex task which is difficult to reproduce. Thus, facial animation for visual speech requires accurately piloting the various articulators, i.e. to synthesize the correct movements and to respect their consistency with the speech signal.

4.3.2. *Visemes for facial animation*

The sounds of language cannot always be visually distinguished from one another, especially those which only differ by the position of a non-visible articulator (for example, voicing or nasality). Several studies have been conducted in order to identify the various articulatory configurations of the phonemes and to group them according to their visual similarities. These studies are based on two complementary methods: visual perception of phonemes which is used to construct classes from the confusion matrices resulting from perceptual tests [BEN 94b, BIN 74, GEN 81, MOU 80, PLA 80] and anatomical measurements on the articulators which are processed by data analysis methods so as to learn partitions of the representation space [ABR 86, BEN 92, FRO 64, PLA 80, TSE 90, ZER 90]. All these results, which are reviewed in Cathiard's PhD dissertation [CAT 94] show that the distinctions which are best transmitted are those corresponding to easily accessible visual features, such as "labial" or "rounded". Moreover, they indicate that the most salient visual classes depend on the language and on the consonantic contexts (for vowels) or vocalic context (for consonants) used in the studies.

Following the definition proposed by Benoît *et al.* [BEN 92], we will call a *viseme* a set of sounds in the language sharing a similar position of the visible speech production organs which correspond to their articulation. Speech being made of transitions, the visemes correspond to instantaneous patterns and each of them correspond to a group of sounds that cannot be visually distinguished. The entire set of visemes covers the "complete inventory" of visible articulatory positions used for a given language. Note carefully that a particular phoneme may belong to several visemes, depending on its context: for example, in French, the pronunciations of [b] in context [by] or in context [bi] result in two distinct positions of the mouth, and so does the pronunciation of [i] in [si] and in [ʃi]. This observation stems from the coarticulation effects which occur in speech production. In fact, the process of speaking cannot be explained as the production of a sequence of discrete units but rather as a set of overlapping articulatory gestures with a number of articulated organs (among which, only a few are visible). The acoustic wave thus obtained is not continuously audible: it is also composed of a significant proportion of silences, during pauses or the holding phase of plosives, during which the speaker may be producing visible gestures so as to anticipate the next sound. In short, some of the speech movements are only audible, some others are only visible and finally some are both audible and visible. Cathiard [CAT 94] has shown that, during a transition [i#y] (with a 160 ms pause between [i] and [y]), subjects are able to perceive the [y] 140 ms before its acoustic beginning and 40 ms are sufficient for establishing the information on the [y] with 100% certainty.

It is clear that coarticulation plays a key role in the processing of visual information by human beings and it is therefore important to render this phenomenon in facial animation: to sum up, there is no one-to-one relationship between a phoneme and a viseme.

4.3.3. *Synchronization issues*

Synchronization issues (evidenced for instance by singers miming a record) is also a key aspect in speech perception. It relates to the inherent synchrony that exists in the transmission of auditory and visual information, at least when the source is located at a "reasonable" distance in front of the sound velocity. Dixon and Spitz [DIX 80] have observed experimentally that subjects are not able to detect an artificial asynchrony between audio and visual speech, as long as the audio signal arrives between 130 ms before and 260 ms after the video image of the speaker's face. In fact, this is an average value, as the sensitivity to the audio/video time lag is very acute for instantaneous events, like a hammer hitting an anvil, for which the previous interval narrows to [−75 ms, +188 ms]. Fast labial phenomena, such as the lip explosions in [p, b, m], fit rather well into this second category. Recent studies have tried to quantify the loss of audiovisual intelligibility entailed from a time lag between audio and visual information in speech. These studies are essentially motivated by problems raised by the videophone for which the image coding/decoding process causes an unavoidable delay of the video signal with respect to the audio signal. Thus, Smeele and Sittig [SME 91] have measured the global intelligibility of phonetically balanced CVC logatomes, presented in an audiovisual form, in an acoustic environment in which other speech sounds had been artificially added. An average intelligibility of 20% correctly identified logatomes was measured for audio-only conditions, versus 65% for audio-visual conditions. However, when the face was presented more than 160 ms after the audio signal, no more benefit was observed with the visual modality over the audio modality alone. Conversely, Smeele *et al.* [SME 92] observed in another experiment, an approximately constant level of intelligibility of about 40% when the audio signal was presented with a time lag between 320 and 1,500 ms after the video signal. In a comparable experiment, Campbell and Dodd [CAM 80] had previously observed that the impact of facial reading on the disambiguation of noisy isolated words was still measurable with time lags of 1,500 ms between audition and vision, whichever modality was ahead of the other one. Conversely, Reisberg *et al.* [REI 87] did not observe any visual benefit on their task of shadowing Kant's text as mentioned above, using a audio/visual lag of ±500 ms. These somehow divergent results confirm the idea that audition and vision mutually influence each other in the perception of speech, even though the extent of this phenomenon is not perfectly clear yet. However, it is important to preserve this synchrony in facial animation for visual speech, which is inherent to audio/visual speech information.

4.3.4. *Source consistency*

The effect of a configurational inconsistency between audio-based and video-based speech information has been demonstrated experimentally by McGurk and MacDonald [MCG 76]. A very strong perceptual illusion is created by presenting the video sequence of a speaker who is articulating a sound, doubled by a different sound on the audio track. Thus, an audiovisual stimulus [ba/ga] is 98% of the time identified as a [da], i.e. an illusory percept which shares phonological features from both syllables but comprises a consonantic phoneme which does not belong to any of the stimuli. For the synthesis of talking faces, we must therefore be careful to avoid such illusions. In fact, as formerly discussed, the phonetic realizations which are the easiest to discriminate on an auditory basis are the most difficult to discriminate visually, and vice versa. Thus, [p], [b] and [m] show identical lip shapes in many languages whereas they are acoustically very different. These three phonemes are generally grouped into the same viseme. On the contrary, automatic speech recognition systems often confuse [p] and [k], whereas these two phonemes turn out to be very different as regards the shape of the speaker's lips. Therefore, for a synthetic face to improve the intelligibility of a speech synthesizer, it is essential that the face movements are consistent with the acoustic wave which they are meant to have produced. Otherwise, contradictory information (similar to the McGurk illusion) may occur during the bimodal integration process by the auditor/spectator which is likely to degrade the intelligibility of the original message significantly. Such a catastrophic effect can inadvertently occur if the movements of the visible articulators are piloted by the acoustic wave, i.e. from the output of a phonetic recognition system.

4.3.5. *Key constraints for the synthesis of visual speech*

The various studies that we have mentioned above have led to interesting results concerning the intrinsic bimodality of speech. Here, we present as a conclusion, a brief summary of some key results, stemming from these experiments and which are relevant to the synthesis of visual speech:

– Even though the acoustic dimension of speech is predominant in spoken communication, the understanding of speech is clearly improved by the image of the speaker's face, especially in adverse acoustic conditions.

– There is no one-to-one relationship between a phoneme and a viseme (the visual shape of the [s] in syllable [si] is very different from that in syllable [sy]). Thus, the constitution of a reference set of visemes must account for coarticulation effects. This is also the case for audiovisual speech synthesis.

– Time lag between both transmission channels must not exceed 200 ms, otherwise the spectator will perceive the asynchrony; furthermore, the additional intelligibility brought about by the visual modality is lost when it exceeds a maximum time lag (example of the videophone).

– Face gestures must be consistent with the acoustic wave, so that the auditor-spectator can identify who is talking when several voices are transmitted simultaneously.

– Face gestures must be consistent with the acoustic wave, so that the auditor/spectator is not subjected to the McGurk illusion [MCG 76], i.e. a bias in the acoustic message caused by inconsistent visual information.

– Face gestures must be consistent so that the auditor/spectator does not fall into a perception mode based on two independent modalities, where the audition does not complement the vision in a synergetic way, but comes across as an independent supplement, as is the case for dubbed video or simultaneous translation, for which the two information sources do not share any natural consistency.

4.4. Synthesis of visual speech

4.4.1. *The structure of an artificial talking head*

Whether it is in 2D or 3D, an artificial talking head is generally represented as a mesh (also called *network* or *lattice*) which defines its volume, and a set of descriptors that characterizes its aspects (called *texture* or *material*). The network is composed of a set of points defined by their 2D or 3D coordinates. These points are linked to their neighbors by ridges, which make up the sides, the set of which form a volume that approximates the face to be modeled. This information can be obtained automatically (laser rotoscopy) or semi-automatically (localization system), but most of the time it is generated manually by a computer graphic designer (3D modeler).

This information then forms the basis on which the animation can take place, by shaping the face as a 3-dimensional network, so as to obtain the right expression at the right time.

4.4.2. *Generating expressions*

In this section, we present several techniques to deform artificial talking heads for animation purposes.

4.4.2.1. *Morphing: interpolation between several face shapes*

Prior to applying the morphing technique, several face shapes must be constructed. Each face must contain the same number of points and their topologies must be identical. For each control point on the face, a new position is interpolated from the coordinates of this point in the different reference faces. In order to vary the influence of one shape on another, a weighting coefficient is used. Thus, an opening/closing movement of the mouth can simply be generated from two extreme shapes, one with the mouth open and one with the mouth closed. The degree of opening of the mouth is governed by the interpolation coefficient between these two shapes. A value of 0 corresponds to the closed-mouth shape, a value of 1 to the open-mouth shape. A coefficient of 0.5 thus corresponds to a half-open mouth.

The main drawback of this technique is the need for a large number of key faces (at least one for each possible movement). However, the *morphing* technique is very common and it is currently implemented in all standard animation software available on the market (3DSMax, SoftImage 3D, Maya, etc.) as well as in real-time rendering engines.

4.4.2.2. *Parametric face*

Parke was the initiator of this deformation technique, which he called "ad hoc parameterization" [PAR 82, PAR 91]. It consists of attributing to a command parameter, a set of operations active on a well-defined collection of points, so as to move them and create a new face shape. For instance, the opening of the jaw is obtained by defining the set of points which belong to the jaw and by applying a rotation to them. The angle of this rotation is directly correlated with the control parameter. The forward movement of the upper lip corresponds to a translation of the points belonging to that lip, proportionally to the command parameter.

The main drawback of this face shaping model is the total connection between the geometric structure of the face and the shaping procedures. Thus, for any new face structure, the definition of the points to be taken into account for a particular parameter has to be completely redefined. In addition, there is no (or very little) propagation of the deformation to the points that are not directly affected by the parameter variation: the movement of each point is entirely characterized by the command parameters.

4.4.2.3. *Functional deformation model*

In order to model the anatomic structure of the face more precisely, Waters [WAT 87] has designed a muscular deformation model. In this approach, characteristic points on the face are defined, which are meant to correspond to cutaneous insertions of the various face muscles. Their movements are directly linked to the action of the command parameters, which in this case are related to the muscle structure. The movement of the points around these nodes is determined by functional calculus. The so-called *zygomatic* parameter mimics the action of this muscle by activating a point located at the corner of the lips. This activation passes on to neighboring points in the network, with some attenuation when the distance to the node increases.

In this approach, the deformation of the skin of the face is modeled by a set of functions. Moreover, some points on the face may not belong to any of the areas of influence, and thus remain static. This is why some models based on mechanics have been proposed, in order to account for the visco-elastic nature of the skin.

4.4.2.4. *Mass-spring network*

In the mass-spring network approach, the nodes of the polygon network (or of the control grid) are defined as masses connected to one another by springs. The skin of the face can be modeled by one [VIA 92], two [LEE 95, WAT 93] or three layers [WAT 90] of masses and springs. The propagation of the skin deformation is obtained by solving the Lagrange equations of classical mechanics. A set of external forces, which generally correspond to muscle forces, is applied on the network. The new facial expression corresponds to the equilibrium point reached by the mass/spring network and calculated by a relaxation method (for instance, the Euler method).

4.5. Animation

Facial animation consists of generating the deformation parameters corresponding to a targeted expression. To the best of our knowledge, there are four methods of facial animation: the analysis of the image of a face, the recovery of a puppeteer's gestures, the analysis of speech signal and generation from a text.

4.5.1. *Analysis of the image of a face*

A first technique consists of placing reflecting or colored marks on an actor's face, at well-defined locations [MAG 89, PAT 91, POU 94, WIL 90]. A video triangulation system tracking the position in space of these marks across time yields trajectories for characteristic points on the face. Commercial products using this technique are currently available (FaceTracker (Motion Analysis), Facetrax (AOA), X-IST (Vierte Art), etc.).

This technique is difficult to implement as any movement of the head with respect to the camera induces a relative movement of the points. As the real movements of the face points may be relatively small, the signal-to-noise ratio can collapse rapidly. Furthermore, the use of marks turns out to be inefficient for measuring the lip movements. This does not provide an accurate measurement of the inner lip geometry, which is highly correlated to the acoustic signal and therefore which is imperative for determining relevant control parameters.

A second method consists of putting make-up on the characteristic areas of the face. Parameter extraction can be done by a color processing algorithm applied to the image [LAL 91], for instance by isolating the areas of interest by *chroma key* techniques. Accurate measurements can then be taken on these areas. This technique makes it possible to detect relevant parameters for lip animation. The LipsInk system, proposed by the Ganymedia company, is based on this principle.

Finally, face analysis systems requiring neither marks nor make-up are being developed for the animation of synthetic faces, but also for other applications such as automatic speech recognition [NAT 96]. For this purpose, many image processing algorithms are developed. Some of them try to recover, from a video recording, relevant information on the speaker's face movements using optical stream techniques [ESS 93a, ESS 93, ESS 95b]. For this purpose, areas for which a movement is detected from one image to the next are first located. Alternatively, other approaches use pattern recognition techniques to locate areas of interest such as the eyes, the mouth, the chin, etc.: homogenous area detection, *a priori* models, contour tracking by curve fitting, statistical methods, etc. [REV 99, SAU 94, SAU 95]. Another method is to have the system learn a set of reference expressions and to classify an unknown image with respect to this set. For this purpose, powerful mathematical tools such as normalized correlations or principal component analysis can be used [TUR 91]. Many research groups currently work along these lines, which leads us to expect significant progress in the years to come.

4.5.2. *The puppeteer*

A puppeteer can animate a face using traditional techniques, the motion of his fingers and his hands being transmitted to a computer, with a data glove system [STU 92, TAR 91]. These methods enable real time facial animation, and therefore a high level of interactivity. They are very common for the animation of live virtual presenters.

4.5.3. *Automatic analysis of the speech signal*

Facial animation from a speech signal [CAS 94, LAV 95, LEW 87, MOR 90, MOR 91, MOR 93] calls for techniques directly related to speech synthesis, recognition and coding. The analysis step consists of extracting relevant information from the speech signal, for animating the face. However, the existence of high variability in the speech signal, the complementarity between audition and vision, the inversion problems and the existence of visible but non-audible gestures (visual anticipation) constitute a number of obstacles hindering a correct prediction of animation control parameters.

4.5.4. *From the text to the phonetic string*

Traditional animation of a talking head requires a phonetic representation of facial expressions. If the speech signal is natural, i.e. if the face's voice is one of an actor, this operation is usually conducted manually – as is the case for traditional animation. This consists of translating phonetically the pronounced text and aligning each phoneme with the corresponding segment of the acoustic waveform. This work can be done automatically with an alignment software derived from automatic speech recognition technologies.

Faces for which the corresponding speech signal is also synthetic [BEN 95, BES 95, COH 93, LEG 96a, LEG 96b, MOR 90, PEA 86, SAI 90, WAT 93b] can recover all necessary information from the text-to-speech system. Indeed, such a system is composed of two modules: a phonetic transcriber, which converts a text (generally input on the keyboard) into a phonetic string, together with duration and other prosodic information, using various levels of analysis (syntactic, semantic, lexical, etc.). In a second step, the synthesizer generates the acoustic wave corresponding to the phonetic string stemming from the previous module. Phonetic information produced by the first module can therefore be exploited in order to predict the movement of the synthetic face.

4.6. Conclusion

The anthropomorphic character of 3D faces makes the general public and the professional extremely demanding as concerns the synchrony and consistency between face gestures and the speech signal. It has therefore become necessary to develop animation techniques and models which account for the specificity and complexity of speech production.

The facial animation of artificial talking heads remains a very complex task and particularly the automatic generation of lip movements and their synchronization with the speech signal. In order to solve this problem, it is important not only to master animation techniques but also to hold a perfect knowledge of the speech production and perception processes.

4.7. References

[ABR 86] ABRY C., BOË L.J., "Laws for lips", *Speech Communication*, 5, p. 97-104, 1986.

[BEN 92] BENOÎT C., LALLOUACHE M.T., MOHAMADI T., ABRY C., "A set of French visemes for visual speech synthesis", *Talking Machines*, p. 485-504, Elsevier Science Publishers B.V., North-Holland, Amsterdam, 1992.

[BEN 94] BENOÎT C., MOHAMADI T., KANDELL S.D., "Effects of phonetic context on audiovisual intelligibility in French", *Journal of Speech and Hearing Research*, 37, p. 1195-1203, 1994.

[BEN 95] BENOÎT C., BESKOW J., COHEN M.M., GRANSTROM B., LE GOFF B., MASSARO D.W., "Text-to-audio-visual speech synthesis over the world", *Proc. of the Speech Maps Workshop*, Grenoble, December 1995.

[BES 95] BESKOW J., "Rule-based visual speech synthesis", *Proc. of the 4th EUROSPEECH Conference* vol. 1, p. 299-302, Madrid, 1995.

[BIN 74] BINNIE C.A., MONTGOMERY A.A., JACKSON P.L., "Auditory and visual contributions to the perception of consonants", *Journal of Speech and Hearing Research*, 17, p. 619-630, 1974.

[BRO 83] BROOKE N., SUMMERFIELD Q., "Analysis, synthesis and perception of visible articulatory movements", *Journal of Phonetics*, 11, p. 63-76, 1983.

[CAM 80] CAMPBELL R., DODD B., "Hearing by eye", *Quarterly Journal of Experimental Psychology*, 32, p. 509-515, 1980.

[CAS 94] CASELLA E., LAVAGETTO F., MIANI R., "A neural approach to lips movements modeling", *Proc. of EUSIPCO-94*, p. 13-16, Edinburgh, September 1994.

[CAT 94] CATHIARD M.-A., La perception visuelle de l'anticipation des gestes vocaliques: cohérence des événements audibles et visibles dans le flux de la parole, Thesis, Psychologie Cognitive, Pierre Mendès France University, Grenoble, 1994.

[COH 90] COHEN M.M., MASSARO D.W., "Synthesis of visible speech", *Behavioral Research Methods*, Instruments and Computers, vol. 22, no. 2, p. 260-263, 1990.

[COH 93] COHEN M.M., MASSARO D.W., "Modeling coarticulation in synthetic visual speech", in M. MAGNENAT-THALMANN and D. THALMANN (eds.), *Models and Techniques in Computer Animation*, Springer-Verlag, Tokyo, 1993.

[COH 94] COHEN M.M., MASSARO D.W., "Development and experimentation with synthetic visible speech", *Behavioral Research Methods, Instruments and Computers*, vol. 26, no. 2, p. 260-265, 1994.

[DIX 80] DIXON N.F., SPITZ L., "The detection of audiovisual desynchrony", *Perception*, p. 719-721, 1980.

[ERB 69] ERBER N.P., "Interaction of audition and vision in the recognition of oral speech stimuli", *Journal of Speech and Hearing Research*, 12, p. 423-425, 1969.

[ERB 75] ERBER N.P., "Auditory-visual perception of speech", *Journal of Speech and Hearing Research*, 40, p. 481-492, 1975.

[ESS 93a] ESSA I.A., PENTLAND A., A vision system for observing and extracting facial action parameters, Technical Report 247, MIT Media Laboratory, Perceptual Computing Group, Cambridge, USA, July 1993.

[ESS 93b] ESSA I.A., DARRELL T., PENTLAND A., Tracking facial motion, Technical Report 272, MIT Media Laboratory, Perceptual Computing Group, Cambridge, USA, 1993.

[ESS 95a] ESSA I.A., BASU S., DARRELL T., PENTLAND A., Modeling, tracking and interactive animation of faces and heads using input from video, Technical Report 370, MIT Media Laboratory, Perceptual Computing Group, Cambridge, USA, 1995.

[ESS 95b] ESSA I.A., Analysis, interpretation and synthesis of facial expressions, PhD Thesis, Massachusetts Institute of Technology, Cambridge, USA, 1995.

[FRO 64] FROMKIN V., "Lip positions in American-English vowels", *Language and Speech* vol. 7, no. 3, p. 215-225, 1964.

[GEN 81] GENTIL M., Etude de la perception de la parole: lecture labiale et sosies labiaux, Technical report, IBM, France, 1981.

[HIL 88] HILL D.R., PEARCE A., WYVILL B., "Animating speech: an automated approach using speech synthesized by rules", *The Visual Computer*, 3, p. 277-289, 1988.

[LAL 91] LALLOUACHE M.T., Un poste visage-parole couleur. Acquisition et traitement automatique des contours des lèvres, Thesis, Institut national polytechnique Grenoble, France, 1991.

[LAV 95] LAVAGETTO F., "Converting speech into lip movements: a multimedia telephone for hard of hearing people", *IEEE Transaction on Rehabilitation Engineering*, vol. 3, no. 1, p. 90-102, 1995.

[LEE 95] LEE Y., TERZOPOULOS D., WATERS K., "Realisitic modeling for facial animation", *Computer Graphics Annual Conference Series*, p. 55-62, 1995.

[LEG 96a] LE GOFF B., BENOÎT C., "A text-to-audiovisual-speech synthesizer for French", *Proc. of the 4th International Conference on Spoken Language Processing*, Philadelphia, USA, October 1996.

[LEG 96b] LE GOFF B., BENOIT C., "Synthèse audiovisuelle de la parole à partir du texte", *Actes des XXIèmes Journées d'Etude sur la Parole*, p. 379-382, Avignon, France, 1996.

[LEW 87] LEWIS J.P., PARKE F.I., "Automated lipsynch and speech synthesis for character animation", *Proc. of Human Factors in Computing Systems and Graphics Interface '87*, p. 143-147, April 1987.

[MAG 89] MAGNO CALDOGNETTO E., VAGGES K., BORGHESE N.A., FERRIGNO G., "Automatic analysis of lips and jaw kinematics in VCV sequences", *Proc. of the EUROSPEECH Conference*, vol. 2, p. 453-456, Paris, France, 1989.

[MAS 90] MASSARO D.W., COHEN M.M., "Perception of synthesized audible and visible speech", *Psychological Science*, 1, p. 55-63, 1990.

[MCG 76] MCGURK H., MACDONALD J., "Hearing lips and seeing voices", *Nature*, 264, p. 746-748, 1976.

[MOH 93] MOHAMADI T., Synthèse à partir du texte de visages parlants: réalisation d'un prototype et mesures d'intelligibilité bimodale, Thesis, Institut national polytechnique de Grenoble, France, 1993.

[MOR 90] MORISHIMA S., AIZAWA K., HARASHIMA H., "A real-time facial action image synthesis system driven by speech and text", *SPIE Visual Communications and Image Processing*, 1360, p. 1151-1157, October 1990.

[MOR 91] MORISHIMA S., HARASHIMA H., "A media conversion from speech to facial image for intelligent man-machine interface", *IEEE Journal on Selected Areas in Communications*, vol. 9, no. 4, 1991.

[MOR 93] MORISHIMA S., HARASHIMA H., "Facial expression synthesis based on natural voice for virtual face-to-face communication with machine", *IEEE Virtual Reality Annual International Symposium*, p. 486-491, Seattle, Washington, September 1993.

[MOU 80] MOURAND-DORNIER L., Le rôle de la lecture labiale dans la reconnaissance de la parole, Medical Thesis, Franche-Comté University, France, 1980.

[NAT 96] STORK D., HENNECKE M. (eds.), *Speechreading by Humans and Machines*, Springer-Verlag, Berlin, NATO-ASI Series no. 150, 1996.

[NEE 56] NEELY K.K., "Effect of visual factors on the intelligibility of speech", *Journal of the Acoustical Society of America*, 28, p. 1275-1277, 1956.

[PAR 82] PARKE F.I., "A parameterized model for facial animation", *IEEE Computer Graphics and Applications*, vol. 2, no. 9, p. 61-70, 1982.

[PAR 90] PARKE F.I., "Parametrized facial animation revisited", *State of the Art in Facial Animation*, vol. 26, p. 44-61, ACM Siggraph'90 Course Notes, 1990.

[PAR 91] PARKE F.I., "Control parametrization for facial animation", in N. MAGNENAT-THALMANN and D. THALMANN (ed.), *Computer Animation '91*, p. 3-14, Springer-Verlag, 1991.

[PAT 91] PATTERSON E.C., LITWINOWICZ P.C., GREENE N., "Facial animation by spatial mapping", in N. MAGNENAT-THALMANN and D. THALMANN (ed.), *Computer Animation '91*, p. 4558, Springer-Verlag, 1991.

[PEA 86] PEARCE A., WYVILL B., WYVILL G., HILL D.R., "Speech and expression: A computer solution to face animation", *Graphics and Vision Interface '86*, p. 136-140, 1986.

[PEL 91] PELACHAUD C., Communication and Coarticulation in Facial Animation, PhD Thesis, Computer and Information Science Department, University of Pennsylvania, Philadelphia, USA, 1991.

[PEL 96] PELACHAUD C., BADLER N.I., STEEDMAN M., "Generating facial expressions for speech", *Cognitive Science*, vol. 20, no. 1, p. 1-46, 1996.

[PER 91] PERRIER P., De l'usage des modèles pour l'étude de la production de la parole, Thesis, Institut national polytechnique de Grenoble, France, 1991.

[PLA 80] PLANT G.L., "Visual identification of Australian vowels and diphthongs", *Journal of Audiology*, 2, p. 83-91, 1980.

[POU 94] POURCEL D., "Le clonage des personnages en images de synthèse", *Proc. of the IMAGINA '94 Conference*, p. 166-172. National Audiovisual Institute, France, February 1994.

[REI 87] REISBERG D., MCLEAN J., GOLDFIELD A., *Easy to Hear but Hard to Understand: A Lip-reading Advantage with Intact Auditory Stimuli*, Lawrence Erlbaum Associates, 1987.

[REV 99] REVERET L., Conception et évaluation d'un système de suivi automatique des gestes labiaux en parole, Thesis, Institut national polytechnique de Grenoble, France, 1999.

[SAI 90] SAINTOURENS M., TRAMUS M.-H., HUITRIC H., NAHAS M., "Creation of a synthetic face speaking in real time with a synthetic voice", *Proc. of the ETRW on Speech Synthesis*, p. 249-252, Autrans, France, 1990.

[SAU 94] SAULNIER A., VIAUD M.L., GELDREICH D., "Analyse et synthèse en temps réel du visage pour la télévirtualité", *Proc. of the IMAGINA '94 Conference*, p. 175-182, National Audiovisual Institute, France, February, 1994.

[SAU 95] SAULNIER A., VIAUD M.L., GELDREICH D., "Real-time facial analysis and synthesis chain", in M. BICHSEL (ed.), *International Workshop on Automatic Face and Gesture Recognition*, p. 86-91, 1995.

[SME 91] SMEELE P.M.T., SITTIG A.C., "The contribution of vision to speech perception", *Proc. of 2nd European Conference on Speech Communication and Technology*, vol. 3, p. 1495-1497, Genova, Italy, September 1991.

[SME 92] SMEELE M.T., SITTIG A.C., VAN HEUVEN V.J., "Intelligibility of audio-visually desynchronised speech: Asymmetrical effect of phoneme position", *Proc. of the International Conference on Spoken Language Processing*, p. 65-68, Banff, Alberta, Canada, October 1992.

[STU 92] STURMAN D.J., Whole-hand input, PhD thesis, Massachusetts Institute of Technology, Cambridge, USA, 1992.

[SUM 54] SUMBY W.H., POLLACK I., "Visual contribution to speech intelligibility in noise", *Journal of the Acoustical Society of America*, 26, p. 212-215, 1954.

[SUM 89] SUMMERFIELD Q., MACLEOD A., MCGRATH M., BROOKE, "Lips, teeth and the benefits of lipreading", in A.W. YOUNG and H.D. ELLIS (eds.), *Handbook of Research on Face Processing*, p. 223-233, Elsevier Science Publisher B.V., North Holland, 1989.

[SUM 79] SUMMERFIELD Q., "Use of visual information for phonetic perception", *Phonetica*, 36, p. 314-331, 1979.

[TAR 91] TARDIF H., "Character animation in real time", *Proc. of SIGGRAPH'91 Conference*, 1991.

[TSE 90] TSEVA A., "Les visèmes vocalique du français: une analyse multidimensionnelle", *Bulletin d'Audiophonologie*, Scientific Annals of Franche-Comté University, 6, p. 381-411, 1990.

[TUR 91] TURK M., PENTLAND A.P., "Face recognition using eigenfaces", *Proc. of CVPR*, p. 586-591, 1991.

[VIA 92] VIAUD M.L., Animation Faciale avec rides d'expression vieillissement et parole, Thesis, Paris XI-Orsay University, France, 1992.

[WAT 90] WATERS K., TERZOPOULOS D., "A physical model of facial tissue and muscle articulation", vol. 26, *State of the Art in Facial Animation*, p. 130-145, ACM Siggraph'90 Course Notes, Dallas, USA, August 1990.

[WAT 87] WATERS K., "A muscle model for animating three-dimensional facial expressions", *Computer Graphics*, vol. 21, no. 4, p. 17-24, July 1987.

[WAT 93] WATERS K., Synthetic muscular contraction on facial tissue derived from computer tomography data, Technical report, Digital Equipment Corporation, Cambridge, USA, April 1993.

[WAT 93b] WATERS K., LEVERGOOD T., DECface: an automatic lip-synchronization algorithm for synthetic faces, Digital Equipment Corp, Cambridge Research Laboratory, Technical Report Series 93/4, September 1993.

[WIL 90] WILLIAMS L., "Electronic mask technology", *State of the Art in Facial Animation*, vol. 26, p. 164-184, ACM Siggraph'90 Course Notes, Dallas, USA, August, 1990.

[ZER 90] ZERLING J.P., Aspects articulatoires de la labialité vocalique en français, Thesis, Institut de phonétique, Strasbourg, France, 1990.

Chapter 5

Computational Auditory Scene Analysis

5.1. Introduction

Until recently, the study of auditory processes has been mainly focused on perceptual qualities such as the pitch, loudness or timbre of a sound produced by a *single source*. Experimentations in psychoacoustics have brought to the fore a relationship between the physical properties of a sound and the feelings which it conveys; the physiological mechanisms underlying this relationship have been sketched. Models of auditory processes have been designed, which are based on the acoustic wave or its spectrum.

Unfortunately, the audio sources around us are rarely active in isolation. We live in a cacophony of superimposed voices, sounds and noise, the resulting spectrum of which is completely different from that of a single source. Each of our ears receives sound waves originating from a multitude of sources. However, it is generally possible to focus one's attention on a specific source and perceive its loudness, its pitch, its timbre, and understand what is being said (when dealing with a speech source), in spite of the presence of competing sounds. Conventional models, which have been designed for isolated sounds, are not powerful enough to account for the perception of multiple sources.

In his day, Helmholtz wondered how we could perceive individual qualities of instruments when they were played together [HEL 77]. However, it is with the works of Bregman that *auditory scene analysis* (or ASA) became a research topic in itself [BRE 90]. For Bregman, the study of the emergence of subjective sources (or

Chapter written by Alain DE CHEVEIGNÉ.

streams) is central, as it logically comes before the determination of their individual qualities. Bregman's ASA is a transposition of visual scene analysis principles into the field of audition.

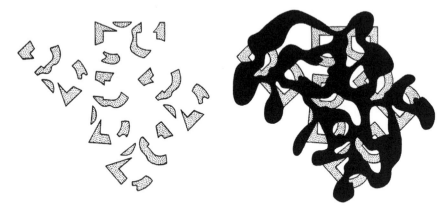

Figure 5.1. *Visual scene analysis. On the left, the fragments are not organized. On the right, the presence of a masking shape enables their grouping on a perceptual basis. ASA searches for similar principles for organizing the auditory world (from [BRE 90])*

Along with the development of computer science and artificial intelligence, attempts have been made to develop *computational auditory scene analysis* (CASA) [BRO 92a, COO 91, ELL 96, LYO 83, MEL 91, WAN 95, WEI 85]. CASA models have a twofold ambition: 1) to describe and explain perceptual processes and 2) to solve practical problems, such as noise removal in a speech recognition system. The influence of computational vision research, in particular the work of Marr [MAR 82] has played a key role.

The concept of the *CASA model* suffers from some ambiguity. For the modeling of perceptual processes, it is not easy to define the border between CASA models and other models, as computational modeling has become quite traditional in several domains. Viewed as a signal processing approach, the specificities and advantages of CASA modeling over other techniques are not completely straightforward. By aiming at being a good auditory model *and* a useful approach, the CASA model runs the risk of reaching neither of these two objectives. Nevertheless, the CASA approach can be fruitful, provided the difference is clearly made between the model and the method, especially when they are evaluated. The insistence in designing a full (and therefore complex) system is a good remedy against the potential reductionism of psychoacoustic models. From a practical viewpoint, applications such as speech recognition need to replicate the noise tolerance ability of the human auditory system. Interesting developments have recently been coming from the CASA approach, in particular the *missing feature theory* [COO 94, COO 97, LIP 97, MOR 98].

5.2. Principles of auditory scene analysis

5.2.1. *Fusion versus segregation: choosing a representation*

In the framework of ASA, the notions of *fusion* and *separation* are often used. Fusion corresponds to situations when some features are attributed to the same audio source (or stream) whereas segregation happens when they are distributed over several sources. To give full meanings to these terms, an internal representation composed of auditory cues must be hypothesized, in which the cues from the various sources can be separated from one another. This may mean a representation of the physical stimulus in the time domain, in the frequency domain or in any one of the various time-frequency representations. Alternatively, a physiological representation can be considered (cochlea-based filter banks, neural coincidence networks, etc.), from which the auditory system is able to extract elements pertaining to each source.

In fact, psychoacousticians use a third type of representation when they describe a stimulus in terms of synthesis parameters (duration, amplitude, frequency or instantaneous phase for each component). This is not exactly a time-frequency representation in the conventional sense, as no representation of this type can provide such an accurate description on the time and frequency axes simultaneously. For example, let us consider a stimulus composed of several sine functions modulated in frequency. In the synthesis operation, the instantaneous frequency is perfectly specified but there is no general method for retrieving these parameters from the stimulus. A time-frequency analysis may provide an approximate estimation, but not a unique exact value that would correspond precisely to the idealistic description of psychoacoustics.

This causes considerable confusion. The *principles of ASA* have been stated by psychoacousticians in terms of synthesis parameters. On the other hand, the CASA model does not have access to this idealistic representation and must deal with what can be extracted from the signal. Many "good ideas" in terms of idealistic representations go flat when they are applied in practice. Revealing these difficulties is one of the merits of the CASA approach.

5.2.2. *Features for simultaneous fusion*

While keeping in mind the abovementioned restrictions, let us consider a stimulus "formed" of a given number of components. We could expect that the auditory system attributes them to a single source, as a sonometer or a speech recognition system would do. Our experience shows that this is not always the case: in general, we "separate the components" of the stimulus and attribute part of them to each source. The following question thus arises: as, in some cases, the components from distinct sources are separable (by segregation), why are they sometimes perceived as being grouped (fusion)? Fusion and segregation are two sides of a same coin. What are the acoustic features that trigger either one?

Harmonicity. A harmonic relationship between components favors their fusion. This is the case when the stimulus is periodic (voiced speech, some musical instruments, etc.). On the contrary, when the stimulus is inharmonic ("polyperiodicity" [MAR 91]), it seems to contain several sources. Concurrent vowels or voices are easier to understand if they follow distinct harmonic series, i.e. if their fundamental frequencies F_0 are different.

Envelope coherence, attack synchronicity. If individual components start simultaneously and their amplitude varies coherently, they tend to be fused. Conversely, an attack asynchrony favors segregation. This is an example of the more general principle known as *common fate*.

Binaural correlation. If the components of a source all have the same binaural relationship, their fusion is favored. A difference in the binaural relationship between a target sound and a masking sound favors the perception of the target sound.

Coherent frequency modulation. This is another example of the common fate principle. If the time-frequency representation is viewed as a picture, the components with coherent modulation should form a pattern and emerge from the static components or from those with incoherent modulation.

All these features have been proposed and implemented with more or less success in some CASA systems.

5.2.3. *Features for sequential fusion*

Similarly to simultaneous fusion, we could imagine that sounds which follow one another over time are always attributed to the same source (fusion). This is not the case: in some situations, the auditory system divides a sequence of sounds into several distinct streams (segregation). Each stream then seems to evolve independently. Each of them can be chosen and "isolated" by attention. The order of the sounds within a given stream can be distinguished, but not from one stream to another. This phenomenon is exploited in Bach's fugues to create the illusion of several melodic lines with a single instrument. Among the features which determine fusion and segregation, let us mention:

– frequency proximity: a sequence of pure sounds with close frequencies tends to fuse into a single stream. The sounds form distinct streams if frequencies are far apart;

– repetitiveness: segregation trends are reinforced by the duration and the repetitiveness of stimuli;

– repetition rate: presenting sound sequences at a fast rhythm favors segregation. A slower rhythm favors fusion;

– timbre similarity: a sequence of sounds with a common timbre tends to favor fusion. Sounds with very different timbres are less likely to fuse and it is difficult to determine their time order.

On the basis of this enumeration, we could expect that, because of its many discontinuities in amplitude, timbre, etc., speech is not perceived as a coherent stream. Paradoxically, this is not the case: a voice keeps its coherence despite these discontinuities.

5.2.4. *Schemes*

The previously mentioned features depend on the signal and underlie what is called *primitive* fusion. The corresponding mechanisms are automatic and unintentional. They do not depend on any training or cognitive context. Situations also exist where the fusion is based on learned *schemes*, on abstract regularities or on the subject's state of mind. The distinction between primitive versus scheme-based fusion is to be put in parallel with the *bottom-up* versus *top-down* processes in artificial intelligence.

5.2.5. *Illusion of continuity, phonemic restoration*

When a short noise is superimposed upon a continuous tone, the tone seems to continue "behind" the noise. The same impression is observed even if the tone is interrupted during the noise, provided the latter is loud enough. This is called the *illusion of continuity*. Similarly, with speech, if a phoneme is replaced by a relatively loud noise, the missing phoneme is perceived as if it were still present. This is the phonemic restoration phenomenon. The "restored" phoneme can vary according to the context (for instance, a stimulus such as "*eel" becomes "wheel", "peel", "meal", etc., according to the semantic context). Quite surprisingly, it is almost impossible to tell which of the restored phrase's phonemes was missing.

5.3. CASA principles

5.3.1. *Design of a representation*

The analogy with visual scene analysis, on which ASA is based, assumes the existence of a "representation", the richness of which is comparable to the 3D space for objects or the 2D space for images (Marr uses the term 2½D to designate the enriched representation stemming from binocular vision and other perception mechanisms related to depth [MAR 82]). As the acoustic wave is of low dimensionality, the CASA model starts by synthesizing an enriched representation.

5.3.1.1. *Cochlear filter*

The conventional CASA model starts with a filter bank. In principle, these filters conform to what is known on cochlear filtering. In practice, a wide variety of filter banks has been proposed, according to the priority decided by the designer on the proximity to a physical model of the cochlea, the conformity with physiological recordings or psychological data, the ease of implementation, etc. Currently, the most popular filter is the *gammatone* filter, which is relatively realistic and easy to implement [COO 91, HOL 88, PAT 92, SLA 93]. The filters of a cochlear filterbank model are generally of constant width (in Hz) up to 1 kHz. Above this frequency, their width is proportional to their central frequency. An additional delay may be added to the output of the channels so as to compensate for the group-delay differences and "align" their impulse responses.

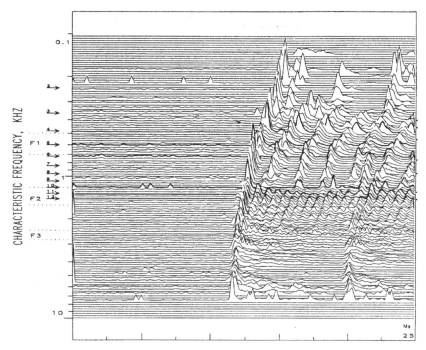

Figure 5.2. *Activity of a group of auditory nerve fibers, for a cat, in response to the synthetic syllable [da]. Cochlear filtering and transduction models try to reproduce this type of response. The progressive delay for low frequency channels (top) due to the propagation time in the cochlea is usually compensated for in the model (from [SHA 85])*

5.3.1.2. *Transduction*

The mechanical vibration of the basilar membrane governs the *probability* of discharge of the auditory nerve fibers which synapse to the inner hair cells.

This process may be modeled with varying degrees of realism:

– since a probability is positive, the transduction shows properties similar to the one of a half-wave rectifier;

– transduction also has compressive properties which can be modeled by a simple instantaneous non-linearity (logarithm, cubic root, etc.), or by an adaptive mechanism: automatic gain control [HOL 90, LYO 82, LYO 84, PAT 92, SEN 85] or hair cell model [MED 86, MED 88];

– in the models by Lyon [LYO 82, LYO 84] and Holdsworth [HOL 90], the gain for each channel varies according to the activity within a time-frequency neighborhood. The physiological basis of this process is unclear, but it has the favorable effect of reinforcing the contrast of the representation in the frequency domain (this is an example of confusion between a model and a method). Some other models go further and incorporate an explicit mechanism of spectral and/or temporal differentiation, an example of which is the LIN (*lateral inhibitory network*) of Shamma [SHA 85];

– non-linear transduction is usually followed by low-pass filtering (time-domain smoothing). Depending on the model, this filtering operation is either light (small time-constant) and represents the loss of synchronization observed physiologically in the high frequencies (between 1 and 5 kHz), or more severe, so as to eliminate the periodic structure of voiced speech and to obtain a stable spectrum over time.

The output of the filter/transduction module can be seen either as a sequence of short-term spectra, or as a set of parallel channels each carrying a filtered version of the signal. This is a high-dimensional representation which is a first step towards a favorable substrate for scene analysis.

5.3.1.3. *Refinement of the time-frequency pattern*

Nevertheless, the output of the filter/transduction module does not have the ideal characteristics of the representation which has been used for synthesis (see section 5.2.1): it tends to lack frequency and/or temporal resolution. The LIN network proposed by Shamma and mentioned above reinforces the spectral contrast [SHA 85]. Deng proposes the cross-correlation between neighboring channels in order to reinforce the formants' representation [DEN 88]. *Synchrony strands* by Cooke lead to a representation which is close to a sum of sine curves and which is well adapted for applying ASA principles (continuity of each strand, common fate, harmonicity, etc.) [COO 91]. These techniques can be interpreted as attempts to extract from the signal a representation close to the ideal representation which is used by psychoacousticians, according to the formulation of ASA's principles.

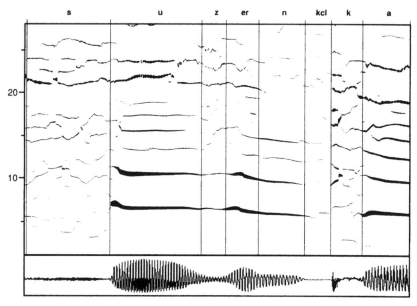

Figure 5.3. *Synchrony strand type representation, in response to a segment of speech (bottom). In the low frequencies, each strand corresponds to one harmonic, whereas in the high frequencies it corresponds to a formant (from [COO 91])*

5.3.1.4. *Additional dimensions*

If the time-domain smoothing is not too severe, the *temporal structure* of each channel stemming from the filter/transduction module can be exploited, leading to an enrichment of the representation with additional dimensions. Inspired by the binaural interaction model put forward by Jeffress, Lyon proposes to calculate the cross-correlation function between the channels from each ear [LYO 83]. Compared to a traditional time-frequency representation, this representation contains an additional dimension: the interaural delay. Maxima can appear at different positions along this dimension, corresponding to the azimuths of the various sources. Lyon samples the representation in terms of sections parallel to the frequency axis in order to isolate a particular source [LYO 83]. Similar attempts have been made since [BOD 96, PAT 96].

Another dimension comes into play if the *autocorrelation* function is calculated for each channel. This idea was originally proposed by Licklider in order to estimate the period in a perceptive pitch model [LIC 59]. In response to a periodic stimulus (as in voiced speech), maxima arise at locations corresponding to the period (and multiples of the period). This principle can be exploited in order to separate concurrent voice correlates. In response to several periodic stimuli (voices), some channels may be dominated by one voice and others by another voice. By selecting channels according to the dominant periods, it is possible to isolate voices. Proposed by Weintraub [WEI 85], this idea was revisited by Mellinger [MEL 91], Meddis and Hewitt [MED 92], Brown [BRO 92a], Lea [LEA 92] and Ellis [ELL 96].

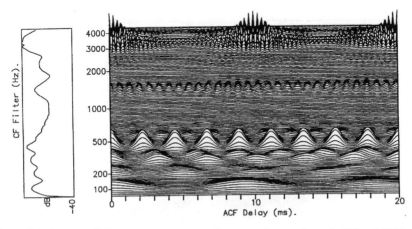

Figure 5.4. *Autocorrelation pattern corresponding to a mixture of vowels ([i] at 100 Hz and [o] at 112 Hz). Each line corresponds to one of the channels of the peripheral filter. Each channel is assigned to a vowel as a function of its dominant periodicity (from [LEA 92])*

Autocorrelation analyzes each channel with a sharp time-resolution, making it possible to resolve periodicity for the speech formants. However, the fine structure also reflects the resonance of the cochlear filters, which does not tell much about the signal. A *time-domain smoothing* operation enables the removal of the fine structure and (hopefully) retains only the modulations reflecting the fundamental period. These modulations can then be evaluated by autocorrelation or by other methods: zero-crossing [COO 91] or Fourier transform [MEY 96, MEY 97]. The *modulation spectrum* of various parameters (physiological features, LPC coefficients, cepstral coefficients, etc), considered to be temporal sequences, has recently created a strong interest, in particular in the field of speech recognition [GRE 96, HER 94, KAN 98, NAD 97].

Among other transformations, let us mention the frequency transition map by Brown and the onset maps by Mellinger, Brown or Ellis, which aim at localizing abrupt time changes that may correspond to the beginning of a sound [BRO 92a, ELL 96, MEL 91].

Each additional dimension enriches the representation. If the acoustic pressure at one ear is dependent on *one* dimension (the time), the set of peripheral channels is dependent on *two* dimensions (time, frequency). When taken into account, the binaural correlation and the autocorrelation (or the modulation spectrum) lead to *four* dimensions: time, frequency, interaural delay and modulation frequency. This "dimensional explosion" is motivated by the hope that cues for concurrent sounds will become separable if the dimensionality is high enough.

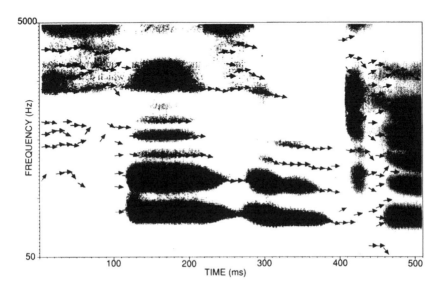

Figure 5.5. *Frequency transition map corresponding to a speech signal (same segment as in Figure 5.3). The arrows indicate the orientation estimated by a time-frequency orientation filter bank (from [BRO 93])*

5.3.1.5. *Basic abstractions*

Most CASA models start with rich, weakly constrained representations (see previous section) and then try to organize information as basic objects, following for example the ASA principles. Synchrony strands proposed by Cooke [COO 91] result from the application of a time continuity constraint to the components in the spectral representation. The principle of harmonicity grouping translates into *periodicity groups* in Cooke and Brown's approach [COO 92] or into *wefts* in Ellis' work [ELL 96]. The principle of attack synchrony is used by Brown to form auditory objects [BRO 92a].

5.3.1.6. *Higher-order organization*

The organization process carries on hierarchically, until the complete partition of the information into sources is obtained. Some models use a purely bottom-up (*data-driven*) process, while other models claim a more complex, top-down strategy calling for artificial intelligence techniques [ELL 96, GOD 97, KAS 97, NAK 97]. The drawback of complex strategies is twofold: they are opaque and they tend to react catastrophically – in the sense that a small disruption of the system's input conditions may produce large changes of its state. However, they are essential for handling the whole set of information sources and hypotheses which take place in the organization of an auditory scene.

5.3.1.7. *Schemes*

Most CASA systems are *data-driven* and rely on ASA principles of the *primitive* type. *Top-down* approaches, relying on *scheme-based* ASA principles, are rare. It is worth mentioning Ellis' proposal to use a speech recognition system in order to guide auditory scene analysis [ELL 97]. When the speech component of the auditory scene is recognized, its contribution to the scene can be determined and the rest of the scene can be analyzed more accurately.

5.3.1.8. *The problem of shared components*

Whatever the richness and dimensionality of the representation, it may be that the assignment of a given element is ambiguous. Strategies vary depending on whether they assign this element to one source only (exclusive allocation), to both sources (duplex assignment) or to none at all. It is also possible to *split* the element according to some continuity criterion in the time or frequency domain [WEI 85]. Such a split can be seen as a failure of the representation, which has not succeeded in partitioning the acoustic information into atomic elements assignable to each source.

5.3.1.9. *The problem of missing components*

Theoretical reasons (that are unfortunately confirmed in practice) tell us that it is impossible to reach perfect separation in each and every situation. For instance, too close frequency components will be confused and assigned to one of the sources at the expense of another. Such masked or uncertain portions will be missing in the representation of the separated sources. Two approaches are possible to address this problem: 1) to re-create the missing information by interpolation or extrapolation from the acoustic or cognitive context [ELL 96, MAS 97]; or 2) to mark the corresponding portion as missing and to ignore it in the subsequent operations, for instance by assigning a zero weight in the pattern recognition step [COO 97, LIP 98, MOR 98].

The first approach, sometimes motivated by an over-literal interpretation of the notion of *phonemic restoration,* may be justified when re-synthesis is intended. The second approach is preferable in speech recognition applications.

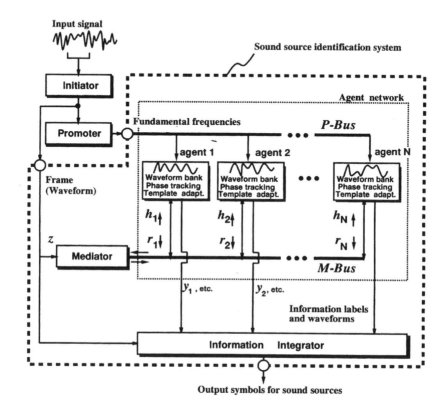

Figure 5.6. *An example of a CASA system structure: the Ipanema system for music analysis. Each agent is dedicated to the tracking of a particular aspect of the signal, under the control of a "mediator" (from [KAS 96])*

5.4. Critique of the CASA approach

The CASA approach is fertile, but it has weaknesses and pitfalls which need to be identified in order to be avoided.

5.4.1. *Limitations of ASA*

ASA is based on the concept that an auditory scene can be treated as a visual scene and that the Gestalt principles can be transposed from the vision domain, provided an adequate representation is chosen and a few adjustments are made to take into account the specificities of the acoustic domain. This idea can however lead to erroneous intuitions.

For example, let us consider harmonicity, which is a key grouping principle in ASA and is widely used in CASA models. ASA would state that spectral or temporal regularities of a harmonic target constitute a pattern which is easy to extract from the background (typically inharmonic or with a different fundamental frequency). Harmonicity would confer to a target some sort of texture which would facilitate its identification. This is not the case: many experiments show that the harmonicity of the *background* (or masking sound) does indeed facilitate segregation, but the harmonicity of the target has hardly any effect [DEC 95, DEC 97b, LEA 92, SUM 92c]. It can also be shown that the target's harmonicity is of limited usefulness in separating concurrent voices in a speech recognition task, and is less useful for the target than for the interference signal [DEC 93b, DEC 94].

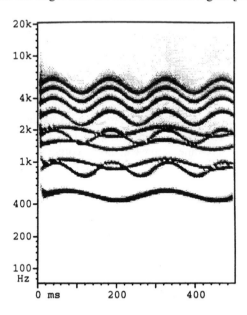

Figure 5.7. *The frequency modulation incoherence between two concurrent sources is exploited by the music analysis system (by [MEL 91]). Psychoacoustic experiments have shown however that this information is not used by the auditory system. This is an example where a CASA system exploits a Gestalt principle (common fate of the frequencies) that may not actually be used by the auditory system*

A second example is the Gestalt principle of *common fate*, which would mean that a spectrum made of components that vary in parallel (coherent frequency modulation) form a particular pattern which is especially easy to distinguish from a static background, or a background that would vary differently. Modulating a target which is not coherent with the background should thus facilitate its identification. Once again, this is not the case: experiments show that frequency modulation has barely no effect other than the instantaneous F_0 differences induced by the modulation [CAR 94, DAR 95, DEM 90, MAR 97, MCA 89, SUM 92b].

One more example: the quality of a target's binaural correlation governs the accuracy with which it can be localized. We may think that this facilitates its segregation, whatever the nature of the background. Here again, this is not the case: segregation depends on the binaural correlation of the *masking sound* and not of the target. A well correlated masking sound is easy to cancel [COL 95, DUR 63]. Curiously, it is not necessary that the correlation be consistent across the various frequency channels [CUL 95].

5.4.2. *The conceptual limits of "separable representation"*

As mentioned above, the purpose of the initial, enriched representation is to allow correlates of different sources to be perceptually separated, by assigning elements of the representation to one source or another. However this goal is not always attained. Many authors have been confronted with the need to split elements of the representation, such as channels of a filterbank, and share them between sources [COO 91, ELL 96, PAR 76, WEI 85]. We may thus wonder whether the separable representation is in itself necessary. For example, the authors of [DEC 97b] have shown that Meddis and Hewitt's model [MED 92], that operates on a separable frequency-delay-time representation, could not explain all the F_0 difference effects on vowel segregation. Conversely, a model that operates on the time structure of nervous discharges in each frequency channel accounts well for segregation phenomena [DEC 97b]. This model performs better, but does not use a separable representation. Another example is the estimation of the fundamental frequencies of simultaneous sounds (for instance, the notes played by instruments which play together), which can be achieved without resorting to a separable representation of the time-frequency or autocorrelation type [DEC 93a, DEC 98].

Separable time-frequency-correlation representations, or the like, are often needed in CASA models. However, they are neither a panacea, nor a must, for auditory organization tasks.

5.4.3. *Neither a model, nor a method?*

The CASA approach offers a fertile and open field for new experiments, ideas, models and methods. This is not risk-free. At best, the CASA specialist is well versed in the auditory sciences (psychoacoustics, physiology, etc.) and yet perfectly in gear with the application domain. At worst, he is neither. It is unfortunately common to see an unrealistic model being defended in the name of "efficiency", or a poorly performing method on the basis that "this is the way the ear works".

Modeling approaches of auditory processes (whether computational or not) are flourishing and it is not always easy to understand the specificity of the CASA model. On the other hand, many techniques exist for source separation, noise reduction, etc. (in particular, *blind separation*), which are not affiliated to the CASA framework. They do not necessarily correspond to perceptual mechanisms, but this does not mean that they are less effective.

5.5. Perspectives

In spite of these weaknesses, the CASA approach contributes to the understanding of perceptual processes and to the design of new concepts in signal processing. Four recent developments are of particular interest.

5.5.1. *Missing feature theory*

There are situations when a CASA system (or similar) does not succeed in restoring some parts of the target signal. The corresponding data are missing. Their replacement by a zero value would alter the meaning of the pattern (for example, in a speech recognition system). Their replacement by an average over time or frequency is hardly a better solution. In some cases, interpolation or extrapolation from the context is justified. However, the optimal solution for a pattern recognition task consists of *ignoring* the missing data by assigning them a zero *weight* [AHM 93, COO 94, COO 96, COO 97, DEC 93b, GRE 96, LIPP 97, MOR 98].

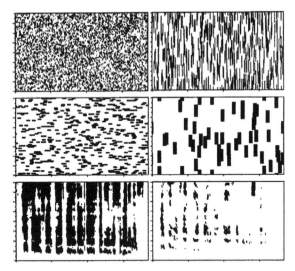

Figure 5.8. *Time-frequency masks used in missing data experiments. Areas in black correspond to available information, while the rest is missing. Recognition rates remain high even with a suppression rate of 80%. It is also possible to carry out training on incomplete data (from [COO 96])*

According to this approach, the CASA module delivers a *reliability map* to the recognition module. The latter module must be in the position to exploit it, which may raise difficulties in practice. For instance, many recognition systems use *cepstral* parameters, the advantage of which is to be orthogonally distributed and to allow the use of HMM (hidden Markov models) with diagonal covariance matrices. A reliability map in the *spectral* domain cannot be directly exploited by such a system. The use of spectral parameters instead of cepstral parameters raises other problems [MOR 96].

The correct use of *missing feature* techniques is certainly a key to the practical application of the CASA approach. They can also be useful in a wider context, for instance, for integrating information from different modalities. For example, an audiovisual recognition system can attribute a small weight to the image when the speaker's face is hidden, or a small weight to the audio modality when the speech sound is masked by noise.

5.5.2. *The cancellation principle*

Conventionally, ASA uses the structure of target sounds (for instance, their periodicity) in order to extract them from a non-structured (or differently-structured) environment. However, it has been observed that this approach is not particularly effective, and that it is not the manner in which the auditory system proceeds. Let us consider the case of two microphones recording two sources with distinct azimuths. A system exploiting the position of the target will yield at best (by beam-forming) a 6 dB signal-to-noise ratio reduction, whereas a system using the knowledge of the position of the interference can theoretically reach an infinite signal-to-noise ratio (even though, in practice, the improvement is limited in the case of reverberation or multiple maskers). Similarly, a system exploiting the target periodicity for reinforcing it will not perform as efficiently as a system exploiting the periodicity of the background to cancel it [DEC 93a, DEC 93b]. The auditory system exploits the background's periodicity rather than that of the target [DEC 95, DEC 97c, LEA 92, SUM 92c]. The cancellation criterion is close to that used by blind separation techniques. Scene analysis by successive cancellation steps is a characteristic of Nakatani's system [NAK 95a, NAK 95b, NAK 97].

Cancellation offers in some cases an infinite rejection (i.e. an infinite improvement of the target/background ratio), but it usually introduces some distortion of the target. For example, the shared (or masked) components are suppressed. Missing feature techniques are useful in this case.

5.5.3. *Multimodal integration*

The development of multimodal speech recognition opens the way for multimodal scene analysis, which would be more than the simple juxtaposition of visual and auditory analysis modules [OKU 99a]. There again, missing feature approaches are promising for the integration of modal data with variable reliability.

5.5.4. *Auditory scene synthesis: transparency measure*

ASA can also be approached from a radically different angle, i.e. from the viewpoint of an audio scene designer. When the audio material is assembled and mixed, it may happen that one of the ingredients is particularly dominant and has a strong masking effect that makes the auditory scene rather confused. Taking into account the various parameters revealed by ASA enables the prediction of the masking power of a source according to its physical characteristics. An *audio transparency* measure would be useful for the designer, to help in choosing the optimal ingredients. Such a measure has been proposed in the future MPEG7 standard for the description of multimedia data [DEC 99b].

5.6. References

[AHM 93] AHMAD S., TRESP V., "Some solutions to the missing feature problem in vision", in S.J. HANSON, J.D. COWAN and C.L. GILES (eds.), *Advances in Neural Information Processing Systems 5*, p. 393-400, San Mateo, Morgan Kaufmann, 1993.

[ASS 90] ASSMANN P.F., SUMMERFIELD Q., "Modeling the perception of concurrent vowels: vowels with different fundamental frequencies", *J. Acoust. Soc. Am.*, 88, p. 680-697, 1990.

[BER 95] BERTHOMMIER F., MEYER G., "Source separation by a functional model of amplitude demodulation", *Proc. of ESCA Eurospeech*, p. 135-138, 1995.

[BOD 96] BODDEN M., RATEIKSHEK K., "Noise-robust speech recognition based on a binaural auditory model", *Proc. of Workshop on the Auditory Basis of Speech Perception*, Keele, p. 291-296, 1996.

[BRE 90] BREGMAN A.S., *Auditory Scene Analysis*, MIT Press, Cambridge, 1990.

[BRO 82] BROKX J.P.L., Nooteboom S.G., "Intonation and the perceptual separation of simultaneous voices", *Journal of Phonetics*, 10, p. 23-36, 1982.

[BRO 92a] BROWN G.J., Computational auditory scene analysis: a representational approach, Sheffield, Department of Computer Science, unpublished PhD Thesis, 1992.

[BRO 92b] BROWN G.J., Cooke M.P., "Computational auditory scene analysis: grouping sound sources using common pitch contours", *Proc. Inst. of Acoust.*, 14, p. 439-446, 1992.

[BRO 93] BROWN G.J., COOKE M., "Physiologically-motivated signal representations for computational auditory scene analysis", in M. COOKE, S. BEET and M. CRAWFORD (eds.), *Visual Representations of Speech Signals*, Chichester, John Wiley and Sons, p. 181-188, 1993.

[CAR 94] CARLYON R., "Further evidence against an across-frequency mechanism specific to the detection of frequency modulation (FM) incoherence between resolved frequency components", *J. Acoust. Soc. Am.*, 95, p. 949-961, 1994.

[CHE 53] CHERRY E.C., "Some experiments on the recognition of speech with one, and with two ears", *J. Acoust. Soc. Am.*, 25, p. 975-979, 1953.

[COL 95] COLBURN H.S., "Computational models of binaural processing", in H. HAWKINS, T. MCMULLIN, A.N. POPPER and R.R. FAY (eds.), *Auditory Computation*, New York, Springer-Verlag, p. 332-400, 1995.

[COO 91] COOKE M.P., Modeling auditory processing and organization, Sheffield, Department of Computer Science, unpublished Thesis, 1991.

[COO 93] COOKE M.P., BROWN G.J., "Computational auditory scene analysis: exploiting principles of perceived continuity", *Speech Comm.*, 13, p. 391-399, 1993.

[COO 94] COOKE M., GREEN P., ANDERSON C., ABBERLEY D., Recognition of occluded speech by hidden markov models, University of Sheffield Department of Computer Science, Technical report, TR-94-05-01, 1994.

[COO 96] COOKE M., MORRIS A., GREEN P., "Recognising occluded speech", *Proc. Workshop on the Auditory Basis of Speech Perception*, Keele, p. 297-300, 1996.

[COO 97] COOKE M., MORRIS A., GREEN P., "Missing data techniques for robust speech recognition", *Proc. ICASSP*, p. 863-866, 1997.

[COO 99] COOKE M., ELLIS D.P.W., "The auditory organization of speech and other sources in listeners and computational models", *Speech Communication*, 35, p. 141-177, 2001.

[CUL 95] CULLING J.F., SUMMERFIELD Q., "Perceptual segregation of concurrent speech sounds: absence of across-frequency grouping by common interaural delay", *J. Acoust. Soc. Am.*, 98, p. 785-797, 1995.

[DAR 95] DARWIN C.J., CARLYON R.P., "Auditory grouping", in B.C.J. MOORE (ed.), *Handbook of Perception and Cognition: Hearing*, New York, Academic Press, p. 387-424, 1995.

[DEC 93a] DE CHEVEIGNÉ A., "Separation of concurrent harmonic sounds: fundamental frequency estimation and a time-domain cancellation model of auditory processing", *J. Acoust. Soc. Am.*, 93, p. 3271-3290, 1993.

[DEC 93b] DE CHEVEIGNÉ A., Time-domain comb filtering for speech separation, ATR Human Information Processing Laboratories, Technical Report, TR-H-016, 1993.

[DEC 94] DE CHEVEIGNÉ A., KAWAHARA H., AIKAWA K., LEA A., "Speech separation for speech recognition", *Journal de Physique*, IV 4, C5-545-C5-548, 1994.

[DEC 95] DE CHEVEIGNÉ A., MCADAMS S., LAROCHE J., ROSENBERG M., "Identification of concurrent harmonic and inharmonic vowels: a test of the theory of harmonic cancellation and enhancement", *J. Acoust. Soc. Am.*, 97, p. 3736-3748, 1995.

[DEC 97a] DE CHEVEIGNÉ A., "Concurrent vowel identification III: a neural model of harmonic interference cancellation", *J. Acoust. Soc. Am.*, 101, p. 2857-2865, 1997.

[DEC 97b] DE CHEVEIGNÉ A., KAWAHARA H., TSUZAKI M., AIKAWA K., "Concurrent vowel identification I: effects of relative level and F0 difference", *J. Acoust. Soc. Am.*, 101, p. 2839-2847, 1997.

[DEC 97c] DE CHEVEIGNÉ A., MCADAMS S., MARIN C., "Concurrent vowel identification II: effects of phase, harmonicity and task", *J. Acoust. Soc. Am.*, 101, p. 2848-2856, 1997.

[DEC 98] DE CHEVEIGNÉ A., "Cancellation model of pitch perception", *J. Acoust. Soc. Am.*, 103, p. 1261-1271, 1998.

[DEC 99a] DE CHEVEIGNÉ A., KAWAHARA H., "Multiple period estimation and pitch perception model", *Speech Communication*, 27, p. 175-185, 1999.

[DEC 99b] DE CHEVEIGNÉ A., SMITH B., A "sound transparency" descriptor ISO/IEC JTC1/SC29/WG11, MPEG99/m5199, 1999.

[DUR 63] DURLACH N.I., "Equalization and cancelation theory of binaural masking-level differences", *J. Acoust. Soc. Am.*, 35, p. 1206-1218, 1963.

[ELL 96] ELLIS D., Prediction-driven computational auditory scene analysis, MIT, unpublished Thesis, 1996.

[ELL 97] ELLIS D.P.W., "Computational auditory scene analysis exploiting speech-recognition knowledge", *Proc. IEEE Workshop on Apps. of Sig. Proc. to Acous. and Audio*, Mohonk, 1997.

[GRE 95] GREEN P.D., COOKE M.P., CRAWFORD M.D., "Auditory scene analysis and hidden Markov model recognition of speech in noise", *Proc. IEEE-ICASSP*, p. 401-404, 1995.

[GRE 97] GREENBERG, "Understanding speech understanding: towards a unified theory of speech perception", *Proc. ESCA Workshop on the Auditory Basis of Speech Perception*, Keele, p. 1-8, 1997.

[HAR 96] HARTMANN W.M., "Pitch, periodicity, and auditory organization", *J. Acoust. Soc. Am.*, 100, p. 3491-3502, 1996.

[HEL 77] HELMHOLTZ H. V., (1877). *On the Sensations of Tone* (English translation A.J. Ellis, 1954), New York, Dover.

[HER 94] HERMANSKY H., MORGAN N., "RASTA processing of speech", *IEEE trans Speech and Audio Process.*, 2, p. 578-589, 1994.

[HOL 88] HOLDSWORTH J., NIMMO-SMITH I., PATTERSON R.D., RICE P., Implementing a gammatone filter bank, MRC Applied Psychology Unit technical report, SVOS Final Report, Appendix C, 1998.

[HOL 90] HOLDSWORTH J., Two dimensional adaptive thresholding, APU AAM-HAP report, Technical Report, vol. 1, Appendix 4, 1990.

[HOL 92] HOLDSWORTH J., SCHWARTZ J.-L., BERTHOMMIER F., PATTERSON R.D., "A multirepresentation model for auditory processing of sounds", in Y. CAZALS, L. DEMANY and K. HORNER (eds.), *Auditory Physiology and Perception*, Oxford, Pergamon Press, p. 447-453, 1992.

[JOR 98] JORIS P.X., YIN T.C.T., "Envelope coding in the lateral superior olive. III. comparison with afferent pathways", *J. Neurophysiol.*, 79, p. 253-269, 1998.

[KAN 98] KANADERA N., HERMANSKY H., ARAI T., "On properties of the modulation spectrum for robust automatic speech recognition", *Proc. IEEE-ICASSP*, p. 613-616, 1998.

[LEA 92] LEA A., Auditory models of vowel perception, Nottingham University, unpublished Thesis, 1992.

[LIC 59] LICKLIDER J.C.R., "Three auditory theories", in S. KOCH (ed.), *Psychology, a Study of a Science*, New York, McGraw-Hill, I, p. 41-144, 1959.

[LIP 97] LIPPMANN R.P., CARLSON B.A., "Using missing feature theory to actively select features for robust speech recognition with interruptions, filtering, and noise", *Proc. ESCA Eurospeech*, KN-37-40, 1997.

[LYO 83] LYON, R.F., "A computational model of binaural localization and separation", in W. RICHARDS (ed.), *Natural Computation*, Cambridge, Mass, MIT Press, p. 319-327, 1983.

[LYO 84] LYON R., "Computational models of neural auditory processing", *Proc. IEEE ICASSP*, 36.1.(1-4), 1984.

[LYO 91] LYON R., "Automatic gain control in cochlear mechanics", in P. DALLOS, C.D. GEISLER, J.W. MATHEWS, M.A. RUGGERO and C.R. STEELE (eds.), *Mechanics and Biophysics of Hearing*, New York, Springer-Verlag, 1991.

[MAR 82] MARR D., "Representing and computing visual information", in P.H. WINSTON and R.H. BROWN (eds.), *Artificial Intelligence: an MIT Perspective*, Cambridge, Mass, MIT Press, p. 17-82, 1982.

[MAR 97] MARIN C., DE CHEVEIGNÉ A., "Rôle de la modulation de fréquence dans la séparation de voyelles", *Proc. Congrès Français d'Acoustique*, p. 527-530, 1997.

[MCA 84] MCADAMS S., Spectral fusion, spectral parsing, and the formation of auditory images, Stanford University, unpublished Thesis, 1984.

[MCA 89] MCADAMS S., "Segregation of concurrent sounds I: effects of frequency modulation coherence", *J. Acoust. Soc. Am.*, 86, p. 2148-2159, 1989.

[MED 88] MEDDIS R., "Simulation of auditory-neural transduction: further studies", *J. Acoust. Soc. Am.*, 83, p. 1056-1063, 1988.

[MED 92] MEDDIS R., HEWITT M.J., "Modeling the identification of concurrent vowels with different fundamental frequencies", *J. Acoust. Soc. Am.*, 91, p. 233-245, 1992.

[MEL 91] MELLINGER D.K., Event formation and separation in musical sound, Stanford Center for Computer Research in Music and Acoustics, unpublished Thesis, 1991.

[MEY 96] MEYER G., BERTHOMMIER F., "Vowel segregation with amplitude modulation maps: a re-evaluation of place and place-time models", *Proc. ESCA Workshop on the Auditory Basis of Speech Perception*, Keele, p. 212-215, 1996.

[MEY 97] MEYER G.F., PLANTE F., BERTHOMMIER F., "Segregation of concurrent speech with the reassigned spectrum", *Proc. IEEE ICASSP*, p. 1203-1206, 1997.

[MOR 98] MORRIS A.C., COOKE M.P., GREEN P.D., "Some solutions to the missing feature problem in data classification, with application to noise robust ASR", *Proc. ICASSP*, p. 737-740, 1998.

[NAD 97] NADEU C., PACHÈS-LEAL P., JUANG B.-H., "Filtering the time sequences of spectral parameters for speech recognition", *Speech Comm.*, 22, p. 315-332, 1997.

[NAK 95a] NAKATANI T., OKUNO H.G., KAWABATA T., "Residue-driven architecture for computational auditory scene analysis", *Proc. IJCAI*, p. 165-172, 1995.

[NAK 95b] NAKATANI T., GOTO M., ITO T., OKUNO H.G., "Multi-agent based binaural sound stream segregation", *Proc. IJCAI Workshop on Computational Auditory Scene Analysis*, p. 84-91, 1995.

[NAK 96] NAKATANI T., GOTO M., OKUNO H. G., "Localization by harmonic structure and its application to harmonic stream segregation", *Proc. IEEE ICASSP*, p. 653-656, 1996.

[NAK 97] NAKATANI T., KASHINO K., OKUNO J.G., "Integration of speech stream and music stream segregations based on a sound ontology", *Proc. IJCAI Workshop on Computational Auditory Scene Analysis*, Nagoya, p. 25-32, 1997.

[OKU 99a] OKUNO H.G., NAKAGAWA Y., KITANO H., "Incorporating visual information into sound source separation", *Proc. International Workshop on Computational Auditory Scene Analysis*, 1999.

[OKU 99b] OKUNO H.G., IKEDA S., NAKATANI T., "Combining independant component analysis and sound stream segregation", *Proc. International Workshop on Computational Auditory Scene Analysis*, 1999.

[PAR 76] PARSONS T.W., "Separation of speech from interfering speech by means of harmonic selection", *J. Acoust. Soc. Am.*, 60, p. 911-918, 1976.

[PAT 92] PATTERSON R.D., ROBINSON K., HOLDSWORTH J., MCKEOWN D., ZHANG C., ALLERHAND M., "Complex sounds and auditory images", in Y. CAZALS, K. HORNER and L. DEMANY (eds.), *Auditory Physiology and Perception*, Oxford, Pergamon Press, p. 429-446, 1992.

[PAT 96] PATTERSON R., ANDERSON T.R., FRANCIS K., "Binaural auditory images and a noiseresistant, binaural auditory spectrogram for speech recognition", *Proc. Workshop on the Auditory Basis of Speech Perception*, Keele, p. 245-252, 1996.

[ROS 97] ROSENTHAL D.F., OKUNO H.G., *Computational Auditory Scene Analysis*, Lawrence Erlbaum, 1997.

[SCH 83] SCHEFFERS M.T.M., Sifting vowels, Gröningen University, Thesis, 1983.

[SEN 85] SENEFF S., Pitch and spectral analysis of speech based on an auditory synchrony model, MIT, unpublished Thesis (Technical Report 504), 1985.

[SHA 85] SHAMMA S.A., "Speech processing in the auditory system I: the representation of speech sounds in the responses of the auditory nerve", *J. Acoust. Soc. Am.*, 78, p. 1612-1621, 1985.

[SLA 93] SLANEY M., An efficient implementation of the Patterson-Holdsworth auditory filter bank, Apple Computer Technical Report, 35, 1993.

[SLA 95] SLANEY M., "A critique of pure audition", *Proc. Computational Auditory Scene Analysis Workshop*, IJCAI, Montreal, 1995.

[SUM 90] SUMMEFIELD Q., LEA A., MARSHALL D., "Modelling auditory scene analysis: strategies for source segregation using autocorrelograms", *Proc. Institute of Acoustics*, 12, p. 507-514, 1990.

[SUM 92a] SUMMEFIELD Q., CULLING J.F., "Auditory segregation of competing voices: absence of effects of FM or AM coherence", *Phil. Trans. R. Soc. Lond.*, B 336, p. 357-366, 1992.

[SUM 92b] SUMMEFIELD Q., "Roles of harmonicity and coherent frequency modulation in auditory grouping", in M.E.H. SCHOUTEN (ed.), *The Auditory Processing of Speech: From Sounds to Words*, Berlin, Mouton de Gruyter, p. 157-166, 1992.

[SUM 92c] SUMMEFIELD Q., CULLING J.F., "Periodicity of maskers not targets determines ease of perceptual segregation using differences in fundamental frequency", *Proc. 124th Meeting of the ASA*, 2317(A), 1992.

[WAN 95] WANG A.L.-C., Instantaneous and frequency-warped signal processing techniques for auditory source separation, CCRMA (Stanford University), unpublished Thesis, 1995.

[WAR 70] WARREN R.M., "Perceptual restoration of missing speech sounds", *Science*, 167, p. 392-393, 1970.

[WAR 72] WARREN R.M., OBUSEK C.J., ACKROFF J.M., "Auditory induction: perceptual synthesis of absent sounds", *Science*, 176, p. 1149-1151, 1972.

[WEI 85] WEINTRAUB M., A theory and computational model of auditory monaural sound separation, Stanford University, unpublished Thesis, 1985.

[YOS 96] YOST W.A., DYE R.H., SHEFT S., "A simulated 'cocktail party' with up to three sound sources", *Perception and Psychophysics*, 58, p. 1026-1036, 1996.

Chapter 6

Principles of Speech Recognition

6.1. Problem definition and approaches to the solution

Human-machine communication can be seen as an exchange of messages coded with methods that enable their transmission through a physical medium [MOR 98]. The process which produces a representation of the messages to be communicated is called coding. Messages are structured into words, themselves represented as sequences of symbols belonging to an alphabet and organized according to a given lexicon. Phrases are built from words using grammar rules and other constraints which constitute *knowledge sources (KSs)* [JUN 96].

Speech understanding and voice dictation systems perform a *decoding process* [BAH 83, DUG 95, JEL 76, JEL 80, JEL 97] using knowledge sources to transform the speech signal carrying the message into representations of this message at various levels of abstraction. The decoding process can produce word sequences or conceptual hypotheses.

As opposed to human-human communication, human-machine communication must produce data structures in a deterministic way. In this case, *deterministic* means that a computational system must generate the same representation each time the same signal is processed. Knowledge sources used by machines for decoding messages are only *models* of those used by humans for the same or similar purposes. One of these models is the *language model* [FED 95, KNE 93, KUH 90]. It models constraints based on which words are acceptable in a given language.

Chapter written by Renato DE MORI and Brigitte BIGI.

One of the most important difficulties in natural language processing by computers originates from the fact that it is virtually impossible to design a grammar G which is able to generate all conceivable sentences from the language and only these sentences. This is due to many factors, the most essential of which is probably the fact that natural languages evolve in a way which cannot be appropriately modeled by known mathematic formalisms. Grammars with a large number of detailed rules are capable of accurately modeling some aspects of a language, but they are too limited for others aspect. These grammars are referred to as *low coverage* grammars. Other grammars can provide a complete coverage but, as they are very general, they can generate sentences which do not belong to a language. An example of such an *over-productive* grammar is the *word pair* grammar, which can generate any pair of words stemming from the language's vocabulary. Overproduction can be controlled by assigning probabilities to the grammar's rules so that undesirable sentences will be generated with a lower probability than the correct ones. Some of these grammars are particularly used in automatic speech recognition, as they can be represented as finite-state stochastic automata. Probability values are associated with the arcs corresponding to transitions between the states of the automaton. For instance, *bigram* probabilities can be associated with word pairs in a stochastic word pair grammar.

Transition arcs in stochastic automata can be substituted by other automata; for example, one for each word corresponding to the various pronunciations of that word. Transition arcs of word automata are labeled with phonemic symbols and the pronunciations are obtained from a *lexical model*. One after the other, each phoneme symbol can be replaced by an *acoustic model* which is also an automaton, the transition arcs of which are associated with probability distributions of observable acoustic parameters or other observable features.

From this hierarchy of components, an integrated network can be obtained and used for generating word or meaning hypotheses corresponding to a given speech signal.

The decoding process must handle ambiguities resulting from distortions introduced by the transmission channel, the various pronunciations and articulations of a word and, quite often, the intrinsic ambiguities which are inherent in the uttered message. Ambiguities can be reduced by exploiting the redundancies in the message. In practice, knowledge is used to transform the input signal into several parameter vector sequences and to obtain from them several levels of symbolic representation. The first level can be based on words, syllables, phonemes or simply acoustic descriptors.

The interpretation of a message is usually obtained as the result of a search process which considers an integrated network as the generator of an observable description of acoustic features:

$$X = x_1 x_2 \cdots x_n \cdots x_N$$

which represents the signal to be analyzed.

The search process aims at finding the optimal state sequence in the integrated network by identifying a path that generates X. Candidate word sequences are obtained by progressively scoring partial paths in the network. These candidates are often called *theories*. Search is constrained by KSs which impose consistency between the components of candidate paths. Redundancy may help to compensate for model and search method limitations.

Figure 6.1. *A simple decoding model*

Modern systems are based on probabilistic scores assigned to the hypothesized candidates. A simple probabilistic system for evaluating the hypotheses consists of a decoder which considers an acoustic observation sequence as the output of an information channel (see Figure 6.1), which receives as input a sequence of symbols representing the speaker's intention. When these symbols are words, they are generally denoted as the following sequence:

$$W = w_1 w_2 \cdots w_k \cdots w_K$$

Word sequence W *is assumed to be coded into a sequence* of acoustic features X. The goal of the recognition process is to reconstruct W based on observation X. If the same X can be produced by several W because the knowledge of the coding process is incomplete or imperfect, then the decoding process may generate erroneous results.

In the case of voice dictation, the decoding process generates word hypotheses by selecting a candidate sequence for which $\Pr(W|X)$ is maximum. If the language model provides the term $\Pr(W)$ and the channel model provides $\Pr(X|W)$, then $\Pr(X,W) = \Pr(X|W)\Pr(W)$ may be calculated.

It must be noted that, as $\Pr(X)$ is equal for all candidates W, the sequence W' for which $\Pr(X,W)$ is maximum is that for which $\Pr(W|X)$ is maximum. $\Pr(X|W)$ is the probability of observing X when W is pronounced. In practice, this probability cannot be evaluated directly from the data. It has to be evaluated with an acoustic model (AM). $\Pr(W)$ is the probability of a word sequence and it is evaluated with a language model (LM).

Conventional voice dictation systems tend not to use semantic knowledge sources to transcribe the acoustic signal. They supply, at their output, the sequence of words W' which maximizes $\Pr(X,W)$.

The model parameters are estimated from data corpora and are generally inaccurate. Probabilities evaluated with various models are combined with specific weights. For instance, the generation of the word sequence hypothesis W' can be based on the following score:

$$W' = \arg\max_W \left\{ \log \Pr(X|W) + \beta \log \Pr(W) \right\} \qquad [6.1]$$

where β is a factor which accounts for the various degrees of modeling inaccuracies.

6.2. Hidden Markov models for acoustic modeling

6.2.1. *Definition*

The most common acoustic models are hidden Markov models (or HMM). A HMM is defined as a pair of discrete periodic stochastic processes $(\mathbf{I}, \mathbf{X})$. Process $\mathbf{I}$ takes its values in a finite set I of *states*, whereas process $\mathbf{X}$ takes its values in an *observation space* X, either discrete or continuous, depending on the nature of the data sequences to be modeled. Both processes satisfy the following equations:

$$\Pr\left(I_t = i \,\middle|\, I_0^{t-1} = i_0^{t-1} \right) = \Pr\left(I_t = i \,\middle|\, I_{t-1} = i_{t-1} \right) \qquad [6.2]$$

$$\Pr\left(X_t = x \,\middle|\, I_0^{t+h} = i_0^{t+h}, X_1^{t-1} \right) = \Pr\left(X_t = x \,\middle|\, I_{t-1}^{t} = i_{t-1}^{t} \right) \qquad [6.3]$$

Equation [6.2] expresses the so-called "first order Markov hypotheses". The states are such that the history before time t has no influence on the future evolution of the process, as long as the current state is specified. Process $\mathbf{I}$ is therefore an homogenous first-order Markov chain.

Equation [6.3] formulates the output independence hypotheses, i.e. neither the evolution of **I** nor the past observations have an influence on the current observation, as long as the last two states are specified. From now on, we will use the more convenient notation i_0^t to denote $I_0^t = i_0^t$, except in the case of ambiguity. Thus, hypotheses from equations [6.2] and [6.3] are rewritten as:

$$\Pr\left(I_t = i \mid i_0^{t-1}\right) = \Pr\left(I_t = i \mid i_{t-1}^t\right) \tag{6.4}$$

$$\Pr\left(X_t = x \mid i_0^{t+h}\right) = \Pr\left(X_t = x \mid i_{t-1}^t\right) \tag{6.5}$$

Moreover, the same operator Pr will be used to denote the probability of discrete events and to denote the value of the probability density function for events in a continuous space.

Random variables of process **X** represent the variability in the acoustic observations for a given speech unit, while process **I** models the various possibilities with which these events can be chained over time.

In equation [6.2], output probabilities at time t are conditioned by the states of process **I** at time instant t-1 and t, i.e. by the transition at time t. It is common to express the conditional probabilities in reference to a state of **I** at instant t only. There is no reason to prefer one formulation over the other, and the resulting models are practically equivalent.

From the previously introduced hypotheses, it is fairly straightforward to derive the following basic properties:

$$\Pr\left(x_s^t \mid i_0^{t+h}, x_1^{s-1}\right) = \Pr(x_s^t \mid i_{s-1}^t)$$
$$\Pr\left(I_t = i \mid i_0^{t-1}, x_1^{t-1}\right) = \Pr(I_t = i \mid i_{t-1}^t)$$
$$\Pr\left(x_{t+1}^{t+h} \mid i_0^t, x_1^t\right) = \Pr(x_{t+1}^{t+h} \mid i_t)$$

where $0 \le s \le t \le T$ and $h > 0$.

6.2.2. Observation probability and model parameters

A model is well defined if it is possible to evaluate the probability density function for any finite sequence X_1^T of observable random variables. These probabilities can be decomposed as:

$$\Pr\left(x_1^T\right) = \sum_{i_0^T \in I^{T+1}} \Pr\left(x_1^T \mid i_0^T\right) \Pr\left(i_0^T\right) \qquad [6.6]$$

where I^{T+1} is the set of all state sequences having a length $T+1$.

Using hypotheses [6.3] and [6.6], the terms involved in the above summation become:

$$\Pr\left(i_0^T\right) = \Pr(i_0) \prod_{t=1}^{T} \Pr\left(i_t \mid i_0^{t-1}\right) = \Pr(i_0) \prod_{t=1}^{T} \Pr\left(i_t \mid i_{t-1}\right)$$

$$\Pr\left(x_1^T \mid i_0^T\right) = \Pr\left(x_1 \mid i_0^T\right) \prod_{t=2}^{T} \Pr\left(x_t \mid x_1^{t-1}, i_0^T\right) = \prod_{t=1}^{T} \Pr\left(x_t \mid i_{t-1}^t\right)$$

With the previous definitions, the probabilities in expression [6.6] can be evaluated with the model parameters as follows (π_{i0} being the probability of starting in state i_0):

$$\Pr(i_0^T) = \pi_{i_0} \prod_{t=1}^{T} a_{i_{t-1}i_t} \qquad [6.7]$$

$$\Pr(x_1^T \mid i_0^T) = \prod_{t=1}^{T} b_{i_{t-1}i_t}(x_t) \qquad [6.8]$$

$$\Pr(x_1^T) = \sum_{i_0^T \in I^{T+1}} \pi_{i_0} \prod_{t=1}^{T} a_{i_{t-1}i_t} b_{i_{t-1}i_t}(x_t) \qquad [6.9]$$

As a consequence, the model parameters are sufficient to evaluate the probability of an observation sequence. However, in HMM applications, the probabilities are evaluated with other equations, corresponding to fast algorithms, which will be described in detail further on.

6.2.3. *HMM as probabilistic automata*

HMM are often viewed as a particular case of probabilistic automata, represented as graphs such as the one depicted in Figure 6.2. The states of these graphs correspond to the states in **I** of the Markov chain, whereas the oriented branches correspond to the transitions, represented by pairs of states i,j with $a_{ij} > 0$.

An observation can be considered as the emission by a system which, at each instant, executes a transition from one state to another, choosing at random a transition according to a probability density which is specific to the state and generates a random vector according to a probability density which is specific to the branch. The number of states in a model and the set of branches between these states define what is usually called the model *topology*.

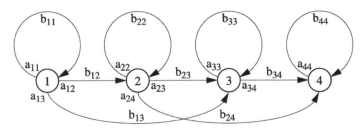

Figure 6.2. *A graphical representation of an HMM. With each arc $i \rightarrow j$ is associated a transition probability a_{ij} and an output density b_{ij}*

In speech recognition applications, *left-right* topologies are often used, i.e. topologies for which $a_{ij} > 0$ for $j < i$, as in the example depicted in Figure 6.2. Moreover, when left-right topologies are used, the first and last states in the model are generally referred to as the *initial* and *final* states. The choice of the first state as the initial state is equivalent to assigning $\pi = [\delta(i,0)]_{i \in I}$ as the probability of the initial state. If F denotes the final state, reached at the end of the observation sequence, the sum in equation [6.9] applies to entire paths, i.e. to the paths which end in F at time T.

6.2.4. Forward and backward coefficients

In the context of use with HMM, partial probabilities are defined in order to be calculated efficiently with iterative algorithms and to enable the extraction of the total emission probability. They are useful quantities for solving decoding problems and for running parameter estimation procedures. This section introduces the definitions of these quantities and the recursive algorithms which are used to compute them.

The *forward* probability $\alpha_t(x^T, i)$ is the probability that the source has emitted the partial sequence x_1^t given that the process I is in state i at time t:

$$\alpha_t\left(x_1^T, i\right) = \begin{cases} \Pr\,(I_0 = i), & t = 0 \\ \Pr\,(I_t = i, x_1^t), & t > 0 \end{cases}$$

This probability can be calculated iteratively as:

$$\alpha_t\left(x_1^T, i\right) = \begin{cases} \pi_i & t = 0 \\ \sum_{j \in I} a_{ji} b_{ji}(x_t)\alpha_{t-1}(x_1^T, j) & t > 0 \end{cases} \qquad [6.10]$$

The *backward* probability $\beta_t(x^T, i)$ is the probability that the source will emit the partial sequence x_{t+1}^T given that the process I is in state i at time t:

$$\beta_t\left(x_1^T, i\right) = \begin{cases} \Pr\,(x_{t+1}^T \mid I_t = i), & t < T \\ 1 & t = T \end{cases}$$

To calculate this quantity, a recurrent formula analogous to [6.10] can be used, except that it applies in a reverse time order as follows:

$$\beta_t\left(x_1^T, i\right) = \begin{cases} 1 & t = T \\ \sum_{j \in I} a_{ij} b_{ij}(x_{t+1})\beta_{t-1}(x_1^T, j) & t < T \end{cases}$$

The *optimal path* probability $v_t(x_1^T, i)$ is the maximum probability of the junction between the partial sequence x_1^t and a state sequence ending in state i at time t:

$$v_t\left(x_1^T, i\right) = \begin{cases} \Pr\,(I_0 = i), & t = 0 \\ \max_{i_0^{t-1}} \Pr\,(i_0^{t-1}, I_t = i, x_1^t), & t > 0 \end{cases}$$

This probability can be calculated as follows:

$$v_t\left(x_1^T, i\right) = \begin{cases} \pi_i & t = 0 \\ \max_{j \in I} a_{ji} b_{ji}(x_1) v_{t-1}(x_1^T, j), & t > 0 \end{cases} \qquad [6.11]$$

The total probability of an observation sequence can be obtained by using one of the following equalities:

$$\Pr\left(x_1^T\right) = \sum_{i \in I} \alpha_T\left(x_1^T, i\right)$$

$$= \sum_{i \in I} \pi_i \beta_0\left(x_1^T, i\right) \qquad [6.12]$$

$$= \sum_{i \in I} \alpha_t\left(x_1^T, i\right)\beta\left(x_1^T, i\right)$$

Coefficient v is used to calculate another version of the score assigned by the model to a sequence of observations. This score corresponds to the probability which contributes most to the summation along the optimal path:

$$\hat{\Pr}\left(x_1^T\right) = \max_{i \in I} v_T(x_1^T, i) \qquad [6.13]$$

In practice, this is the quantity generally used to make the recognition decision. If the model has initial and final states, the final probabilities can be obtained by selecting only the value of the proper coefficient in the final state.

All these algorithms have an $O(MT)$ complexity, where M is the number of non-zero transitions and T is the length of the input sequence. M can be equal to N^2 at most, if N is the number of states in the model. However, it is usually much smaller, as the transition matrix is generally sparse.

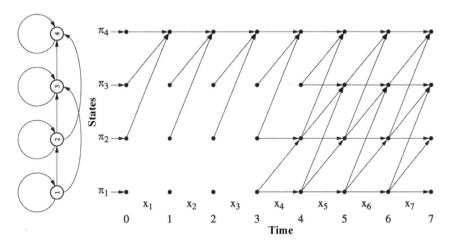

Figure 6.3. *A state-time lattice. Each column in the lattice represents the various states at a given instant. The arcs of the lattice correspond to the model transitions between successive instants. They are valued as the product of the transition probability by the emission probability*

The probabilities defined above are calculated with the support of a lattice, which corresponds to the unfolding of the original graph along the time axis (see Figure 6.3). The states in this lattice correspond to pairs (t,i), where t is the time index and i is a state of the model. The branches represent the transitions; they are the elementary components of the possible paths in the model between the initial and final instants. Each interval between two columns corresponds to a particular element in the observed sequence. Each column in the lattice keeps the values of one of the preceding probabilities for the partial sequence ending in different states in the model. For a given observation sequence x_1^T, each branch $(t-1,i) \rightarrow (t,j)$ is assigned a "weight" value obtained as $a_{ij}b_{ij}(x_t)$. Thus, each path can be assigned a score corresponding to the product of the weights of each branch along the path. This score corresponds to the emission probability for the observed sequence along the considered path:

$$q\left(x_{t+1}^{t+h}, i_t^{t+h}, \vartheta\right) = \prod_{s=t+1}^{t+h} a_{i_{s-1}i_s} b_{i_{s-1}i_s}(x_t) = \Pr\left(x_{t+1}^{t+h}, i_{t+1}^{t+h} \mid I_t = i_t\right)$$

where ϑ represents the current situation of the model parameters. The calculation is carried out column-wise, in synchrony with the natural order of the observations. For each time segment, the scores for each state in a given column are updated with a recursive formula which uses the score values from the adjacent column, the transition probabilities of the model and the values of the output densities for the current observation.

In Figure 6.3, a state is marked in the lattice together with the elements that contribute to its score for α and β. The light gray arrows indicate the branches for which the scores are summed to obtain the probability, whereas the darker arrows show the paths contributing to the probability. The probability corresponds to the path with the highest score among the light gray arrows.

6.3. Observation probabilities

The product of Q independent densities forms an output density:

$$b(x) = \prod_{h=1}^{Q} b^h(x^h)$$

x_1 can be a vector of popular acoustic features called cepstral coefficients at time t, x_2 a vector resulting from the time derivatives of the elements of x_1. Hidden

Markov models with such densities are called *discrete* HMM. The first recognition systems based on HMM used these models for their simplicity and their efficiency.

In practice, the calculation of a discrete probability is based on the direct access to a table. However, coding speech events with a small number of symbols is an important simplification which entails a considerable loss of information. For this reason, more recent implementations model the sequence of real valued vectors resulting from the extraction of acoustic parameters directly. Even if this approach improves the model resolution, it requires a number of hypotheses on the type of output densities, as they must belong to an adequate parametric family so that their representation with a limited number of parameters is feasible. One of the most obvious choices for a continuous parametric density is the multivariate Gaussian density:

$$N_{\mu,\Sigma}(x) = \frac{1}{\sqrt{(2\pi)^D \det(\Sigma)}} \exp\left(-\frac{1}{2}(x-\mu)^* \Sigma^{-1}(x-\mu)\right)$$

where D is the dimension of the observation space, i.e. the length of a vector. The parameters of the density $N_{\mu,\Sigma}$ are the mean vector μ and the symmetric covariance matrix Σ.

Gaussian (or normal) densities are a frequent choice for modeling random variables ranging around a mean value and resulting from several unknown factors. Matrix Σ expresses the scattering of the data around mean value μ.

Weighted sums of Gaussian densities are popular to approximate densities more finely than a single Gaussian does:

$$M(x) = \sum_{k=1}^{K} w_k N_{\mu_k,\Sigma_k}(x)$$

Mixture densities are flexible enough to approach arbitrarily any density, provided a sufficient number of components is used. Statistical parameters such as the mean can be adapted to the speaker or to a new context of use [ANA 97, GAU 94, LEG 95].

6.4. Composition of speech unit models

Automatic speech recognition using stochastic models is achieved by a probabilistic decoding process which consists of choosing, from amongst the wide range of possible linguistic events, the event that corresponds to the observed data with the highest probability. This means that the model must be able to assign to each possible event a score for the hypothesis that this event has generated the observed data.

In practice, an observation does not correspond to a single word but to a sequence of words. If the language is composed of a limited set of sentences, it is conceivable to have a global model for each sentence. However, when the set of valid sentences is very large, or even infinite, a different approach must be considered. Dealing with a large number of models creates difficulties for both the training and the recognition phases. If the target vocabulary is composed of several thousand words and if a specific model for each word is necessary, it would require several occurrences of each word in the training corpus in order to train the entire set of models. It is practically impossible to collect these samples. Furthermore, with such an approach, the recognition language cannot be extended to include words that do not occur in the training data set. These problems can be solved by representing the meaningful linguistic events as the concatenation of elementary speech units belonging to a list of reasonable size. Specific models are built for these units only and the more complex ones are modeled as the composition of more basic units included in the complex one. Other types of models are described for example in [BAH 93, LEE 96, MAR 94, SCH 94].

This structure is particularly well suited for HMM-based automatic speech recognition, for the following reasons. First of all, speech production is based on the emission of sequences of units stemming from a limited set of elements: the phonemes of the language. Secondly, the HMM graph structure greatly facilitates the construction of composed models. In English, for example, there are approximately 40 phonemes. Suppose that an HMM is available for each of them, and that their topology distinguishes between initial and final states. In order to evaluate the probability that a given speech signal corresponds to the pronunciation of the word "*one*", it is possible to use the models corresponding to the phonemic transcription [w ah n] and to construct a word model by simply concatenating the phoneme models, as described in Figure 6.4.

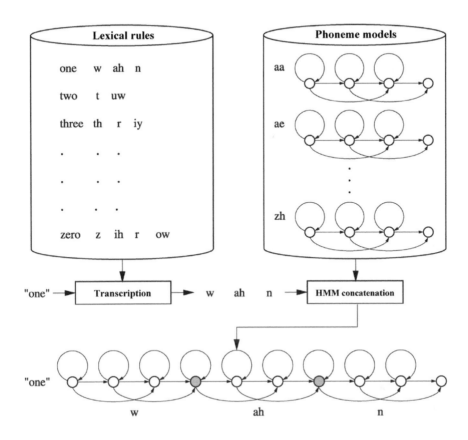

Figure 6.4. *Construction of a word model by concatenating and linking phoneme models*

It then becomes clear that if a complete set of elementary models is available, a specific model can be constructed for each word in the language, as soon as its phonemic transcription is known. Model composition also simplifies the training phase. A set of examples is necessary to train a model. When the elementary units are phonemes, it is not possible to collect isolated examples, as only full words or entire sentences can be pronounced in a natural way. Therefore, training phoneme models with relevant data requires the segmentation of the training data, i.e. the specification of the label and the position of each phoneme in the recording. This task is at best tedious but sometimes purely unfeasible, when the training corpus is very large. However, when using concatenated speech unit models, the processing of the training data is made easier.

HMM training consists of an iterative procedure for which, at each step, new values of the model parameters are calculated depending on the previous values and the training data. An iteration of the process can be decomposed in two steps: during the first step, all training pronunciations are used to update the set of intermediate values, each of them corresponding to one element in the model. Then, these "cumulated" values are used to calculate a new set of parameters. This enables the use of composition for training elementary models from continuous speech sentences. Only the phonemic transcription needs to be known for each sentence. From there, a composed HMM representing the whole sentence can be constructed and the accumulators for each of its components can be updated. Once the training data are processed, the parameters of each model are reestimated on the basis of all phoneme occurrences in the training set. It is worth noting that, in this procedure, the accumulators for a phoneme model are also influenced by the training data of the other models, with a contribution that depends on the extent to which all these data correspond to the current model.

The following may seem surprising, but there is experimental evidence that, despite this fact, it is possible to initialize the training phase with a set of "neutral" models (for example, models which are all identical) and obtain correct convergence towards a selective set of phoneme models. Details on the training algorithms can be found in [MOR 98], Chapters 5 and 6.

6.5. The Viterbi algorithm

The algorithm used to calculate probability values v introduced in section 6.2.4 is known as the Viterbi algorithm. It can be viewed as an application of dynamic programming in order to find the path with the optimal score in an oriented lattice. Taking into account the fact that some transitions are void, the algorithm is based on recursive equation [6.1]. Once the whole observation sequence x_1^T has been processed, the path with the maximum score can be found by calculating $\hat{\Pr}(x_1^T) = \max_{i \in I}(v_t(x_1^T, i))$. When applying the recursion formula, it is possible to keep track of the state index corresponding to the maximum value, by storing a matrix of backtracking pointers $\Phi_t(x_1^T, i)$. From these backtracking pointers, the most likely path in the state-time lattice can be traced back. The path is specified by a sequence of state-time pairs $\left[\hat{i}_l, \hat{t}_l\right]_{l=0}^L$, the length of which $(L+1)$ can be larger than the length of the observation sequence $T+1$, as several states can be visited at the same time if the model has void transitions.

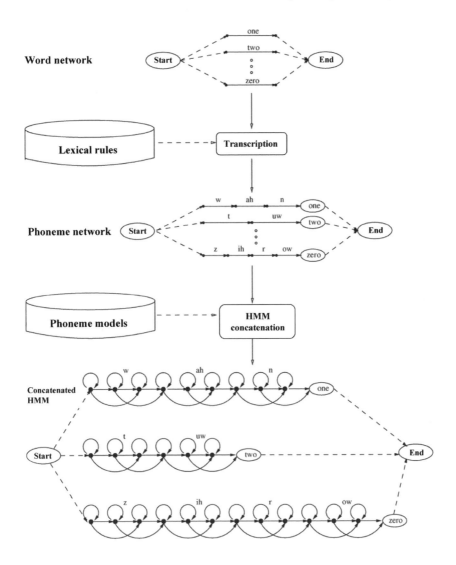

Figure 6.5. *Construction of a composed model for the recognition of a one-digit utterance. The dotted lines in the graph represent void transitions used to connect the models*

The optimal path corresponds to the alignment of the speech segments with the model states. In composed models, different states may correspond to different phonemes and therefore the time alignment provides a localization of the phonemes in the utterance, synchronized on the input and output states for each elementary model. The Viterbi algorithm is more efficient when applied to composed models when elementary models are connected to one another by void transitions.

Imagine that the recognition vocabulary is composed of 11 digits in English ("oh", "zero", "one", ..., "nine") and that the words are pronounced in isolation. In order to decide on the recognized word in the input utterance, one solution could be to calculate the emission probability for the observed input data with each digit model and then to choose the model yielding the maximum value. A more efficient solution consists of constructing a composed model as in Figure 6.5, executing the Viterbi algorithm and retrieving the most probable path. Thanks to the topology of the composed model, each complete path corresponds uniquely to the pronunciation of a digit. It is therefore possible to decide on the recognized digit by simply examining the sequence of states visited along the best path.

Given the importance of the Viterbi algorithm in speech recognition applications, a description of the basic procedures which are required to implement a Viterbi-based HMM decoder is proposed in Table 6.1. The sequence of observation x_1^T is assumed to be given, as well as the number N of states in the model.

The two main functions are *viterbi* and *backtrack*. The first is used to fill matrix $[v_t(i)]$ with the scores of the partial paths and stores the information needed by the backtracking procedure in both matrices $[\Phi_t(i)]$ and $[\tau_t(i)]$. Here, the state index varies from 1 to N, while the time index ranges from 0 to T. The value $\Phi_t(i)$ represents the last state visited for the best path ending in that state at time t, while $\tau_t(i)$ is used to find out if the last transition in the path is empty. The *viterbi* function calls, at each instant, both procedures *expand_empty_trans* and *expand_full_trans* which update a column in the lattice while keeping track of the transitions, whether they are void or active.

6.6. Language models

The goal of the language model is to evaluate the probability $\Pr(W_1^T)$ for a sequence of words $W_1^T = w_1, ..., w_t, ..., w_T$ which can be written as:

$$\Pr(W_1^T) = \prod_{t=1}^{T} \Pr(w_t | h_t) \qquad [6.14]$$

function **viterbi**	function **backtrack**
begin	begin
$\quad t := 0$;	$\quad t := T$; $j := 1$;
$\quad$ for $i := 1$ to N do begin	$\quad$ for $i := 2$ to N do begin
$\quad\quad v_t(i) := \pi_i$;	$\quad\quad$ if $v_t(j) < v_t(i)$
$\quad\quad \Phi_t(i) := nil$;	$\quad\quad$ then $j := i$;
$\quad\quad \tau(i) := nil$;	$\quad$ end
$\quad$ end	$\quad l := -1$;
$\quad$ expand_empty_trans;	$\quad$ while $j \neq nil$ do begin
$\quad$ for $t := 1$ to T do begin	$\quad\quad l := l + 1$;
$\quad\quad$ for $i := 1$ to N do	$\quad\quad \widehat{i_l} := j$; $\widehat{t_l} := t$;
$\quad\quad\quad v_t(i) := 0$;	$\quad\quad j := \Phi_t(\widehat{i_l})$; $t := \tau_t(\widehat{i_l})$;
$\quad\quad$ expand_full_trans;	$\quad$ end
$\quad\quad$ expand_empty_trans;	$\quad$ reverse $(\widehat{i_0^l})$;
$\quad$ end	$\quad$ reverse $(\widehat{t_0^l})$;
$\quad \widehat{v} := v_T(1)$	$\quad$ return $(\widehat{i_0^l}, \widehat{t_0^l})$
$\quad$ for $i := 2$ to N do begin	end
$\quad\quad$ if $\widehat{v} < v_T(i)$ then $\widehat{v} := v_T(i)$	
$\quad$ end	
$\quad$ return $\widehat{v}$	
end	
procedure **expand_empty_trans**	procedure **expand_full_trans**
begin	begin
$\quad$ for $i := 1$ to N do push(i);	$\quad$ for $i := 1$ to N do begin
$\quad i :=$ pop();	$\quad\quad$ for $j := 1$ to N do begin
$\quad$ while $i \neq nil$ do begin	$\quad\quad\quad \widehat{v} := v_{t-1}(i)a_{ij}b_{ij}(x_t)$;
$\quad\quad$ for $j := 1$ to N do begin	$\quad\quad\quad$ if $v_t(j) < \widehat{v}$ then begin
$\quad\quad\quad \widehat{v} := v_t(i)a_{ij}^{\Sigma}$;	$\quad\quad\quad\quad v_t(j) := \widehat{v}$;
$\quad\quad\quad$ if $v_t(j) < \widehat{v}$ then begin	$\quad\quad\quad\quad \Phi_t(j) := i$;
$\quad\quad\quad\quad v_t(j) := \widehat{v}$	$\quad\quad\quad\quad \tau_t(j) := t - 1$;
$\quad\quad\quad\quad \Phi_t(j) := i$;	$\quad\quad\quad$ end
$\quad\quad\quad\quad \tau_t(j) := t$;	$\quad\quad$ end
$\quad\quad\quad\quad$ push_unique();	$\quad$ end
$\quad\quad\quad$ end	end
$\quad\quad$ end	
$\quad$ end	
$\quad i :=$ pop();	
$\quad$ end	
end	

Table 6.1. *Description of the basic functions required to implement
a recognition system based on the Viterbi algorithm*

where $h_t = w_1, ..., w_{t-1}$ can be considered as the *history* of the word w_t. With the increasing length of the word sequence h_t, probabilities $\Pr(w_t|h_t)$ are becoming more and more difficult to estimate from data because it would require an extremely large set of examples. A simplification of the history h_t is therefore introduced so as to reduce its cardinality. An *n-gram* model approximates the history of a word as the *n*-1 previous words, namely:

$$h_t \approx w_{t-n+1} \cdots w_{t-1} \qquad [6.15]$$

The n-gram approximation reduces the volume of data required to estimate statistically the probability $\Pr(W_1^T)$ considerably [BRO 92, DEL 94, LAU 93, MER 94, ROS 94]. Of course, such an approximation degrades the accuracy of the model. However, even a trigram (3-gram) model requires a very large quantity of textual data (corpus) to estimate the model efficiently. For instance, a trigram model for a 1,000 word vocabulary requires the estimation of approximately 10^9 probabilities.

Another important aspect that illustrates the difficulty in estimating n-grams, and which is inherent to corpora is the lack of coverage and the poor quality of the training data. Experimentally, some correct word sequences are rare and only show a few occurrences, unless the corpus is very large.

6.6.1. *Perplexity as an evaluation measure for language models*

In general, language models for speech recognition are evaluated according to their impact on the recognition accuracy. Nevertheless, they can be evaluated separately, by considering for instance their ability to predict the sequence of words in a text [JEL 90]. The most common measure is the *perplexity* measure. However, while perplexity is usually a good indication of the language model quality, its correlation with the recognition rate is not straightforward. In reality, speech recognition performance is influenced by the acoustic similarity of words, which is not taken into account in the expression of perplexity.

If L is the size of vocabulary V and if the emission of words is independent of the emission of their history, the measure of information (or *entropy*) is defined in the following way:

$$H(S) = -\left\{ \sum_{i=1}^{L} \Pr(w_i) \log(\Pr(w_i)) \right\}$$

A source with entropy H delivers the same information content as a source emitting words with equal probability stemming from a vocabulary V' of size L', such that:

$$L' = 2^H$$

If the source emits sequences of length n, then:

$$H(S) = -\lim_{n \to \infty} \frac{1}{n} \left\{ \sum_{W_1^n} \Pr(W_1^n) \log(\Pr(W_1^n)) \right\}$$

where the sum extends to all possible sequences of words W_1^n.

If the source is ergodic and if its statistical properties do not vary over time, then all the sequences of same length have the same probability and the entropy becomes:

$$H(S) = -\lim_{n \to \infty} \frac{1}{n} \log(\Pr(W_1^n))$$

If the entropy is now estimated on a corpus of n words, the previous equality becomes:

$$H(S) = -\frac{1}{n} \log(\Pr(W_1^n))$$

If the probabilities are evaluated with a model, a logprob quantity is usually preferred:

$$LP(W_1^n) = -\frac{1}{n} \log\{\Pr'(W_1^n)\} = -\frac{1}{n} \log\left\{ \Pr\left(w_1 \prod_{i=2}^{n} \Pr(w_i|w_{i-1}) \right) \right\} \qquad [6.16]$$

The perplexity of a language model derived from a corpus is defined as follows:

$$PP = 2^{LP(W_1^n)} = \Pr(W_1^n)^{-1/n} \qquad [6.17]$$

In order to evaluate a language model, the probability of the model is estimated on a training set and the perplexity is evaluated with that model on a text corpus which is completely distinct from the training corpus. This illustrates the fact that perplexity is strongly dependent on the training data.

Another important aspect of language models is the size of the vocabulary. In fact, an increase in the vocabulary size causes a considerable growth of the number of n-grams and their estimation requires a much larger volume of data and computing resources. This can be avoided by limiting the vocabulary to the k most frequent words in the corpus. Nevertheless, this choice creates out-of-vocabulary (OOV) words. Thus, the estimation of the language model and the test corpus for perplexity evaluation are usually restricted to a limited vocabulary, involving only the words in V for the n-grams.

6.6.2. *Probability estimation in the language model*

A simple formulation of the n-gram estimation is introduced for more complex models. Let us consider a set of *training texts* $W = w_1...w_T$ composed of words belonging to a finite vocabulary V. Based on the Markovian hypothesis which limits h_t to the n-1 previous words, the set of texts can be learned, without any information loss, on the basis of:

$$S = h_n w_n, h_{n+1} w_{n+1},..., h_T w_T$$

Given h, a sample S_h can be extracted from S by taking the subsequence of all the n-grams in S starting with h, yielding:

$$S_h = h w_{h_1},..., h w_{h_m}$$

where $\{h_1, h_2,..., h_m\} \subseteq \{n, n+1,..., T\}$ can be viewed as an m word occurrence realization extracted independently according to the probability distribution $\Pr(w|h)$. For the other words, the sample is composed of IID (independent and identically distributed) random variables. In the rest of this chapter, for the sake of simplicity, the context h will be considered as known and will be omitted from the notations. Thus, the goal is to estimate the discrete probability distribution $\Pr(w)$, defined on w, from a training sample of IID random variables:

$$S = w_1,..., w_m$$

It must be noted that w refers to a word in the vocabulary whereas w_i denotes the ith word in the sequence. The distribution $\Pr(w)$ is assumed as belonging to $\{\Pr(w; \theta), \theta \in \Theta\}$, where θ is a vector of parameters which specifies the distribution and Θ is the parameter space. Two simple and general distributions are considered: 1) *discrete* and 2) *symmetric discrete*.

6.6.2.1. *The discrete distribution*

The parameter space of the discrete distribution is the simplex:

$$\Theta = \left\{\theta = [\theta_w]_{w \in V} : \theta_w \geq 0, \forall w \in V, \sum_{w \in V} \theta_w = 1\right\}$$

which assigns directly a parameter to each word in the vocabulary. In fact:

$$\Pr(w;\theta) = \theta_w \qquad [6.18]$$

This distribution belongs to the *exponential family*, and such is the case for the corresponding probability in S, which is the multinomial distribution, i.e.:

$$\Pr(S;\theta) = \frac{m!}{\prod_{w\in V} c(w)!} \prod_{w\in V} \theta_w^{c(w)} \quad \text{where} \quad c(w) = \sum_{i=1}^{m} \delta(w_i = w) \qquad [6.19]$$

with $\delta(e) = 1$ if e is true and 0 otherwise. The multinomial distribution of equation [6.19] belongs to the exponential family, and the vector $[c(w)]_{w\in V}$ represents a sufficient number of statistics in S.

6.6.2.2. Symmetric discrete distribution

If, for the sake of symmetry, we assume that words which have the same frequency in S must have the same probability, a slightly different parametric distribution can be defined:

$$\Pr(w;\theta) = \frac{\theta_r}{n_r} \quad \text{where} \quad r = c(w)$$

where θ_r represents the total probability over all words occurring r times in S, whereas n_r denotes the number of different words occurring r times in S. Such a model requires fewer parameters than the partitions of w. Therefore, the parameter space Θ for the symmetric distribution is the simplex with dimensions equal to the size of the partitions.

The probability of S with symmetric distribution is:

$$\Pr(S;\theta) = \frac{m!}{\prod_{w\in V} c(w)!} \prod_{w\in V} \left(\frac{\theta_{c(w)}}{n_{c(w)}}\right)^{c(w)} = \frac{m!}{\prod_{w\in V} c(w)!} \prod_{r\geq 0} \left(\frac{\theta_r}{n_r}\right)^{rn_r} \qquad [6.20]$$

Given the parametric model $\Pr(w;\theta)$, the problem is to find the optimal value for its vector parameters. The corresponding problem of estimation is a traditional problem in parametric statistics and it can be approached in different ways. Two distinct estimation criteria are particularly popular: the maximum likelihood and the Bayesian approach.

6.6.3. *Maximum likelihood estimation*

The language model probabilities are estimated with the help of a criterion which considers the parameter θ as an unknown quantity which has to be determined. The optimal estimation is defined as the one that maximizes the likelihood of observing the sample S. Let us denote as $S = w_1,..., w_m$ some IID random variables. For these variables, the maximum likelihood (ML) estimation is defined as:

$$\theta^{ML} = \arg\max_{\theta\in\Theta} \Pr(S;\theta) = \arg\max_{\theta\in\Theta} \prod_{i=1}^{m} \Pr(w_i;\theta) \qquad [6.21]$$

EXAMPLE 6.1.– The maximum likelihood estimation for the discrete distribution can be readily obtained by maximizing the logarithm of the probability rather than the probability itself. Taking this into account when calculating Θ, and suppressing a constant factor, the following Lagrangian is obtained:

$$L(\theta;\lambda) = \sum_{w\in V} c(w)\log\theta_w + \lambda\left(1 - \sum_{w\in V}\theta_w\right)$$

By deriving this quantity with respect to θ_w ($w\in V$) and equaling the result to zero, we obtain:

$$\frac{\partial L}{\partial\theta_w} = \frac{c(w)}{\theta_w} - \lambda = 0 \qquad \forall w\in V$$

Concerning w, a reorganization of the terms followed by a summation yields:

$$\sum_{w\in V} c(w) = \left(\sum_{w\in V}\theta_w\right)\lambda = \lambda \qquad [6.22]$$

By substituting λ, the equation becomes:

$$\theta_w^{ML} = \frac{c(w)}{\sum_{w\in V} c(w)} = \frac{c(w)}{m} \qquad [6.23]$$

EXAMPLE 6.2.– If the maximization is applied to the logarithm of the symmetric probability distribution, θ is written:

$$\theta_r^{ML} = \frac{n_r r}{m}$$

and it will produce the same probabilities as the discrete distribution.

6.6.4. *Bayesian estimation*

Bayesian estimation is conceptually distinct from the approach described in the previous section and can be applied to adapt a language model from one domain to another. In fact, the parameter vector θ can be considered as a random variable for which an *a priori* distribution can be assumed. In this case, the estimation problem can be understood as finding an estimation of θ, given a training sample of IID random variables $S = w_1,..., w_m$ and an *a priori* distribution $\Pr(\theta)$ for the parameters θ. Applying Bayes rules, the *a posteriori* distribution of θ is:

$$\Pr(\theta|S) = \frac{\Pr(S|\theta)\Pr(\theta)}{\Pr(S)} \tag{6.24}$$

The posterior distribution of θ combines the *a priori* existing knowledge on θ with the empirical data sample. Both expressions of the probabilities, namely $\Pr(S|\theta)$ and $\Pr(S;\theta)$, correspond to the same distribution.

The estimation of θ can be derived from the posterior distribution in several cases. Two of them are detailed here:

1) The *maximum a posteriori* (MAP) criterion tends to optimize θ so as to maximize the *a posteriori* probability. By eliminating the constant factor $\Pr(S)$ in equation [6.24], the MAP criterion appears as a generalization of the ML criterion.

$$\theta^{MAP} = \arg\max_{\theta \in \Theta} \Pr(\theta|S) = \arg\max_{\theta \in \Theta} \Pr(S|\theta)\Pr(\theta) \tag{6.25}$$

The simplest type of a priori knowledge, which corresponds though to the weakest level of *a priori* knowledge on θ, is the uniform distribution:

$$\Pr(\theta) = s(\theta) = \frac{1}{vol(\Theta)} \tag{6.26}$$

where $vol(\Theta)$ denotes the volume of the parameter space Θ. With this particular distribution, the MAP estimate is equivalent to the ML estimate.

2) The Bayesian criterion (B) considers θ with regard to its posterior distribution:

$$\theta^B = E[\theta|S] = \int_\Theta \theta \Pr(\theta|S)\, d\theta = \frac{\int_\Theta \theta \Pr(S|\theta) \Pr(\theta)\, d\theta}{\int_\Theta \Pr(S|\theta)\, d\theta} \qquad [6.27]$$

A sufficient condition for the posterior distribution of equation [6.24] to be calculable is that S has sufficient statistics, which is verified if $\Pr(w;\theta)$ is of the exponential family. In fact, the existence of sufficient statistics implies the existence of distributions $\Pr(\theta)$ for which the posterior distribution is of the same family as the prior distribution.

The Dirichlet distribution has such a property:

$$\Pr(\theta|S') = \frac{(m'+k-1)!}{c'(1)!\dots c'(k)!} \prod_{w \in V} \theta_w^{c'(w)}$$

where $c'(w) = \sum_{i=-m'+1}^{0} \delta(w_i = w)$ are sufficient statistics evaluated on $S' = w_{-m'+1}, \cdots, w_0$.

Applying Bayesian criterion [6.24] with this prior distribution, the following ratio is obtained:

$$\theta_w^B = \frac{\int_\Theta \theta_w \prod_{z \in V} \theta_w^{c(z)+c'(z)}\, d\theta}{\int_\Theta \prod_{z \in V} \theta_w^{c(z)+c'(z)}\, d\theta} = \frac{c(w)+c'(w)+1}{m+m'+k} \qquad \forall w \in V$$

The estimated MAP is then similar to the ML of the multinomial distribution:

$$\theta^{MAP} = \arg\max_{\theta \in \Theta} \prod_{w \in V} \theta_w^{c(w)+c'(w)} = \left[\frac{c(w)+c'(w)}{m+m'} \right]_{w \in V}$$

6.7. Conclusion

An n-gram language model generates a representation which corresponds to a finite-state stochastic automaton which contains all the n-grams in parallel. If each word is represented by its corresponding hidden Markov model, the result is an integrated network which constitutes the knowledge source used for generating word hypotheses from the continuous speech signal [JEL 97].

A fundamental issue for an efficient use of the methods presented in this chapter is robustness. Considerable effort is deployed to improve recognition rates when training conditions (speakers, background noise, channel distortions) are different from the test conditions. One of the most efficient compensation approaches for robustness is obtained by transforming the speech signal. The adaptation of acoustic and language model parameters is another way to achieve robustness efficiently. The adapted parameters are those which characterize probability distributions, such as means and variances. In large vocabulary continuous speech recognition, tens of thousands of acoustic Gaussian parameters have to be adapted, which requires in principle a huge volume of adaptation data. Interesting methods have been proposed recently to achieve adaptation with a limited data set. The solution consists of applying the same transformation to a set of parameters. For the language model, adaptation consists of using recognized words to predict other words and increase their probability as they have a semantic relationship with the observed words. The search for improved robustness remains a challenge for the future development of speech recognition systems.

6.8. References

[ANA 97] ANASTASAKOS T., MC DONOUGH J., MAKHOUL J., "Speaker adaptive training: a maximum likelihood approach to speaker normalisation", *Proc. of the IEEE International Conference on Acoustics, Speech and Signal Processing*, p. 1043-1046, Munich, Germany, 1997.

[BAH 83] BAHL L.R., JELINEK F.J., MERCER R.L., "A maximum likelihood approach to continuous speech recognition", *IEEE Transactions on Pattern Analysis and Machine Intelligence*, vol. 5, no. 2, p. 179-190, 1983.

[BAH 93] BAHL L.R., BELLEGARDA J.R., DE SOUZA P.V., GOPALAKRISHNAN P. S., NAHAMOO D., PICHENY M.A., "Multonic Markov word models for large vocabulary continuous speech recognition", *IEEE Transactions on Speech and Audio Processing*, vol. 1, no. 3, p. 334-344, July 1993.

[BRO 92] BROWN P.F., DELLA PIETRA V.J., LAI J.C., DE SOUZA P., MERCER R.L., "Class-based *n*-gram models of natural language", *Computational Linguistics*, vol. 18, no. 4, p. 467-479, 1992.

[DEL 94] DELLA PIETRA S.A., DELLA PIETRA V.J., Statistical modeling using maximum entropy, Research report, IBM, 1994.

[DUG 95] DUGAST C., AUBERT X., KNESER R., "The Philips large-vocabulary recognition system for American English, French and German", *Eurospeech*, Madrid, Spain, 1995.

[FED 95] FEDERICO M., CETTOLO M., BRUGNARA F., ANTONIOL G., "Language modeling for efficient beam-search", *Computer Speech and Language*, 9, p. 353-379, 1995.

[GAU 94] GAUVAIN J.L., LEE C.H., "Maximum *a posteriori* estimation for multivariate Gaussian mixture observations of Markov chains", *IEEE Transactions on Speech and Audio Processing*, 2, p. 291-298, 1994.

[JEL 76] JELINEK F.J., "Continuous speech recognition by statistical methods", *Proceedings of the Institute of Electrical and Electronic Engineers (IEEE)*, 4, p. 532-556, 1976.

[JEL 80] JELINEK F.J., MERCER R.L., "Interpolated estimation of Markov source parameters from sparse data", in *Pattern Recognition in Practice*, p. 381-397, Amsterdam, Netherlands, 1980.

[JEL 90] JELINEK F.J., "Self-organized language modeling for speech recognition", in Alex WEIBEL and Kay-Fu LEE (eds.), *Readings in Speech Recognition*, p. 450-505, Morgan Kaufmann, Los Altos, CA, 1990.

[JEL 97] JELINEK F.J., *Statistical Methods for Speech Recognition*, The MIT Press, 1997.

[JUN 96] JUNQUA J.C., HATON J.P., *Robustness in Automatic Speech Recognition*, Kluwer Academic, 1996.

[KNE 93] KNESER R., STEINBISS V., "On the dynamic adaptation of stochastic language models", *Proc. of the IEEE International Conference on Acoustics, Speech and Signal Processing*, vol. II, p. 586-588, Minneapolis, MN, 1993.

[KUH 90] KUHN R., DE MORI R., "A cache-based natural language model for speech recognition", *IEEE Trans. Pattern Anal. Machine Intell.*, vol. 12, no. 6, p. 570-582, 1990.

[LAU 93] LAU R., ROSENFELD R., ROUKOS S., "Trigger-based language models: a maximum entropy approach", *Proc. of the IEEE International Conference on Acoustics, Speech and Signal Processing*, vol. 2, p. 45-48, Minneapolis, MN, 1993.

[LEE 96] LEE C.H., SOONG F.K., PALIWAL K.K., *Automatic Speech and Speaker Recognition: Advanced Topics*, Kluwer, 1996.

[LEG 95] LEGGETER C.J., WOODLAND P.C., "Maximum likelihood linear regression for speaker adaptation of continuous density hidden Markov models", *Computer Speech and Language*, 9, p. 171-185, 1995.

[MAR 94] MARI J.F., HATON J.P., "Automatic word recognition based on second-order hidden Markov models", *Proc. of the International Conference on Spoken Language Processing*, p. 274-277, Yokohama, Japan, September 1994.

[MER 94] MERIALDO B., "Tagging English text with a probabilistic model", *Computational Linguistics*, vol. 20, no. 2, p. 155-172, 1994.

[MOR 98] DE MORI R., *Spoken Dialogues with Computers*, Academic Press, 1998.

[ROS 94] ROSENFELD R., Adaptive statistical language modeling: a maximum entropy approach, PhD Thesis, School of Computer Science, Carnegie Mellon University, Pittsburgh, PA, 1994.

[SCH 94] SCHUKAT-TALAMAZZINI E.G., KUHN T., NIEMANN H., "Speech recognition for spoken dialog systems", in DE MORI R. NIEMANN H. and HAHNRIEDER G. (eds.), *Progress and Prospects of Speech Research and Technology*, Saint Augustin, Germany, 1994.

Chapter 7

Speech Recognition Systems

7.1. Introduction

Speech recognition deals mainly with the transcription of a voice signal into a sequence of words. Most systems rely on a statistic modeling of the speech generation process (see Chapter 6). Under this approach, the message results from a linguistic model which provides an estimation of $\Pr(W)$ for any sequence of words $W = (w_1, w_2, \ldots)$ and an acoustic channel, which encodes the message W into the signal X, and which is modeled as a probability density function $f(X|W)$. Speech recognition thus consists of maximizing the *a posteriori* probability of W, which is equivalent to maximizing the product $\Pr(W)f(X|W)$. In that way, speech recognition can be decomposed into three problems: the development of the linguistic model $\Pr(W)$, the development of the acoustic model $f(X|W)$ and the development of a decoder exploiting these two models.

This chapter mostly focuses on large vocabulary continuous speech recognition (LVCSR). The range of applications for this type of technology extends well beyond automatic text dictation. LVCSR can, for instance, be used for voice access to databases, content-based indexing of audiovisual documents, the transcription of conversations and spoken language translation. Progress made in speech recognition also benefits other technologies which rely on similar basics, such as speaker recognition and language identification.

Chapter written by Jean-Luc GAUVAIN and Lori LAMEL.

The main modules involved in a speech recognition system are depicted in Figure 7.1. These modules are:

   −   the main knowledge sources (speech and text corpora, pronunciation lexicons);

   −   the acoustic feature extraction module (acoustic analysis);

   −   the acoustic and linguistic models, the parameters of which are estimated during the training phase;

   −   the decoder which uses these models.

Statistical models are usually estimated on large speech and text corpora, the preparation of which requires a substantial effort.

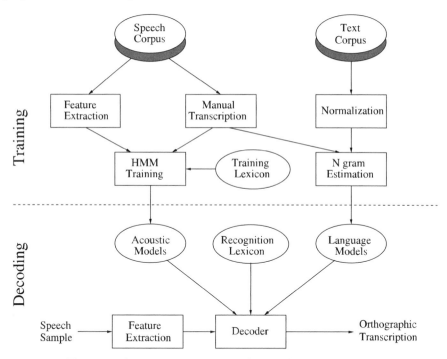

Figure 7.1. *Synoptic representation of a speech recognition system including the training and the decoding processes*

In this chapter, we will review the main modules of a speech recognition system and then we will discuss the applicative constraints together with a few application domains. We will then conclude this chapter with a few research perspectives and challenges.

7.2. Linguistic model

Language models (LM) represent the constraints related to the regularities of natural language [ROS 00]. Grammatical constraints can be described as context-free grammars (which can be built manually for small vocabularies) or as statistical models. The most common approach relies on n-gram models which capture the syntactic and semantic constraints of the language by assuming that the probability of a word sequence $W = (w_1, w_2, ..., w_k)$ can be approximated as $\prod_{i=1}^{k} \Pr(w_i | w_{i-n+1}, ..., w_{i-2}, w_{i-1})$, assuming that the left context h_i of each word can be reduced to its n-1 previous words. For a trigram model, this is written as (see Chapter 6):

$$\Pr(w_i | h_i) \approx \Pr(w_i | w_{i-2}, w_{i-1})$$

The n-gram probabilities can be estimated by applying the maximum likelihood criterion, which corresponds to the frequency of occurrence of each n-word sequence in the training corpus (texts or speech transcriptions). For instance, in the case of a trigram model, the probability is expressed as:

$$\Pr(w_i | w_{i-2}, w_{i-1}) \approx \frac{C(w_{i-2}, w_{i-1}, w_i)}{C(w_{i-2}, w_{i-1})}$$

where $C(.)$ denotes the number of occurrences of the n-gram in the training data. When training data are not in sufficient quantity, the statistics for rare (or unobserved) n-grams are generally smoothed using a back-off mechanism exploiting lower order statistics [KAT 87]. Thus, if there are not enough data to provide a reliable probability estimate from the trigram counts, the probability of the rare trigrams are estimated from the bigram probabilities:

$$\Pr(w_i | h_i) \approx \Pr(w_i | w_{i-1}) B(w_{i-1}, w_{i-2})$$

where $B(w_{i-1}, w_{i-2})$ is a back-off coefficient required to guarantee that the sum of probabilities for a given context is always equal to 1. This back-off mechanism offers an additional advantage as the size of the model can be arbitrarily reduced by increasing the minimum number of observations needed to include an n-gram probability in the model. This property can also be exploited to reduce the computational load during the decoding process. Word bigram, trigram and 4-gram models are the most widely used language models, even though marginal improvements have been reported with larger contexts (5-grams or higher) [BAH 95, HIN 96, LJO 95, WOO 95] or with word class n-grams [SAN 96]. Language models are often evaluated and compared in terms of test set likelihood or, equivalently, in terms of their perplexity, defined as $Px = \Pr(text|LM)^{-(1/L)}$ (see Chapter 6), where L is the number of words in the text.

Given a corpus of texts (or transcriptions), it can seem relatively easy to build an n-gram language model. Most steps are simple and use word counting and sorting procedures. Decisions with a significant impact cover the definition of the words (including possibly compound words or acronyms) and the vocabulary selection, as well as the back-off strategy and the corresponding estimation technique. The normalization of the texts on which the statistics are carried out must be done very carefully. One benefit of normalization is to reduce the lexical variability and therefore to increase the lexical coverage of the texts for a given vocabulary size. Normalization rules are usually specific to each language. Typical normalization operations are the expansion of numerical expressions, the processing of acronyms, isolated letters, capitalization and punctuation. Text preparation requires the segmentation and the tagging of the various entities (sentences, paragraphs, articles, tables, etc.), but also the detection of corrupted texts and more generally of any text that is not usable for probability estimation. It is also necessary to correct the most frequent spelling mistakes. Some normalization processes are necessary in order to reduce the vocabulary size, such as the processing of hyphens and apostrophes, or the decomposition of compound words in German. For instance, the number 1991, which is written *neunzehnhunderteinundneunzig* in German, can be decomposed as *neunzehn hundert ein und neunzig*. For dictation systems, it is essential to convert the texts into a form which is closer to read language, taking into account the probability of the variants estimated by comparing the reference texts with some transcriptions of these texts when they are read. For instance, the number 150 can be pronounced *"one hundred fifty"* or *"one hundred and fifty"* and 1/8 can be pronounced *"one eighth"* or *"an eighth"* [GAU 95].

The vocabulary used by the recognition system is generally chosen to maximize the coverage of development data. On average, each out-of-vocabulary word causes 1.5 to 2 recognition errors [PAL 95]. A trivial solution to reduce this error rate is to increase the size of the vocabulary [GAU 95].

It is often difficult to find the best compromise between the need for a large amount of data (in order to accurately estimate the language model parameters) and the importance of having data which accurately represent the language of the application. Frequently we must deal with data of different origins in very unbalanced proportions. One solution is to construct a language model for each data source and then to interpolate those models. The interpolating coefficients can then be estimated with the EM (*Expectation-Maximization*) algorithm [DEM 77] by maximizing the likelihood of development data. An alternative consists of simply fusing the *n*-gram counts and constructing a model from these counts. If some data sources are more representative than others, the *n*-gram counts can be weighted accordingly but this approach may pose estimation problems, particularly with respect to the back-off coefficients. The main steps for developing a language model are depicted in Figure 7.2.

The use of word classes can yield improved model robustness with regard to the training data, by reducing the number of parameters to be estimated. This technique is particularly efficient when some *a priori* knowledge about the user's phraseology is available.

In most recognition systems, the language model is static. In a dialog system, it can also be dynamically chosen among a finite number of models, as a function of the dialog state. Adaptation techniques can increase the model representativity with respect to the processed data. For text dictation, various approaches have been proposed to adapt the language models using words already recognized in the document, such as a simple *cache* model [JEL 91, ROS 94], a *trigger* model [ROS 92] or a *theme concordance* model [SEK 96]. The *cache* model is based on the idea that words that have just appeared in a dictated document are likely to reoccur. The advantage of such a model for short documents is naturally very limited. The *trigger* model tries to overcome this difficulty by increasing the probability for words that were observed to often occur simultaneously with the trigger word, in other documents. The *theme concordance* model relies on key words occurring in processed data to search for documents on the same topic, from which new sub-language models are built and used to re-decode the current document. Despite the growing interest for adaptive language models, only a few marginal improvements have been obtained, compared to a well-trained static n-gram model.

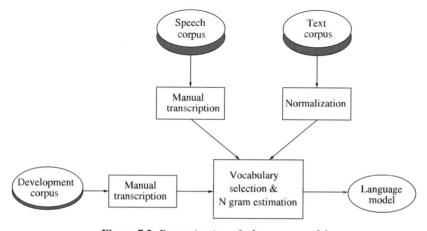

Figure 7.2. *Determination of a language model*

7.3. Lexical representation

The pronunciation dictionary constitutes the link between the acoustic model and the linguistic model. Developing such a dictionary requires, on the one hand, the definition and selection of the vocabulary entries and, on the other hand, the description of each lexical entry in terms of elementary acoustic units. The performance level of the recognition system is naturally dependent on the quality and the consistency of the pronunciations corresponding to each lexical entry [LAM 96]. For large vocabulary recognition systems, the most common acoustic unit is the phoneme. Each language has a specific set of phonemes. For instance, English can be described with approximately 45 phonemes, German requires 49, French 35 (see Table 7.1) and Spanish only 26. Pronunciation dictionaries are usually limited to the most common pronunciations and typically do not explicitly represent allophones, as most allophonic variants can be predicted by pronunciation rules. Moreover, there is often a *continuum* between several allophones of a same phoneme, which makes the choice difficult and subjective. Using a phonemic representation leaves the task of predicting the observed variants to the acoustic model.

Pronunciation dictionaries generally require both language and system expertise. A few studies have been conducted with the aim of automatically deducing the phonemic forms from data [COH 89, FOS 96, RIL 96, TAJ 95], but, to our knowledge, these promising methods did not yield any qualitative improvement, even when the pronunciations were obtained from manually transcribed data [RIL 99].

Fast speakers tend to severely reduce unstressed syllables (occasionally, they are not pronounced at all), especially in long words with several of them. Even though long words are generally easy to recognize, an incorrect phonemic transcription for a long word can cause an error on a neighboring word, which, if short, may even be omitted. In order to reduce this type of error, alternative pronunciations must be allowed, with the elision of some phonemes in unstressed syllables. In French, pronunciation variants stem from two major sources: the schwa and the liaison (see Figure 7.3). The schwa elision has an impact on the number of syllables and, therefore, on the prosody.

The use of compound words in the pronunciation lexicon makes it possible to represent reduced forms for word sequences such as, in French, *"je ne sais pas"* (*"I don't know"*), which can be pronounced /ʃepa/, or *"je voudrais"* (*"I'd like"*), said as /ʒdʀɛ/. These reductions met in spontaneous speech can be modeled with phonological rules instead of being explicitly included in the lexicon [OSH 75]. They are then accounted for in the phonetic graph used by the decoder, both in the training phase and in the recognition phase [COH 89, GIA 91, LAM 92]. The use of these rules during the training phase provides a better estimation of the acoustic model parameters by reducing the number of phonemic transcription errors. The same mechanism can also be used for handling liaisons and consonantic reductions at the end of words, in French [GAU 94c].

When dealing with spontaneous or conversational speech, it can be worth increasing the number of phonemes by adding extra-linguistic units such as breathing noise, laughter or hesitations.

IPA symbol	example	IPA symbol	example
p	pont	i	lit
b	bon	e	blé
t	ton	ɛ	sel
d	don	a	patte
k	cou	o	saule
g	gond	ɔ	sol
f	fou	u	fou
v	vin	y	suc
s	sot	ø	feu
z	zèbre	œ	leur
ʃ	chat	ə	petit
ʒ	jour	œ̃	brun
r	rond	ã	chant
l	la	ɔ̃	bon
j	yole	m	motte
ɥ	lui	n	note
w	oui	ɲ	digne

Table 7.1a. *Phone set for French*

Consonants		Vowels	
p	pen, copy, happen	ɪ	kit, bid, hymn, minute
b	back, baby, job	e	dress, bed, head, many
t	tea, tight, button	æ	trap, bad
d	day, ladder, odd	ɒ	lot, odd, wash

k	key, clock, school	ʌ	strut, mud, love, blood
g	get, giggle, ghost	ʊ	foot, good, put
tʃ	church, match, nature	iː	fleece, sea, machine
dʒ	judge, age, soldier	eɪ	face, day, break
f	fat, coffee, rough, photo	aɪ	price, high, try
v	view, heavy, move	ɔɪ	choice, boy
θ	thing, author, path	uː	goose, two, blue, group
ð	this, other, smooth	əʊ	goat, show, no
s	soon, cease, sister	aʊ	mouth, now
z	zero, music, roses, buzz	ɪə	near, here, weary
ʃ	ship, sure, national	eə	square. fair, various
ʒ	pleasure, vision	aː	start, father
h	hot, whole, ahead	ɔː	thought, law, north, war
m	more, hammer, sum	ʊə	poor, jury, cure
n	nice, know, funny, sun	ɜː	nurse, stir, learn, refer
ŋ	ring, anger, thanks, sung	ə	about, common, standard
l	light, valley, feel	i	happy, radiate. glorious
r	right, wrong, sorry, arrange	u	thank you, influence, situation
j	yet, use, beauty, few	n̩	suddenly, cotton
w	wet, one, when, queen	l̩	middle, metal
ʔ	(glottal stop) department, football	ˈ	(stress mark)

Table 7.1b. *Phone set for English (www.phon.ucl.ac.uk/wells)*

7.4. Acoustic modeling

Most speech recognition systems use Hidden Markov Models (HMM) for the acoustic modeling [YOU 96]. Others use segmental models [GLA 99, OST 92, ZUE 89] or neural networks [ABB 99, HOC 94] for computing acoustic likelihoods. However, all systems operate in the HMM framework for combining the acoustic and linguistic information in a single network which represents the language of the application. Acoustic models must account for the various sources of variability, whether they originate from the linguistic context, the speaker, the acoustic environment or the recording conditions.

sont	sɔ̃ / sɔ̃t(V)
les	le(C) / lez(V)
contenu	kɔ̃t{ə}ny
était	[eɛ]tɛ / [eɛ]tɛt(V)
vingt-deux	vœ̃t{ə}dø / vœ̃{n}dø / vœ̃t{ə}døz(V) / vœ̃{n}døz(V)
autres	otrə / otr(V) / otrəz(V) / ot
décembre	desɑ̃brə / desɑ̃br(V) / desɑ̃b(C)
squatter	skwate / skwater(V) / skwat[œɛ]r
désertions	dezɛr[st]jɔ̃
Budapest	bydapɛst / bydapɛstə (C)
Morgan	mɔrgɑ̃ / mɔrgan
Wonder	wɔndœr / wɔndɛr / vɔ̃dɛr

Figure 7.3. *Examples of pronunciation variants for French: optional phones are between braces {} and alternative choices between square brackets []. The parentheses () specify contextual constraints, where V denotes a vowel and C a consonant*

7.4.1. *Feature extraction*

For speech recognition purposes, the speech signal is sampled at a frequency ranging between 8 and 16 kHz (8 kHz for telephone speech and 16 kHz for broadband signals). Features are then extracted from the signal with two objectives: to reduce its variability while preserving as much linguistic information as possible and to reduce the volume of information to be processed by the recognition system (see Chapter 1).

Most systems use short-term cepstrum coefficients derived from a Fourier Transform or a Linear Prediction Coding scheme. The advantage of cepstrum coefficients over other features is twofold: on the one hand, they form a compact representation of the short-term spectrum and on the other hand, they are almost uncorrelated with one another. The complexity of the acoustic models is therefore lower.

The most common approaches consist in deriving the cepstrum coefficients from a MFCC analysis [DAV 80] or from a PLP analysis [HER 90]. In both cases, a short-term power spectrum (20 to 30 ms) is estimated on a MEL[1] scale, at a typical rate of 10 ms. The voice signal X is thus converted into a sequence of acoustic vectors each of them representing a 10 ms interval:

$$X = (\mathbf{x}_1, \mathbf{x}_2, \cdots, \mathbf{x}_t)$$

The MFCC coefficients are obtained as a Discrete Cosine Transform applied to the logarithm of the power spectrum, whereas the PLP coefficients are obtained as the Inverse Fourier Transform of the cubic root of the power spectrum, followed by a linear prediction analysis. It is usual to retain only the first 12 cepstrum coefficients to which the logarithm of the normalized energy is generally added so as to form a 13-dimensional feature vector. Both feature sets (MFCC and PLP) have been used with comparable success. The PLP analysis appears to be slightly more robust to noisy conditions [KER 96, WOO 96].

The normalization of the cepstrum mean over the duration of a whole utterance provides a reduced dependency of the models regarding the recording conditions and the transmission channel. Depending on the operating conditions of the recognition system, the mean cepstrum can be calculated over the entire recording (for instance, over a whole sentence) or by using only the N previous acoustic vectors (N being in the order of 100, i.e. 1 second of speech). In order to reduce the difficulties stemming from the independence hypothesis assumed by the HMM, it is usual to augment the dimension of the acoustic features by appending the cepstrum coefficients with their first and second derivatives.

---

1 The MEL scale is an approximation of the frequency resolution of the human auditory system (see Chapter 1).

7.4.2. *Acoustic-phonetic models*

HMM are used in most speech recognition systems to model sequences of acoustic vectors. The popularity of these models is due to their simplicity and to the existence of efficient estimation schemes (see Chapter 6). The most widely used acoustic units in LVCSR systems are allophones, namely contextual phonemic units, which are modeled as a 3- to 5-state Markov chain (see Figure 7.4), each being associated with a probability density function on the acoustic vectors. More precisely, given an N-state HMM, the stochastic process is described by the joint probability density function $f(X, \mathbf{s})$ of the speech signal represented as a sequence of acoustic vectors $X = (x_1, x_2, ..., x_t)$ and the state sequence $\mathbf{s} = (s_1, s_2, ..., s_t)$:

$$f(X, \mathbf{s}) = \pi_{s_0} \prod_{t=1}^{T} a_{s_{t-1}s_t} f(\mathbf{x}_t | s_t)$$

where π_{s_0} is the initial state probability, a_{ij} the transition probability from state i to state j and $f(\cdot | s)$ the probability density function corresponding to state s (see Chapter 6).

The use of phonemic units makes it possible to describe the vocabulary as a pronunciation dictionary. It is of course possible to recognize speech without explicitly using a pronunciation dictionary, for instance by using word models (which were very popular in the 1980s) or more elementary units such as *fenones* [BAH 88]. Compared to word models, the use of phonemic units results in a reduction of the total number of parameters. They also make it possible to model coarticulation effects across words, while facilitating vocabulary modifications. *Fenones* are well adapted to the automatic learning of pronunciation, but they are difficult to use when dealing with *a priori* phonetic knowledge.

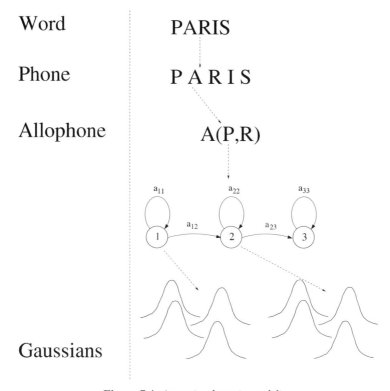

Word   PARIS

Phone   P A R I S

Allophone   A(P,R)

Gaussians

Figure 7.4. *Acoustic-phonetic modeling*

Today, contextual phone models (triphones or pentaphones) are the most commonly used acoustic units. A list of the 15 most frequent triphones in French is given in Table 7.2. Compared to larger units (such as diphones, di-syllables or syllables), contextual phone models offer a wider range of contextual dependencies, with a back-off possibility towards shorter contexts. Several types of contexts have been investigated, including 1-phone contexts (left context or right context), 2-phone contexts (triphones and generalized triphones [LEE 88]) for which the position of the phoneme in the word can be additionally distinguished, 4-phone contexts (pentaphone [WOO 95]) or any combination of these contexts. The choice of the appropriate context usually arises from a compromise between resolution and robustness, and depends naturally on the available training data. A key point is to adapt the number of model parameters to the quantity of training data. A very efficient technique for limiting the number of model parameters without affecting the resolution consists of taking advantage of the similarity across some of the HMM states for a given phoneme in similar contexts, by tying the state distributions. This approach is used in most systems, with minor implementation variants and denomination differences (*senones* [HWA 92], *genones* [DIG 94], *PELs* [BAK 92], *tied-states* [YOU 93]). Of course, this parameter sharing principle reduces the model size. It can be used at different modeling levels (allophone, HMM state or Gaussian component) [TAK 95, YOU 92] but the flexibility is particularly profitable at the Gaussian component level, as a substantial reduction can be obtained without increasing the error rate.

triphone	% occurrence	triphone	% occurrence	triphone	% occurrence
a(p,r)	0.4075	r(t,ə)	0.2831	ã (r,s)	0.1830
j(s,ɔ̃)	0.4053	s(a,j)	0.2552	r(f,ã)	0.1777
ə(d,l)	0.3790	y(s,r)	0.2257	v(a,ε)	0.1768
u(p,r)	0.3144	d(.,ə)	0.1901	r(p,ɔ̃)	0.1758
l(ə,a)	0.2909	ε(s,t)	0.1899	i(.,l)	0.1750

Table 7.2. *List of the most frequent triphones in French.*
The phonetic context is given between parentheses

In practice, both agglomerative (bottom-up) and splitting (top-down) classification techniques yield comparable results. Quite frequently, a decision tree is used to supervise the classification process, particularly for modeling phonemic contexts that have not been observed in the training data [HWA 93, YOU 94]. The type of questions that are used in the tree branches concerns the position of the phoneme and the properties of the neighboring phonemes [ODE 92]. Figure 7.3 exemplifies the most frequent questions for the French language (out of a complete list of 150 questions).

The most classically used probability density function in HMM speech models is the Gaussian Mixture Model (GMM), defined as:

$$f(\mathbf{x}_t|s) = \sum_{k=1}^{K} \omega_k N(\mathbf{x}_t|\mathbf{m}_{sk}, \Sigma_{sk})$$

where ω_k is the weight of Gaussian component k in the mixture, m_k its mean vector and Σ_k it covariance matrix (usually assumed diagonal). The main advantage of continuous observation densities versus discrete or semi-discrete densities comes from the fact that the number of parameters can be easily adapted to the amount of available training data. As a consequence, it is possible to achieve a very accurate modeling of very frequent states without requiring any smoothing technique for modeling the distribution in rare states. On the contrary, discrete or semi-continuous distributions use an *a priori* fixed number of parameters, which does not offer a great precision, unless parameter smoothing and sharing techniques are used in addition. Of course, the choice between one option or the other also depends on the applicative constraints, in particular, the memory capacity and the computational resources.

Question	% application	Question	% application
voiced-consonant[+1]	5.7	low vowel[+1]	2.1
phone-r[+1]	5.2	phone-i[+1]	2.0
sonorant[+1]	4.5	phone-a[+1]	2.0
front vowel[+1]	4.3	vowel[+1]	1.9
unvoiced[-1]	4.0	back vowel[+1]	1.9
unvoiced[+1]	3.3	front vowel[-1]	1.6
phone-r[-1]	2.8	strident[+1]	1.5
sonorant[-1]	2.5	coronal[+1]	1.4
silence[+1]	2.4	phone-a[-1]	1.4
non-speech[-1]	2.3	open vowel[+1]	1.4

Table 7.3. *List of the most frequent questions used to group the states of the phonetic models by means of a decision tree. Notations [+1] and [-1] indicate whether the question addresses the right context or the left context*

The training of acoustic models requires the estimation of the HMM parameters (the transitions probabilities) and of the probability density functions for each state, i.e. the mean vectors, the covariance matrices and the Gaussian mixture weights. The standard method is based on the maximum likelihood (ML) criterion which consists of determining the parameter values that maximize the training data likelihood. This optimization is usually achieved by means of the EM algorithm [DEM 77], which is an iterative algorithm starting from initial values. The states of the model are aligned on the training data and the parameters are obtained by means of the Baum-Welch re-estimation formulae [JUA 85, LIP 82, BAU 70]. This algorithm guarantees that the training data likelihood increases at each iteration. During the alignment step, each acoustic vector can be associated with several states (when using the forward-backward algorithm) or with a single one (Viterbi algorithm). This second solution yields a slightly lower likelihood but, in practice, the accuracy of the models is comparable, especially for large amounts of training data. It is worth noting that the EM algorithm does not guarantee convergence to the absolute maximum likelihood. It must also be underlined that the ML estimate is rarely the one that yields the best system performance. Some practical implementation details turn out to be essential, in particular the initialization step prior to the iterative procedure and the use of *a priori* constraints on the parameters. Figure 7.5 depicts the main steps in the development of acoustic models.

As the objective of the training procedure is to determine the best models for representing the observed data (and the *a priori* knowledge), the accuracy of the recognition system depends greatly on the data representativity. A few approaches for reducing this dependency are described in the forthcoming sections.

Speaker independence is obtained by estimating the acoustic model parameters by means of large corpora containing utterances from at least 100 distinct speakers. Of course, some key differences exist between speakers, whether they are based on anatomical properties or social background, and it can be useful to model these differences explicitly. For instance, better results are achieved when male and female speakers are modeled separately. These models are generally obtained by adapting initial gender-independent models using a Bayesian adaptation technique [GAU 94b], but they can also be estimated separately when sufficient data are available.

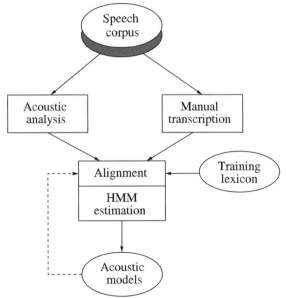

Figure 7.5. *Training procedure for acoustic models*

7.4.3. *Adaptation techniques*

Error rates in speech recognition systems increase when the field conditions differ from the training conditions. A few methods which can reduce the effect of this type of mismatch are described in this section. These approaches are mainly applicable to HMM models with continuous probability density functions. Similar techniques have also been designed for discrete and semi-continuous distributions.

The adaptation of acoustic models enables the compensation of discrepancies between training and field conditions (whether they stem from the acoustic environment, the microphone or from the transmission channel). It also makes it possible to adapt to the specificities of the speaker, for instance non-native speakers. Conventional approaches are based on additive and convolutive compensation techniques, using a noise model and adaptation techniques which do not require any *a priori* knowledge on the type of difference to compensate for. In fact, there is almost never any *a priori* knowledge available to the system concerning the acoustic channel, the noise or the speaker characteristics. Therefore, model adaptation must be carried out on the sole basis of information contained in the data already processed or to be processed, adding optionally, the transcriptions generated by the system from data that have just been processed.

Adaptation techniques may also be used to compensate for the lack of training data representative of the application. Thus, a small quantity of representative data can be used to adapt multi-purpose models which have been previously estimated on a large amount of data. Adaptation is usually carried out in a supervised mode, using an exact transcription of the adaptation data.

The most commonly used techniques for the adaptation of acoustic models are model composition [GAL 92], Bayesian adaptation [GAU 94b] and transformation methods such as linear regression [LEG 95]. Bayesian adaptation can be viewed as a means of introducing probabilistic constraints on the model parameters during the training procedure. HMM models are still estimated with the EM algorithm but on the basis of their maximum likelihood *a posteriori* (MAP) instead of the training data likelihood [GAU 94b]. When the *a priori* distribution of the model parameters is determined from existing models, this approach is referred to as an adaptation procedure. For instance, speaker-independent models can be adapted to yield gender-dependent models or speaker-specific models. The same approach can be used to adapt models to particular acoustic conditions.

Linear transform techniques applied to model parameters are particularly efficient for unsupervised adaptation. These transforms are generally estimated by maximizing the likelihood of the adaptation data. They are almost always applied to the mean vectors of the Gaussian components but they can also be used to adapt the covariance matrices [LEG 95]. The same transform can be applied to the whole set of Gaussians within the acoustic model or, alternatively, distinct transforms can be applied to disjoint sets of Gaussians, depending on the quantity of available adaptation data. The number of classes can be adjusted dynamically, as a function of the data, after having organized the set of Gaussians as a tree-structure. Using a single transform considerably reduces the number of parameters to estimate, which makes the method robust to transcription errors. This is particularly appealing in the case of unsupervised adaptation.

Using the same type of linear transforms in the process of speaker-independent models training makes it possible to estimate simultaneously the transforms specific to each speaker in the training corpus and the parameters of the acoustic HMM models [ANA 96]. The resulting models can then be more efficiently adapted to a new speaker.

Model composition can be used to compensate for an additive noise by combining a noise model with a noise-free speech model [GAL 92]. As opposed to blind adaptation (such as the linear transform approach presented above), this approach allows advantageously to directly model the acoustic channel. For practical reasons, the noise spectral density is usually assumed to be Gaussian and the noisy speech model is assumed to have the same structure and the same number of parameters as the noise-free model (which is typically a HMM model with GMM state density functions). Several approaches have been put forward for model composition, e.g. the log-normal approximation, numerical integration and data generation [GAL 95]. These three approaches are based on approximations, when estimating the model parameters corresponding to the first and second derivatives of the cepstral features.

For a given application, the choice of adaptation technique depends on the nature of the discrepancy between the training and operational data, as well as on the quantity of available adaptation data. Model composition is essentially used to compensate differences stemming from an additive noise, while Bayesian adaptation and linear regressions are general tools which can compensate for differences related to speakers or recording conditions.

Adaptation data can be a subset of training data, for instance to adapt models from one task to another, or to build specific models (adapted to a particular speaker, to a class of speakers, to an accent, etc.) for which adaptation data can be restricted to the data to be decoded. In the first case, adaptation is supervised when the transcription of the training data is available. In the second case, only unsupervised adaptation techniques can be considered in general.

In practice, acoustic models are made more robust to speaker variability and acoustic conditions by normalizing the average cepstrum and its variance. Another technique called *vocal-tract length normalization* consists of reducing the inter-speaker variability by linearly modifying the frequency axis of the signal to be recognized [AND 94]. The method evaluates several ratio values (typically 13 values between 0.88 and 1.12) [LEE 96] and chooses the ratio value with highest probability. As for the linear regression technique, this transform can also be used during the training phase.

7.5. Decoder

One of the key issues in LVCSR is the design of efficient decoding algorithms, capable of exploring the immense search space that results from the combination of the acoustic and linguistic models. More precisely, the goal of the decoder is to determine the sequence of words with maximum probability, given the lexicon and both the acoustic and linguistic models.

In practice, the objective is often reduced to finding the sequence of words corresponding to the most likely state sequence in the Markov model, i.e. the best path in a network (the search space) in which each node is associated with a state in the Markov model and with an acoustic observation (a cepstral vector corresponding to 10 ms of signal). Moreover, an exhaustive search is usually not tractable and therefore techniques have been developed in order to reduce the complexity by limiting the search to a fraction of the entire space. The most common technique for small and medium vocabularies is the frame-synchronous Viterbi decoding with beam-search [NEY 84] which calls for a dynamic programming technique (see Chapter 6).

This approach is also used for large vocabularies, provided that the following techniques are used in conjunction, so as to reduce significantly the required computational power: phonetic trees dependent on the previous words [NEY 92] possibly in conjunction with a dynamic decoding [ODE 94], fast evaluation of a list of candidate words (fast-match) [BAH 92, GIL 90], multi-pass decoding techniques such as the forward-backward search [AUS 91] or the progressive search [MUR 93] and re-evaluation of the N-best solutions [SCH 92]. An alternative to the frame-synchronous search is an asynchronous decoder using the A* algorithm (stack decoder) [BAH 83, GOP 95, PAU 92]. Dynamic decoders (frame-synchronous or A*-based) achieve an on-line combination of the acoustic and linguistic models so as to minimize the memory required by the decoding procedure. They also allow for the use of more complex models than static decoders do, but they need to call on more elaborate pruning strategies in order to remain sufficiently efficient. This type of decoder is very attractive for applications that require real-time computation. Static decoders usually call for more memory resources than dynamic decoders do. This is why they are only used with small language models (less than 1 million parameters). However, it has been demonstrated that it is possible, thanks to optimization techniques applied to probabilistic automata, to achieve static decoding with much larger language models [MOH 98].

Many systems use multi-pass decoders in order to reduce the memory and computation resources, especially when real-time decoding is not necessary [AUS 91, GAU 94a, MUR 93, RIC 95, WOO 95]. In a multi-pass decoder, knowledge sources are introduced progressively, which makes it possible to reduce the complexity of each pass and to often obtain a decoding operation which is globally faster than a one-pass decoder [NGU 99]. This decoding paradigm requires a specific interface between the successive passes, in order to avoid any loss of information resulting in decoding errors (which add on to those stemming from the models). The information is usually represented as a word lattice or as a list of the most likely hypotheses. Word lattices are graphs in which the nodes correspond to signal frames and the arcs correspond to a word, with an acoustic score and a linguistic score. Multi-pass decoders are generally not well suited to real-time applications as no hypothesis can be produced before the last pass. However, when a slight delay is acceptable, it is possible to process the signal block-wise. This type of decoding also offers a means of easily adding on knowledge sources to the decoding process without having to integrate them to the Markov framework.

7.6. Applicative aspects

In this section we briefly review some aspects which have to be taken into account when deploying a recognition system. We will not however approach integration aspects and ergonomical issues, which are considered outside the scope of this chapter.

7.6.1. *Efficiency: speed and memory*

A fast decoding is essential for the deployment of practical applications. Corresponding constraints influence the choice of the model structure and size. Speaker-independent systems are based on GMM acoustic distributions: as a result, acoustic likelihood computation can represent up to 30% and sometimes up to 50% of the decoding time. This is due to the large number of states which are necessary to model phonetic contexts and due to the large number of Gaussian components which are necessary to represent inter-speaker variability (typically, 16 to 32 Gaussians per state).

A frequent technique for speeding up the likelihood computation is the use of vector quantization in the cepstral vector space. This technique consists of preparing a short list of Gaussians to be considered for each state in the Markov model and each region of the space thus quantized [BOC 93]. The number of Gaussians to be considered can be reduced to a fraction (30% to 70%) of the total number of

Gaussians corresponding to the states occupied during the decoding process, with only a negligible increase in the error rate.

Sharing parameters between several phonetic models is a very efficient approach for reducing the total number of parameters. This technique can be applied at different levels in the model (allophones, states and Gaussian components) [TAK 95]. In section 7.4, it has been underlined that sharing phonetic models and tying state densities is a way to bypass the estimation problem, when dealing with insufficient quantities of training data. For larger training corpora, the optimal way to share parameters (in terms of accuracy) may require a large number of free parameters (from 5,000 to 30,000 states). Below this number, the system performance may not be optimal. Flexibility is more important when this technique is applied to the Gaussian components as it becomes possible to drastically reduce the number of parameters without sacrificing the accuracy of the models. This is exemplified in the marginal distribution clustering technique for acoustic subspaces [MAK 97, TAK 95]. This technique provides a considerable model size reduction with a negligible loss of accuracy. In its baseline version, this approach can be understood as a scalar quantization of the parameters [TAK 95].

As far as the decoding step is concerned, the optimal solution often derives from a compromise between the model complexity and the number of hypotheses that are evaluated. Generally, the best models have more parameters but, as they are more accurate, the size of the search space may be reduced without too much penalizing the decoding performance. Many techniques have been proposed to reduce decoding errors and to limit their impact on the recognition rate. Of course, constraints on the computational resources must be taken into account when developing the acoustic and linguistic models. For each operating point, the model complexity must be adapted to the decoding strategy. One of the most attractive decoding strategies for real-time implementations is the one-pass frame-synchronous dynamic decoding strategy which uses phonetic trees in order to represent the entire set of active solutions according to their linguistic context [AUB 99, NEY 92, ODE 94].

The number of parameters of an n-gram model for large vocabulary applications may be too large (several tenth of millions) with respect to practical constraints. One of the properties of these models is that they offer the possibility to reduce arbitrarily the number of parameters for a given order n, by using the lower order (n-1)-gram model for rare n-grams. Other techniques have been proposed to reduce the size of the n-gram models, with a negligible effect on the error rate [SET 96, STO 98]. Another solution to this problem consists of using a cache memory for the probability values which have been recently used by the decoder, while keeping the entire model on the disk [RAV 96].

7.6.2. *Portability: languages and applications*

Providing portability, i.e. the capacity to migrate a technology from one domain to another or from one language to another, is a major challenge for the developers of speech recognition systems. Whereas a given technology has been used successfully for a variety of applications and languages, its porting to a new domain or to a new language still requires a substantial effort, as both the acoustic and linguistic models and the pronunciation dictionary must be adapted. This model adaptation requires sufficient quantities of properly annotated data, the cost of which remains high (in terms of manpower and time).

Acoustic models estimated on a sufficiently large and diverse corpus (for instance, 10 hours of speech uttered by 100 speakers) can be used for a range of applications, provided the appropriate normalization and compensation techniques are applied. This is not the case for the language model and the pronunciation dictionary, which both are very domain-dependent.

For some applications, such as text dictation in a specific domain, large quantities of text are available, which can be used to build the language model. For other tasks, in particular for dialog systems, representative text data may be very limited or even inexistent, and in such cases, a data collection campaign is unavoidable. Such data collection campaigns are carried out in real–life conditions with a recognition system whose models are regularly updated in order to improve the system and to reduce the collection costs [GLA 99].

Building the pronunciation dictionary is often the most expensive adaptation step. Even though grapheme-to-phoneme conversion algorithms exist for many languages, these algorithms are almost exclusively developed for speech synthesis and are therefore poorly adapted to the needs of speech recognition. Semi-automatic approaches are frequently used for building this dictionary. Initial pronunciations are obtained by means of grapheme-to-phoneme conversion rules, then pronunciation variants are generated from data-derived rules, and finally, some pronunciations are checked manually, especially for words of foreign origin [LAM 96].

Developing a speech recognition system for a new language is not conceptually different from developing a system for a new application in the same language. Of course, the language-dependent components (phoneme set, phonological rules, etc.) have to be replaced. The concept of a word, which is language-dependent, has a significant impact on the system design. Taking into account the specificities of a language, we can of course improve the recognition system performance. For instance, in tonal languages such as Mandarin, it is useful to model explicitly the evolution of the fundamental frequency.

On the lexical level, a fixed-size vocabulary may have a very different coverage from one language to another. In particular, strongly flexional languages require larger vocabularies in order to obtain a coverage ratio comparable to other languages. By comparing the number of distinct words in newspaper articles for English, French, German and Italian, it can be noted that German texts contain twice as many words as French texts and four times as many as English and Italian texts [LAM 95][2]. As a consequence, 5,000 words are necessary in English, 20,000 in French and 65,000 in German in order to provide a 95% lexical coverage for newspaper articles.

7.6.3. Confidence measures

The estimation of a confidence measure, i.e. an *a posteriori* probability associated with the recognized word or sentence, is essential for a vast majority of applications [CHA 97, GIL 97, SIU 97, WEI 97, WES 98]. At the sentence level, the goal is to obtain an estimate of $\Pr(W|X)$ corresponding to the hypothesis $W = (w_1, w_2, ...)$. The hypothesis is often made that this probability can be replaced by $\Pr(W|X, \lambda)$, assuming that the models used by the recognition system (acoustic model, language model and lexicon, here denoted as λ) are correct and that the decoder is error-free in the sense that it identifies the most probable solution. This approximation can be further developed by using simpler models than those used by the decoder, so as to reduce the number of operations required to estimate this measure. For instance, the language model can be substituted by a simpler phonotactic model [GAU 96a]. Most applications require a confidence measure at the word level, corresponding to an estimation of $\Pr(w_i|X)$, i.e. the probability of the ith word in the hypothesized word sequence (or, more precisely, of $\Pr(w_i|X, \lambda)$).

An estimation of the latter probability can be efficiently obtained by means of the forward-backward algorithm (see Chapter 6) applied to a word lattice produced by the recognition system together with the recognition hypothesis [WES 98]. As this estimation relies on incorrect models, it is common practice to use additional cues such as the word and phoneme durations, the speaking rate or the signal-to-noise ratio, in order to obtain a more accurate estimation of $\Pr(w_i|X)$. These cues can be combined by means of a logistic regression [GIL 97], a generalized additive model [SIU 97] or a neural network [WEI 97].

---

2 The newspaper articles compared in this study are the *Wall Street Journal* (American English, 37 Mwords) [LDC 94], *Le Monde* (French, 38 Mwords) [GAU 94c], the *Frankfurter Rundschau* (German, 36 Mwords) [ACL 02] and the *Il Sole 24 Ore* (Italian, 26 Mwords) [FED 95], where the total number of words for each corpus is given in Mwords (million words).

The parameters for this module must be estimated on development data, by maximizing a function related to the confidence measure quality such as the mutual information between the confidence measure and the correctness of the hypothesis. As concerns the appropriate set of cues, it is specific to each application.

7.6.4. *Beyond words*

Apart from the uttered words, other attributes can be identified in the audio signal. This additional information may be linguistic (punctuation, semantic marking) or acoustic (speaker's identity, acoustic conditions, speech turns, confidence measure, etc.).

For what concerns acoustic attributes, the same modeling techniques have been applied successfully to gender classification, speaker identity recognition and acoustic characterization. Concerning continuous audio streams, it is advantageous to slice the data into homogenous acoustic segments, to identify and then remove segments containing no speech and finally to group speech segments from the same speaker. This information can be used to segment transcriptions and to facilitate their indexing by an information retrieval system. Semantic marking can be achieved through techniques developed for written language processing, provided these techniques are adapted to spoken language and to the specificities of automatic transcription.

7.7. Systems

Speech dictation is one of the most natural applications of automatic speech recognition technology. It has been the focus of many developments and there exist today inexpensive products for a large number of languages and professional domains. The most distinctive property of this application is that the processed speech is produced with the explicit intention to be transcribed by a machine.

A second type of application concerns the transcription and indexing of audio data, such as radio and television programs, audio- and video-conferences, recording of meetings, or any other audio document that needs to be indexed. These data can be qualified as "found data" in the sense that they have not been produced with the aim of being recognized by a machine. Generally, the language level, the recording conditions and the acoustic environment are far less controlled than those for dictation systems.

The third family of applications which we consider in this chapter is that of dialog systems. Most of these are used for voice access to databases or for the automatic routing of telephone calls.

7.7.1. *Text dictation*

Today, a wide variety of software products are available for continuous speech dictation of texts, either as office automation applications for the general public or as dedicated products profiled for particular professional needs (doctors, lawyers, etc.). The speech signal processed by these systems has generally been produced by a known speaker and, because of the nature of the task, the level of language is very close to that of written language. Sometimes, the speaker is actually reading an already written document. Recording conditions are generally based on a head-set microphone as its proximity to the speaker's mouth (a few centimeters) guarantees a very low noise level.

The first generation of such products required that the user utter isolated words (a pause of at least 300 ms between each word) and necessitated a long training session during which the user was requested to read a text. Isolated word speech input was greatly simplifying the system's task and the error correction process. Today, most systems use speaker-independent acoustic models which are adapted in a transparent manner during the use of the system and continuous speech is the favored speaking mode.

Dictation offers a simple way to evaluate the recognition performance by comparing the orthographic transcription of the text output by the system with the reference one, i.e., the transcription of the spoken text. For this reason, this task has been used as a benchmark for the evaluation of large vocabulary recognition systems, in particular in the framework of the NIST (National Institute for Science and Technology) evaluation campaigns. The metrics used to evaluate a speech recognition system is the percentage of erroneous words, which is obtained by summing three types of errors: substitutions, insertions and deletions. The error rate is defined as:

$$\frac{n_{subs} + n_{ins} + n_{del}}{n_{words\ in\ ref}}$$

and is calculated by minimizing a cost function. Typically, the costs associated with the three types of errors are respectively 1, 0.75 and 0.75. It is worth noting that the above definition can yield an error rate higher than 100%.

In American English, state-of-the-art speaker-independent systems with a head-set microphone achieve around 7% error rate [PAL 95, PAL 96]. This error rate is measured on the NAB (*North American Business News*) corpus of read texts distributed by the LDC (Linguistic Data Consortium) for a 65,000 word vocabulary. On the same data, recorded with a remote microphone in a noisy environment (55 dBA, 15 dB Signal-to-Noise Ratio), the error rate doubles to approximately 14%, with a noise compensation module. The error rate for spontaneous dictation is also in the order of 14% and it increases up to 20% for texts read over the telephone. Similar results are reported for other languages such as French or German [DOL 07, YOU 97].

7.7.2. *Audio document indexing*

Speech recognition is a key technology for indexing audio and audiovisual documents, such as radio and television programs or meeting and teleconference recordings. Audio transcription provides ways of accessing the content of audiovisual documents [GAR 98], by reducing the required time for identifying recordings in large multimedia databases or audiovisual streams [ABB 99, DEJ 99, HAU 95]. The synoptic representation of such a system is represented in Figure 7.6. Usual techniques for text indexing can be applied to automatic transcriptions of audiovisual documents in spite of the existence of errors in the transcriptions [GAU 00b, JOU 99].

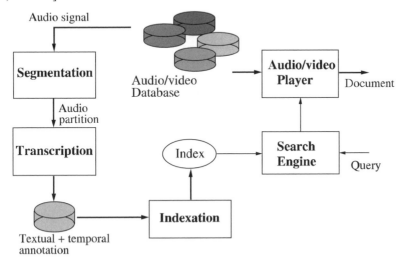

Figure 7.6. *Synoptic representation of an audio indexing system*

Audiovisual documents are especially difficult to transcribe as they contain segments of various acoustic and linguistic types. Transitions between segments can be either progressive or fast. For example, abrupt speaker changes can occur, but also soft transitions in the presence of a musical background. The speaker diversity is very wide and the same speaker may be speaking in very different acoustic conditions across the same document. All speech styles can be observed, from spontaneous to well-prepared speech. This type of non-homogenous audio stream is quite different from the signal which is observed at the input of a dictation system.

The first step consists of segmenting the audio stream (acoustic condition changes, speaker turns, etc.) and identifying the characteristics of each segment (speech/non-speech detection, telephone versus broadband speech, speaker gender, presence of music, speaker's identity, etc.). This step leads to a set of segments with acoustic attributes such as speaker gender, signal bandwidth, etc., which the recognition system can use to choose the most adequate acoustic models for decoding the speech contained in the segment.

In order to obtain acoustic models that are robust with regard to the various acoustic conditions, these models are estimated on large quantities of audio data (from a few dozen to a few hundred hours) representative of the acoustic diversity of the documents. Similarly, the linguistic models are estimated on large text corpora, such as newspaper articles, electronic news bulletins, commercial transcriptions of audio documents and also detailed transcriptions containing hesitations and breathing noises.

For these systems, the decoding is often achieved in two or three passes. The first pass produces a hypothesis which is used in the subsequent phases, so as to adapt in an unsupervised manner the acoustic models to each speaker in the document.

In recent years, significant progress has been made in the transcription of radio and TV broadcast news programs [PAL 97, PAL 98, PAL 99]. The average error rate is in the order of 20%, but a wide variation is observed according to the segment type: from 8% for speech segments uttered by the presenters, up to 50% for some commentaries taking place in difficult acoustic conditions.

The same technology can be applied to other types of problems, such as the transcription of teleconferences or, more generally, the transcription of telephone conversations (hotlines, call centers, etc). Each of these tasks poses a set of problems specific to the signal acquisition, the speaker population, their speaking style, the linguistic content, etc. The closest corpora to these types of data, on which results have been published, are the conversational speech corpora Switchboard and CallHome [GOD 92]. The error rates reported for these data range between 30% and 40% [YOU 98], which is substantially higher than the error rates achieved on information programs. The CallHome data are particularly difficult to process, as they consist of conversations between people who know each other and who speak in a very familiar manner.

7.7.3. *Dialog systems*

The design of systems able to conduct an oral dialog leads us well beyond the transcription of speech into text. These systems call for components that can handle the comprehension and the generation of natural speech, as well as dialog management. The recording conditions and the possible association of speech with other communication modalities, such as touch screens, are also important aspects to be taken into account. Problems posed by speech recognition in dialog systems are analyzed in [BER 99, GIB 97, DYB 98, GAU 96a, GAU 96b, GLA 99, GOL 99, GOR 97, PEC 93].

Telephone services, for which voice is the natural communication mode, are excellent applications for dialog systems; they have therefore been a set target for many developments. The most basic systems simply mimic the interfaces of the DTMF services, with very constrained dialogs and comprehension functionalities reduced to a minimum: for instance, call routing systems and automatic telephone switchboards. Other application domains are transport timetable information systems (by air or by train), stock exchange rates and weather forecasts. As the whole interaction is vocal, the dialog strategy and the response generation are essential aspects, especially for mixed-initiative dialog systems, in which the user is free to change the direction of the dialog at his own will.

Information kiosks and multimedia interfaces (especially on the Internet) represent another class of potential applications: cash dispensers, direction terminals or tourist information systems. For most multimedia interfaces, the input modalities are restricted to a touch screen and a keyboard; speech can be used as an alternative modality or as a complement to the existing ones, in order to improve the user-friendliness and the efficiency of the services.

These two classes of applications share many common points, but significant differences exist, especially with respect to the dialog strategy and the speech input conditions. Indeed, dialog plays a much more important role in telephone servers, as several exchanges between the user and the machines are often necessary to obtain the information. For instance, it is preferable, in a dialog situation, to ask the user to state more precisely his or her request rather than enumerating all the possible answers, whereas, in the case of multimedia applications, it may be more efficient to visually present the user with all the solutions and let him or her point out the right choice. Regarding the speech signal input, the differences are also significant. The telephone signal has a reduced bandwidth and it can be corrupted with variability stemming from the transmission network. For multimedia interfaces, variability is essentially set off by the distance between the microphone and the user's mouth, which makes the system rather sensitive to the surrounding acoustic conditions. This is all the more crucial as kiosks are installed in public places.

These applications enforce a few constraints on recognition systems. The main demands are for the system to operate in real-time, in a speaker-independent mode and to be able to handle spontaneous speech. Real-time response means that the delay between the instant when the speaker stops talking and the instant when the system answers must be short. An acceptable delay is about 0.5 seconds. This generally means that the recognition system must process the utterance during the acquisition phase, which somehow disqualifies multi-pass decoders and non-causal cepstrum and energy normalization schemes.

Linguistic models used in dialog systems explicitly represent extra-linguistic events such as hesitations, breathing noise, laughter, etc. It is also common practice to use compound words for word sequences which are subject to strong coarticulation effects or to phonemic reduction. Language models can be obtained either manually from some *a priori* knowledge specific to the application, or automatically for *n*-gram like statistical models. The first solution is generally used for very directive systems, whereas the second is better adapted for mixed initiative systems.

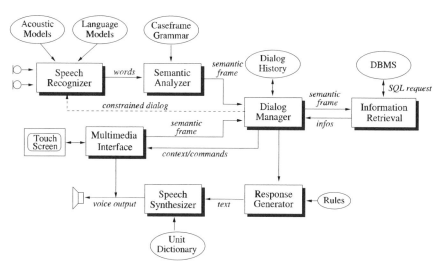

Figure 7.7. *Architecture of a dialog system*

The architecture of a dialog system is represented in Figure 7.7. The main components are the speech recognition system, the semantic analysis module and the dialog management engine. The latter controls the module in charge of generating requests to the database, as well as the response generator. For multimedia interfaces, the voice interface can be combined with other modalities such as a touch screen or a keyboard.

Several approaches can be considered to connect the recognition system and the semantic analysis module. In most cases, a bottom-up approach is used, where the recognition result is passed on to the understanding module without any information feed-back to the recognition module. The recognition result can consist of a single hypothesis (a word sequence), a list of ordered N-best hypotheses or a word lattice. The choice between these options depends on the nature of the language model used in the recognition system and on the understanding method used subsequently (rule-based or statistical approach).

A confidence measure can be associated with each word in the hypothesis so that unsure words can either be rejected by the recognition system, ignored by the comprehension module or trigger a confirmation dialog. This rejection mechanism has strong implications on the user interaction (request for repetition) and it tends to cause longer dialogs. However, by rejecting uncertain hypotheses, the system error rate becomes substantially lower, which reduces the user disturbance.

Aspect	*Measures*
Speech recognizer	word error, content word error, confidence measures
Semantic analyzer *Dialog*	semantic frame error, slot error response
System	global measure (success, #turns, time, waiting time etc.)
Subjective	questionnaires

Table 7.4. *Some assessment metrics for spoken dialog systems*

Whereas it is relatively simple to evaluate a speech recognition system, the evaluation of a dialog system is considerably more difficult, because of the interactive nature of the dialog and the subjective human perception of a system efficiency [BER 99, GIB 97, MAD 92]. It is also important to evaluate not only the various system components, but also the global system efficiency by means of objective and subjective measures [LAM 00a, MIN 98].

Table 7.4 summarizes some of the measures used to evaluate the various modules, as well as the dialog system as a whole. Beyond the rate of incorrect words, it can be useful to measure the error rate on words which bear some meaning for the application. Measures frequently used for semantic analysis are the error rates on the semantic schemes and on the attributes of these schemes, while the dialog management engine is evaluated in terms of response relevance. Objective measures for the global system evaluation are the success rate, the average number of speech turns per call, the number of repetitions, the dialog duration, the waiting time and the number of calls. In the case of multimodal systems, the efficiency of speech interaction can be compared to other modalities, such as touching a screen and typing on a keyboard (for user requests) or to a presentation of the system results in the form of texts or graphs (for system responses).

Evaluation plays an essential role during the system development phase. Several types of evaluation procedures can be set up, each of them having its strong points and its costs. Of course, an automatic evaluation has a lower cost than an evaluation which requires significant human intervention. Automatic evaluation is possible for some of the system components, in particular, speech recognition and literal comprehension (i.e. without taking the context into account), but it is not possible as of today to evaluate the dialog management engine automatically.

Depending on the complexity of the application and on the audio signal quality, the error rate for recognition systems is very variable, ranging from less than 5% for a limited vocabulary and a head-set microphone, up to 25% or more for a telephone system with a vocabulary of several thousand words. It is sometimes difficult to compare results across applications and systems, as the transcription conventions and hypothesis normalization can be very different.

7.8. Perspectives

Over the past decades, the capacities of speech recognition systems have significantly increased, whether it is in terms of vocabulary size, speaking mode or training constraints. Noticeable progress has been made in the handling of non-homogenous data, such as radio and TV programs, which contain changes in speakers, acoustic conditions, topics and sometimes even languages. This progress stems from a more robust acoustic analysis, training procedures which better benefit from very large audio and text corpora, audio stream segmentation algorithms, unsupervised techniques for the adaptation of acoustic models and more efficient decoders able to use higher order linguistic models [GAU 00a].

In spite of these numerous advances, the level of performance of a machine remains well below that of a human transcriber [DES 96, EBE 95, LIP 97, VAN 95]. The ratio between the machine and the human error rates lies somewhere between 5 and 10, depending on the type of data. To bridge this gap, it is necessary to improve the models at all levels: acoustic, lexical, syntactic and semantic. In particular, the automatic processing of conversational speech remains a true challenge.

It is well known that there can be a very large difference between error rates from one speaker to another (up to a factor 20 or more). This stems from a variety of factors [FIS 96], in particular the speaking style and rate. For moderate speaking rates (120-160 words per minute), there is no correlation between the speaking rate and the error rate, whereas for speaking rates above 180 words per minute, the correlation is high [PAL 95]. The reduction of this difference requires the adaptation of pronunciation rules so as to predict pronunciation variants as a function of the observed pronunciations for a given speaker. A speaker who utters a word in a particular manner is of course likely to utter similar words in the same manner. The same applies to some phonological rules which occur across words. Currently, there is no system which takes this type of dependency into account.

As far as linguistic modeling is concerned, the techniques that have been investigated in order to account for long term grammatical agreements have not been successful. These techniques would be particularly useful for processing strongly flexional languages for which n-gram models are clearly not the optimal solution. The adaptation of linguistic models is a challenge for radio and TV broadcast news transcription systems, for which it is particularly important to keep linguistic models up-to-date. New topics can come up suddenly and stay active for a very variable lapse of time. The existence of contemporaneous data sources, such as Internet news sites, enables some automatic language model update.

The design and development of application-independent systems is another major challenge for the field. From a large transcribed speech corpus, it is possible to develop acoustic models for a variety of applications; this is not the case for language models, for which a wide coverage of the application domain is essential. With the current technology, porting a given system to a new application or a new language requires the existence of a sufficient quantity of transcribed data. The development of training techniques which only require a little supervision is also a promising research direction [KEM 99, LAM 00b].

Even though speech recognition is far from being a solved problem, a wide variety of applications are now within reach with the current technology. The two most promising domains concern information servers and automatic indexing of audio documents.

7.9. References

[ABB 99] ABBERLEY D., KIRBY D., RENALS S., ROBINSON T., "The THISL Broadcast News Retrieval System", *Proc. ESCA ETRW on Accessing Information in Spoken Audio*, p. 14-19, Cambridge, UK, April 1999.

[ACL 02] ACL-ECI CDROM, distributed by Elsnet and LDC, 2002.

[AND 94] ANDREOUM A., KAMM T., COHEN J., "Experiments in Vocal Tract Normalisation", *Proc. CAIP Workshop: Frontiers in Speech Recognition II*, 1994.

[ANA 96] ANASTASAKOS T., MCDONOUGH J., SCHWARTZ R., MAKHOUL J., "A Compact Model for Speaker Adaptation Training", *Proc. ICSLP'96*, p. 1137-1140, 1996.

[AUB 99] AUBERT X., "One Pass Cross Word Decoding for Large Vocabularies Based on a Lexical Tree Search Organization", *Proc. ESCA Eurospeech'99*, 4, p. 1559-1562, Budapest, Hungary, September 1999.

[AUS 91] AUSTIN S., SCHWARTZ R., PLACEWAY P., "The Forward-Backward Search Strategy for Real-Time Speech Recognition", *Proc. IEEE ICASSP-91*, p. 697-700, Toronto, Canada, May 1991.

[BAH 83] BAHL L.R., JELINEK F., MERCER R.L., "A Maximum Likelihood Approach to Continuous Speech Recognition", *IEEE Trans. Pattern Analysis & Machine Intelligence*, vol. 5, no. 2, p. 179-190, March 1983.

[BAH 88] BAHL L.R., BROWN P., DE SOUZA P., MERCER R.L., PICHENY M., "Acoustic Markov Models used in the Tangora Speech Recognition System", *Proc. IEEE ICASSP-88*,1, p. 497-500, New York, April 1988.

[BAH 92] BAHL L.R., DE SOUZA P.V., GOPALAKRISHNAN P.S., NAHAMOO D., PICHENY M., "A Fast Match for Continuous Speech Recognition Using Allophonic Models", *Proc. IEEE ICASSP-92*, CA, 1, p. 17-21, San Francisco, CA, March 1992.

[BAH 95] BAHL L.R., *et al.*, "The IBM Large Vocabulary Continuous Speech Recognizer for the ARPA NAB News Task", *Proc. ARPA Spoken Language Systems Technology Workshop*, p. 121-126, Austin, January 1995.

[BAK 92] BAKER J., *et al.*, "Large Vocabulary Recognition of Wall Street Journal Sentences at Dragon Systems", *Proc. DARPA Speech & Natural Language Workshop*, p. 387-392, Harriman, New York, February 1992.

[BAU 70] BAUM L.E., PETRIE T., SOULES G., WEISS N., "A Maximization Technique Occurring in the Statistical Analysis of Probabilistic Functions of Markov Chains", *Ann. Math. Stat.*, 41, p. 164-171, 1970.

[BER 99] BERNSEN N.O., DYBKJAER L., HEID U., "Current Practice in the Development and Evaluation of Spoken Language Dialogue Systems", *Proc. ESCA Eurospeech'99*, p. 11471150, Budapest, Hungary, September 1999.

[BOC 93] BOCCHIERI E., "Vector quantization for efficient computation of continuous density likelihoods", *Proc. IEEE ICASSP-93*, 2, p. 692-695, Minneapolis, May 1993.

[CHA 97] CHASE L., "Word and acoustic confidence annotation for large vocabulary speech recognition", *Proc. ESCA Eurospeech'97*, p. 815-818, Rhodes, Greece, September 1997.

[COH 89] COHEN M., Phonological Structures for Speech Recognition, PhD Thesis, University of California, Berkeley, 1989.

[DAV 80] DAVIS S., MERMELSTEIN P., "Comparison of Parametric Representations of Monosyllabic Word Recognition in Continuously Spoken Sentences", *IEEE Trans. Acoustics, Speech, and Signal Processing*, vol. 28, no. 4, p. 357-366, 1980.

[DEJ 99] DE JONG F., GAUVAIN J.L., DEB HARTOG J., NETTER K., "Olive: Speech Based Video Retrieval", *Proc. CBMI'99*, Toulouse, October 1999.

[DEM 77] DEMPSTER A.P., LAIRD M.M., RUBIN D.B., "Maximum Likelihood from Incomplete Data via the EM Algorithm", *Journal of the Royal Statistical Society Series B (methodological)*, 39, p. 1-38, 1977.

[DES 96] DESHMUKH N., GANAPATHIRAJU A., DUNCAN R.J., PICONE J., "Human Speech Recognition Performance on the 1995 CSR Hub-3 Corpus", *Proc. ARPA Speech Recognition Workshop*, p. 129-134, Harriman, New York, February 1996.

[DIG 94] DIGALAKIS V., MURVEIT H., "Genones: Optimization the Degree of Tying in a Large Vocabulary HMM-based Speech Recognizer", *Proc. IEEE ICASSP-94*, 1, p. 537-540, Adelaide, Australia, April 1994.

[DOL 97] DOLMAZON J.M., *et al.* "Organisation de la première campagne AUPELF pour l'évaluation des systèmes de dictée vocale", *Journées Scientifiques et Techniques du Réseau Francophone d'Ingénierie de la Langue de l'AUPELF-UREF*, p. 13-18, Avignon, France, April 1997.

[DRE 97] DRENTH E.W., RÜBER B., "Context-dependent probability adaptation in speech understanding", *Computer Speech & Language*, vol. 11, no. 3, p. 225-252, July 1997.

[DYB 98] DYBKJAER L., *et al.*, "The DISC Approach to Spoken Language Dialogue Systems Development and Evaluation", *Proceedings of the First International Conference on Language Resources and Evaluation (LREC)*, p. 185-189, Granada, Spain, May 1998.

[EBE 95] EBEL W.J., PICONE J., "Human Speech Recognition Performance on the 1994 CSR Spoke 10 Corpus", *Proc., ARPA Spoken Language Systems Technology Workshop*, p. 53-59, Austin, January 1995.

[FED 95] FEDERICO M., CETTOLO M., BRUGNARA F., ANTONIOL G., "Language Modeling for Efficient Beam-Search", *Computer Speech & Language*, vol. 9, no. 4, p. 353-379, October 1995.

[FIS 96] FISHER W.M., "Factors Affecting Recognition Error Rate", *Proc. ARPA Speech Recognition Workshop*, p. 47-52, Harriman, New York, February 1996.

[FOS 96] FOSLER E., *et al.*, "Automatic Learning of Word Pronunciation from Data", *Proc. ICSLP'96*, Addendum, p. 28-29, Philadelphia, October 1996.

[GAL 92] GALES M.J.F., YOUNG S.J., "An Improved Approach to Hidden Markov Model Decomposition of Speech and Noise", *Proc. IEEE ICASSP-92*, p. 233-236, San Francisco, March 1992.

[GAL 95] GALES M.J.F., YOUNG S.J., "Robust Continuous Speech Recognition using Parallel Model Combination", *Computer Speech & Language*, vol. 9, no. 4, p. 289-307, October 1995.

[GAR 98] GAROFOLO J.S., VOORHEES E.M., AUZANNE C.G.P., STANFORD V.M., LUND B.A., "1998 TREC-7 Spoken Document Retrieval Track Overview and Results", *Proc. 7th Text Retrieval Conference TREC-7*, NIST Special Publication 500-242, p. 79-90, Gaithersburg, MD, November 1998.

[GAU 94a] GAUVAIN J.L., LAMEL L.F., ADDA G., ADDA-DECKER M., "The LIMSI Nov93 WSJ System", *Proc. ARPA Spoken Language Technology Workshop*, p. 125-128, Princeton, March 1994.

[GAU 94b] GAUVAIN J.L., LEE C.H., "Maximum *A Posteriori* Estimation for Multivariate Gaussian Mixture Observations of Markov Chains", *IEEE Trans. Speech & Audio Processing*, vol. 2, no. 2, p. 291-298, April 1994.

[GAU 94c] GAUVAIN J.L., LAMEL L.F., ADDA G., ADDA-DECKER M., "Speaker-Independent Continuous Speech Dictation", *Speech Communication*, vol. 15, no. 1-2, p. 21-37, October 1994.

[GAU 95] GAUVAIN J.L., LAMEL L.F., ADDA-DECKER M., "Developments in Continuous Speech Dictation using the ARPA WSJ Task", *Proc. IEEE ICASSP-95*, p. 65-68, Detroit, May 1995.

[GAU 96a] GAUVAIN J.L., GANGOLF J.J., LAMEL L., "Speech Recognition for an Information Kiosk", *Proc. ICSLP'96*, p. 849-852, Philadelphia, October 1996.

[GAU 96b] GAUVAIN J.L., LAMEL L., "Large Vocabulary Continuous Speech Recognition: from Laboratory Systems towards Real-World Applications", *Institute of Electronics, Information and Communication Engineers*, J79-D-II, p. 2005-2021, December 1996.

[GAU 00a] GAUVAIN J.L., LAMEL L., "Large Vocabulary Continuous Speech Recognition: Advances and Applications", *Proceedings of the IEEE*, vol. 88, no. 8, p. 1181-1200, August 2000.

[GAU 00b] GAUVAIN J.L., LAMEL L., BARRAS C., ADDA G., KERCADIO Y., "The LIMSI SDR system for TREC-9", *Proc. of the Text Retrieval Conference, TREC-9*, Gaithersburg, November 2000.

[GIA 91] GIACHIN E., ROSENBERG A.E., LEE C.H., "Word Juncture Modeling using Phonological Rules for HMM-based Continuous Speech Recognition", *Computer Speech & Language*, 5, p. 155-168, 1991.

[GIB 97] GIBBON D., MOORE R., WINSKI R. (eds.), *Handbook of Standards and Resources for Spoken Language Systems*, Mouton de Gruyter, Berlin, New York, 1997.

[GIL 90] GILLICK L., ROTH R., "A Rapid Match Algorithm for Continuous Speech Recognition", *Proc. DARPA Speech & Natural Language Workshop*, p. 170-172, Hidden Valley, June 1990.

[GIL 97] GILLICK L., ITO Y., YOUNG J., "A Probabilistic Approach to Confidence Measure Estimation and Evaluation", *Proc. IEEE ICASSP-97*, p. 879-882, Munich, April 1997.

[GLA 99] GLASS J.R., HAZEN T.J., HETHERINGTON I.L., "Real-time Telephone-based Speech Recognition in the Jupiter Domain", *Proc. IEEE ICASSP-99*, 1, p. 61-64, Phoenix, March 1999.

[GOD 92] GODFREY J., HOLLIMAN E., MCDANIEL J., "SWITCHBOARD: Telephone Speech Corpus for Research and Development", *Proc. IEEE ICASSP-92*, p. 517-520, San Francisco, March 1992.

[GOL 99] GOLDSCHEN A., LOEH D., "The Role of the Darpa Communicator Architecture as a Human Computer Interface for Distributed Simulations", *Proc. 1999 Simulation Interoperability Standards Organization (SISO) Spring Simulation Interoperability Workshop (SIW)*, Orlando, 14-19 March, 1999.

[GOP 95] GOPALAKRISHNAN P.S., BAHL L.R., MERCER R.L., "A Tree Search Strategy for Large Vocabulary Continuous Speech Recognition", *Proc. IEEE ICASSP-95*, 1, p. 572-575, Detroit, May 1995.

[GOR 97] GORIN A.L., RICCARDI G., WRIGHT J.H., "How May I Help You?", *Speech Communication*, vol. 23, no. 1-2, p. 113-127, October 1997.

[HAU 95] HAUPTMANN A.G., WITBROCK M., CHRISTEL M., "News-on-Demand – An Application of Informedia Technology", *Digital Libraries Magazine*, September 1995.

[HER 90] HERMANSKY H., "Perceptual linear predictive (PLP) analysis of speech", *J. Acoust. Soc. America*, vol. 84, no. 4, p. 1738-1752, 1990.

[HIN 96] HINDLE D.M., LJOLJE A., RILEY M.D., "Recent Improvements to the AT&T Speechto-Text (STT) System", *Proc. ARPA Speech Recognition Workshop*, February 1996.

[HOC 94] HOCHBERG M.M., RENALS S.J., ROBINSON A.J., KERSHAW D., "Large vocabulary continuous speech recognition using a hybrid connectionist-HMM system", *Proc. ICSLP'94*, Yokohama, Japan, p. 1499-1502, September 1994.

[HWA 92] HWANG M., HUANG X., "Subphonetic Modeling with Markov States – Senone", *Proc. IEEE ICASSP-92*, San Francisco, 1, p. 33-36, March 1992.

[HWA 93] HWANG M.Y., HUANG X., ALLEVA F., "Predicting Unseen Triphones with Senones", *Proc. IEEE ICASSP-93*, II, p. 311-314, Minneapolis, April 1993.

[JEL 91] JELINEK F., MERIALDO B., ROUKOS S., STRAUSS M., "A Dynamic Language Model for Speech Recognition", *Proc. DARPA Speech & Natural Language Workshop*, p. 293-295, Pacific Grove, February 1991.

[JOU 99] JOURLIN P., JOHNSON S.E., SPÄRCK JONES K., WOODLAND P.C., "General Query Expansion Techniques for Spoken Document Retrieval", *Proc. SIGIR'99*, August 1999.

[JUA 85] JUANG B.-H., "Maximum-Likelihood Estimation for Mixture Multivariate Stochastic Observations of Markov Chains", *AT&T Technical Journal*, vol. 64, no. 6, 1985.

[KAT 87] KATZ S.M., "Estimation of Probabilities from Sparse Data for the Language Model Component of a Speech Recognizer", *IEEE Trans. Acoustics, Speech & Signal Processing*, vol. 35, no. 3, p. 400-401, March 1987.

[KEM 99] KEMP T., WAIBEL A., "Unsupervised Training of a Speech Recognizer: Recent Experiments", *Proc. ESCA Eurospeech'99*, Budapest, Hungary, 6, p. 2725-2728, September 1999.

[KER 96] KERSHAW D., ROBINSON A.J., RENALS S.J., "The 1995 Abbot Hybrid Connectionist-HMM Large-vocabulary Recognition System", *Proc. ARPA Speech Recognition Workshop*, p. 93-98, Harriman, New York, February 1996.

[LAM 92] LAMEL L.F., GAUVAIN J.L., "Continuous Speech Recognition at LIMSI", *Proc. ARPA Workshop on Continuous Speech Recognition*, Stanford, p. 59-64, September 1992.

[LAM 95] LAMEL L.F., DEMORI R., "Speech Recognition of European Languages", *Proc. IEEE Automatic Speech Recognition Workshop*, Snowbird, Utah, p. 51-54, December 1995.

[LAM 96] LAMEL L.F., ADDA G., "On Designing Pronunciation Lexicons for Large Vocabulary, Continuous Speech Recognition", *Proc. ICSLP'96*, 1, p. 6-9, Philadelphia, October 1996.

[LAM 00a] LAMEL L., *et al.*, "The Limsi Arise System", *Speech Communication*, vol. 31, no. 4, p. 339-354, August 2000.

[LAM 00b] LAMEL L., GAUVAIN J.L., ADDA G., "Lightly Supervised Acoustic Model Training", *Proc. ISCA ITRW ASR2000*, p. 150-154, Paris, France, September 2000.

[LDC 94] LDC, CSR corpus. Language model training data, NIST Speech Disc 22-1 and 22-2, LDC, August 1994.

[LEE 88] LEE K.-F., "Large-vocabulary speaker-independent continuous speech recognition: The SPHINX system", PhD Thesis, Carnegie-Mellon University, 1988.

[LEE 96] LEE L., ROSE R.C., "Speaker Normalisation Using EFficient Frequency Warping Procedures", *Proc. IEEE ICASSP-96*, 1, p. 353-356, Atlanta, May 1996.

[LEG 95] LEGGETTER C.J., WOODLAND P.C., "Maximum Likelihood Linear Regression for Speaker Adaptation of Continuous Density Hidden Markov Models", *Computer Speech and Language*, 9, p. 171-185, 1995.

[LIP 82] LIPORACE L.R., "Maximum Likelihood Estimation for Multivariate Observations of Markov Sources", *IEEE Transactions on Information Theory*, vol. 28, no. 5, p. 729-734, 1982.

[LIP 97] LIPPMANN R.P., "Speech Recognition by Machines and Humans", *Speech Communication*, vol. 22, no. 1, p. 1-15, July 1997.

[LJO 95] LJOLJE A., RILEY M.D., HINDLE D.M., PEREIRA F., "The AT&T 60,000 Word Speech-To-Text System", *Proc. ARPA Spoken Language Systems Technology Workshop*, p. 162-165, Austin, January 1995.

[MAD 92] MADCOW, "Multi-site Data Collection for a Spoken Language Corpus", *Proc. DARPA Speech & Natural Language Workshop*, Harriman, New York, p. 7-14, February 1992.

[MAK 97] MAK B., BOCCHIERI E., "Subspace Distribution Clustering for Continuous Observation Density Hidden Markov Models", *Proc. Eurospeech'97*, p. 107-110, Rhodes, Geece, September 1997.

[MIN 98] MINKER W., "Evaluation Methodologies for Interactive Speech Systems", *LREC'98*, Granada, p. 199-206, May 1998.

[MOH 98] MOHRI M., RILEY M., HINDLE D., LJOLIE A., PEREIRA F., "Full Expansion of Context-Dependent Networks in Large Vocabulary Speech Recognition", *Proc. IEEE ICASSP-98*, p. 665-668, Seattle, May 1998.

[MUR 93] MURVEIT H., BUTZBERGER J., DIGALAKIS V., WEINTRAUB M., "Large-Vocabulary Dictation using SRI's Decipher Speech Recognition System: Progressive Search Techniques", *Proc. IEEE ICASSP-93*, Minneapolis, MN, pp. II-319-322, April 1993.

[NEY 84] NEY H., "The Use of a One-Stage Dynamic Programming Algorithm for Connected Word Recognition", *IEEE Trans. Acoustics, Speech and Signal Processing*, vol. 32, no. 2, p. 263-271, April 1984.

[NEY 92] NEY H., HAEB-UMBACH R., TRAN B.H., OERDER M., "Improvements in Beam Search for 10000-Word Continuous Speech Recognition", *Proc. IEEE ICASSP-92*, I, p. 9-12, San Francisco, March 1992.

[NGU 99] NGUYEN L., SCHWARTZ R., "Single-Tree Method for Grammar-Directed Search", *Proc. IEEE ICASSP-99*, 2, p. 613-616, Phoenix, March 1999.

[ODE 92] ODELL J.J., The Use of Decision Trees with Context Sensitive Phoneme Modelling, MPhil Thesis, Cambridge University Engineering Dept, 1992.

[ODE 94] ODELL J.J., VALTCHEV V., WOODLAND P.C., YOUNG S.J., "A One Pass Decoder Design for Large Vocabulary Recognition", *Proc. ARPA Human Language Technology Workshop*, p. 405-410, Princeton, March 1994.

[OSH 75] OSHIKA B.T., ZUE V.W., WEEKS R.V., NEU H., AURBACH J., "The Role of Phonological Rules in Speech Understanding Research", *IEEE Trans. Acoustics, Speech, Signal Processing*, 23, p. 104-112, 1975.

[OST 92] OSTENDORF M., KANNAN A., KIMBALL O., ROHLICEK J.R., "Continuous Word Recognition Based on the Stochastic Segment Model", *Proc. ARPA Workshop on Continuous Speech Recognition*, p. 53-58, Stanford, September 1992.

[PAL 95] PALLETT D.S., *et al.*, "1994 Benchmark Tests for the ARPA Spoken Language Pro-gram", *Proc. ARPA Spoken Language Systems Technology Workshop*, Austin, p. 5-36, January 1995.

[PAL 96] PALLETT D.S., *et al.*, "1995 Hub-3 Multiple Microphone Corpus Benchmark Tests", *Proc. ARPA Speech Recognition Workshop*, Harriman, New York, February 1996.

[PAL 97] PALLETT D.S., FISCUS J.G., PRZYBOCKI M.A., "1996 Preliminary Broadcast News Benchmark Test", *Proc. DARPA Speech Recognition Workshop*, p. 22-46, Chantilly, February 1997.

[PAL 98] PALLETT D.S., FISCUS J.G., MARTIN A.F., PRZYBOCKI M.A., "1997 Broadcast News Benchmark Test Results: English and Non-English", *Proc. DARPA Broadcast News Transcription & Understanding Workshop*, p. 5-11, Landsdowne, VA, February 1998.

[PAL 99] PALLETT D.S., FISCUS J.G., GAROFOLO J.S., MARTIN A.F., PRZYBOCKI M.A., "1998 Broadcast News Benchmark Test Results: English and Non-English Word Error Rate Performance Measures", *Proc. DARPA Broadcast News Workshop*, p. 5-12, Herndon, VA, March 1999.

[PAU 92] PAUL D.B., "An Efficient A* Stack Decoder Algorithm for Continuous Speech Recognition with a Stochastic Language Model", *Proc. IEEE ICASSP-92*, p. 405-409, San Francisco, March 1992.

[PEC 93] PECKHAM J., "A New Generation of Spoken Dialogue Systems: Results and Lessons from the SUNDIAL Project", *Proc. ESCA Eurospeech'93*, p. 33-40, Berlin, September 1993.

[RAV 96] RAVISHANKAR M.K., Efficient Algorithms for Speech Recognition, PhD Thesis, Carnegie Mellon University, 1996.

[RIC 95] RICHARDSON F., Ostendorf M., Rohlicek J.R., "Lattice-Based Search Strategies for Large Vocabulary Recognition", *Proc. IEEE ICASSP-95*, 1, p. 576-579, Detroit, May 1995.

[RIL 96] RILEY M.D., Ljojle A., "Automatic Generation of Detailed Pronunciation Lexicons", in *Automatic Speech and Speaker Recognition*, Kluwer Academic Pubs, p. 285-301, 1996.

[RIL 99] RILEY M.D., *et al.*, "Stochastic Pronunciation Modelling from Hand-labelled Phonetic Corpora", in *Automatic Speech and Speaker Recognition, Speech Communication*, vol. 29, no. 2-4, p. 209-224, 1999.

[ROS 92] ROSENFELD R., HUANG X., "Improvements in Stochastic Language Modeling", *Proc. DARPA Workshop on Speech & Natural Language*, p. 107-111, Harriman, New York, February 1992.

[ROS 94] ROSENFELD R., Adaptive Statistical Language Modeling, PhD Thesis, Carnegie Mellon University, 1994. (see also Tech. rep. CMU-CS-94-138)

[ROS 00] ROSENFELD R., *Adaptive Statistical Language Modeling, Proceedings of the IEEE*, vol. 88, no. 8, p. 1270-1278, August 2000.

[SAN 96] SANKAR A., *et al.*, "Noise-Resistant Feature Extraction and Model Training for Robust Speech Recognition", *Proc. ARPA Speech Recognition Workshop*, p. 117-122, Harriman, New York, February 1996.

[SCH 92] SCHWARTZ R., AUSTIN S., KUBALA F., MAKHOUL J., "New Uses for N-Best Sentence Hypothesis, within the BYBLOS Speech Recognition System", *Proc. IEEE ICASSP-92*,I, p. 1-4, San Francisco, March 1992.

[SEK 96] SEKINE S., GRISHMAN R., "NYU Language Modeling Experiments for the 1995 CSR Evaluation", *Proc. ARPA Speech Recognition Workshop*, p. 123-128, Harriman, New York, February 1996.

[SEY 96] SEYMORE K., ROSENFELD R., "Scalable Backoff Language Models", *Proc. ICSLP'96*, 1, p. 232-235, Philadelphia, October 1996.

[SIU 97] SIU M., GISH H., "Evaluation of Word Confidence for Speech Recognition Systems", *Computer Speech & Language*, vol. 13, no. 4, p. 299-318, October 1999.

[STO 98] STOLCKE A., "Entropy-based Pruning of Backoff Language Models", *Proc. DARPA Broadcast News Transcription & Understanding Workshop*, p. 270-274, Landsdowne, VA, February 1998.

[TAJ 95] TAJCHMAN G., FOSLER E., JURAFSKY D., "Building Multiple Pronunciation Models for Novel Words Using Exploratory Computational Phonology", *Proc. ESCA Eurospeech'95*, 3, p. 2247-2250, Madrid, Spain, September 1995.

[TAK 95] TAKAHASHI S., SAGAYAMA S., "Four-level Tied Structure for Efficient Representation of Acoustic Modeling", *Proc. IEEE ICASSP-95*, p. 520-523, Detroit, May 1995.

[VAN 95] VAN LEEUWEN D.A., VAN DEN BERG L.G., STEENEKEN H.J.M., "Human Benchmarks for Speaker Independent Large Vocabulary Recognition Performance", *Proc. ESCA Eurospeech'95*, p. 1461-1464, Madrid, Spain, September 1995.

[WEI 97] WEINTRAUB M., BEAUFAYS F., RIVLIN Z., KONIG Y., STOLCKE A., "Neural-Network based Measures of Confidence for Word Recognition", *Proc. ICASSP-97*, p. 887-890, Munich, April 1997.

[WES 98] WESSEL F., MACHEREY K., SCHLÜTER R., "Using Word Probabilities as Confidence Measures", *Proc. IEEE ICASSP-98*, p. 225-228, May 1998.

[WOO 95] WOODLAND P.C., LEGGETTER C.J., ODELL J.J., VALTCHEV V., YOUNG S.J., "The Development of the 1994 HTK Large Vocabulary Speech Recognition System", *Proc. ARPA Spoken Language Systems Technology Workshop*, p. 104-109, Austin, January 1995.

[WOO 96] WOODLAND P.C., GALES M.J.F., PYE D., VALTCHEV V., "The HTK Large Vocabulary Recognition System for the 1995 ARPA H3 Task", *Proc. ARPA Speech Recognition Workshop*, p. 99-104, Harriman, New York, February 1996.

[YOU 92] YOUNG S.J., "The General Use of Tying in Phoneme-Based HMM Speech Recognisers", *Proc. IEEE ICASSP-92*, San Francisco, p. 569-572, March 1992.

[YOU 93] YOUNG S.J., WOODLAND P.C., "The Use of State Tying in Continuous Speech Recognition", *Proc. ESCA Eurospeech'93*, 3, p. 2203-2206, Berlin, September 1993.

[YOU 94] YOUNG S.J., ODELL J.J., WOODLAND P.C., "Tree-Based State Tying for High Accuracy Acoustic Modeling", *Proc. ARPA Human Language Technology Workshop*, p. 307-312, Princeton, March 1994.

[YOU 96] YOUNG S.J., "A Review of Large-Vocabulary Continuous Speech Recognition", *IEEE Signal Processing Magazine*, vol. 13, no. 5, p. 45-57, September 1996.

[YOU 97] YOUNG S.J., *et al.*, "Multilingual Large Vocabulary Speech Recognition: the European SQALE Project", *Computer Speech & Language*, vol. 11, no. 1, p. 73-89, January 1997.

[YOU 98] YOUNG S.J., CHASE L., "Speech Recognition Evaluation: a Review of the U.S. CSR and LVCSR programmes", *Computer Speech & Language*, vol. 12, no. 4, p. 263-279, October 1998.

[ZUE 89] ZUE V., GLASS J., PHILLIPS M., SENEFF S., "The MIT Summit Speech Recognition System: A Progress Report", *Proc. DARPA Speech & Natural Language Workshop*, p. 179189, Philadelphia, February 1989.

Chapter 8

Language Identification

8.1. Introduction

When listening to their native language, speech and hearing-enabled humans can identify the language being spoken immediately. For non-native but familiar languages, we are able to do almost as well. This is the capacity we would like automatic devices to share. Human capabilities are certainly impressive on short speech fragments, however the number of native and familiar languages generally remains very limited compared to the number of distinct spoken languages in the world. Furthermore the risk of error, even if low, is not negligible.

Technically speaking, language identification (LId) involves identifying the language being spoken by an unknown speaker using a given speech sample. Automatic LId appears as one of the expected capacities of "smart" speech-to-text or speech-to-speech systems in a multilingual context. Automatic LId raises many challenging research issues which contribute to progress towards multilingual speech technologies [SCH 06, NAV 06, GEO 04, ZIS 99]. Major questions remain in the domain of automatic language identification. What levels of information contribute most effectively to high accuracy language identification? What are the best acoustic parameters to catch language-specific information? Which algorithms, methods and approaches are the most promising for LId? Which language-specific resources are required for spoken language modeling? What is the correlation between signal duration and the confidence limits of language identification? Related questions of interest also concern human performance: how do humans proceed in identifying languages? Can insight into human cognition be helpful for automatic devices and

---

Chapter written by Martine ADDA-DECKER.

vice versa? Another important question concerns the definition of what is to be considered a language. Beyond the LId issue itself, automatic processing of speech from multiple languages produces baseline resources which enable or enhance large scale comparative studies in phonetics, phonemics and prosody. Although this is not the main goal of LId research, it is an important interdisciplinary byproduct, with the potential of fruitful interactions between researchers in linguistics and speech scientists (e.g. [GEN 07]). For spoken minority languages that are only scarcely described and documented, automatic processing tools may help linguists to elaborate phonemic, morphological and text processing systems.

Automatic LId has been an active research domain for about 30 years, with the pioneering works of Leonard and Doddington (1974). These relied on an acoustic filter-bank approach, and with the work of House and Neuburg (1977). The former relied on an acoustic filter-bank approach [LEO 74], whereas the latter made the first contribution to LId using language-specific phonotactic constraints [HOU 77]. Different sources of information are known to contribute to human language identification, the most important ones being acoustic, phonetic, phonemic and phonotactic levels, but prosody, lexical and morphological knowledge should also be included. These levels are not of the same importance to LId, nor are they equally easy to capture in computationally-tractable models. Acoustic-phonemic and phonotactic approaches have taken advantage of decades of research, first by linguists to describe languages using compact phonetic and phonemic systems [IPA 99, LAD 96, SCH 97, VAL 99], and, more recently, by computer speech scientists developing phone models for automatic speech recognition with appropriate language resources. Hence, in the early years of LId research, acoustic-phonemic and phonotactic modeling became the most popular approach to LId [LAM 94, LAM 95, DAL 96, ZIS 94, ZIS 96a]. Other sources of information, such as prosody or morphology, can be used as a complementary rather than a stand-alone approach.

Until recently, research in speech processing has primarily addressed the world's major languages, for which standardized writing systems exist. For these, the prolific production of various types of language resources already allows for language-specific modeling beyond the mere acoustic level. An important issue is the seamless extension of automatic LId devices to additional languages. The relative paucity of language-specific resources for most spoken languages has motivated research into acoustic-based approaches, as the lack of resources restricts the range of LId modeling options. Advances in these approaches have included both the bootstrapping of LId devices to more languages and a reduction in the related costs in terms of human effort. However, it is worth noting that there are widespread efforts to build language resources in an increasing number of spoken languages, so as to enable a variety of multilingual speech technologies, including automatic transcription and indexing of audio documents, speech translation and synthesis. Research into multilingual speech processing has been supported by the European Commission for almost 30 years now, as dealing with linguistic diversity (there are more than 20 official languages now) is one of Europe's greatest challenges [MAR 98, MAR 05]. Resources are collected and distributed by the European Language Resource Distribution Agency (ELRA/ELDA) [ELD]. In the USA the Defense Advanced Research Project Agency (DARPA) has been massively encouraging multilingual research for more than 15 years and the Linguistic Data Consortium (LDC) now has an impressive catalog of multilingual resources (speech, text and lexica) in at least 80 languages [LDC 07].

Given the progress achieved in automatic LId, there is a growing interest in more subtle LId problems, such as dialect and accent identification [ZIS 96b, KUM 96], as witnessed by recent studies on dialectical forms of Mandarin, Spanish, Arabic, different South African languages and regional accents of English and French [TOR 04, BAR 99, NIE 06]. However, in multilingual contexts, speakers do not necessarily communicate in their native languages. A speech signal may include code switching, code mixing and possibly non-native speech. The definition of language identification, as stated at the beginning of this chapter, may then become more complex: the speech signal conveys information about the language being spoken and the speaker's native language. Should LId systems only identify the language being spoken or should they also be able to give some information about the speaker's accent (related to his/her native language)? Until recently LId has mainly addressed the problem of identifying a spoken language, where speakers use their native language for communication. Research in automatic language identification may also include the challenges of non-native speech to increase the usability of speech systems in multilingual environments [TEI 97, WAN 99, WIT 99, COM 01, BAR 07].

A range of applications could benefit from automatic language identification: telephone companies need to quickly identify the language of a foreign caller to route his/her call to operators who speak that language. Automatic language identification could also be useful in governmental intelligence and monitoring applications. For multilingual translation services, systems need to identify the language being spoken prior to translation. Multilingual audio indexing systems may be enhanced, by integrating dialect and accent identification capacities beyond LId itself.

8.2. Language characteristics

The number of distinct spoken languages is very high, even though the exact number is difficult to establish. Depending on the source the number of distinct languages is estimated to vary between 4,000 and 8,000 [CRY 97, COM 90, GOR 05, KIR 06, LAV]. The reasons for this lack of an exact figure stems from variable definitions of what should be counted as a language. On a diachronic axis, there is the question of living vs extinct languages [HOM 06]. Latin or ancient Greek, although extinct in the sense that there are no longer native speakers, remain alive through educational usage, as well as their literary and historical legacy. An important number of minority languages that are still used, however, are in real danger of extinction today. From the synchrony viewpoint, the distinction between language and dialect is not clear-cut. A living language can be considered as a shared and collectively agreed code or competence [CHO 77] of a community, a population or a social group at a given time and place. A spoken language then corresponds to the oral performance of a set of speakers issued from such a population, community or group. Using this definition of spoken language, the number of existing languages obviously depends on the granularity applied to achieve homogenous populations, as well as time and space divisions. Dialects of a given language are supposed to imply, beyond potential pronunciation shifts, some changes in lexical and grammatical levels. However, they are supposed to retain a mutual intelligibility. Different language/dialect classifications help to explain differences in enumerating the number of languages.

Whatever the exact number of distinct spoken languages, the vast majority of the world's population speaks a very limited set of them. Zipf's law [ZIP 49], which applies to word occurrences in large corpora, also seems to apply to language populations: about 95% of all the speakers make use of only 5% of the world's languages. Another important fact concerning spoken language technologies in general and automatic language identification in particular is that only 5-10% of the world's spoken languages have a corresponding writing system [DAN 96]: most languages are exclusively spoken. This point is also worth bearing in mind when discussing different approaches to LId that rely on high-level language-specific resources. Table 8.1 shows the most widely spoken languages in the world, ranked by the number of native speakers [CRY 93]. The numbers of official language speakers

Language	# L1		# L1+L2		Family
	Rank	Speakers (M)	Rank	Speakers (M)	
Chinese	1	1,000	2	1,000	Sino-Tibetan
English	2	350	1	1,400	Indo-European
Spanish	3	250	4	280	Indo-European
Hindi	4	200	3	700	Indo-European
Arabic	5	150	7	170	Afro-Asiatic
Bengali	6	150	10	150	Indo-European
Russian	7	150	5	270	Indo-European
Portuguese	8	135	8	160	Indo-European
Japanese	9	120	10	120	Japanese
German	10	100	11	100	Indo-European
French	11	70	6	220	Indo-European

Table 8.1. *Most frequent languages ranked according to the number of native speakers (L1), including secondary speakers (L1+L2), expressed in millions (M), after [CRY 93]*

(L1+L2) are also given. It is interesting to note that Chinese (including Mandarin, Xiang, Hakka, Gan and Minbei) ranks first due to the huge number of native speakers, whereas English becomes number one if official language populations are added. French, whilst holding rank 11 for native speakers, progresses to rank 6 when secondary speakers are included. For Bengali, the opposite tendency is observed. The field of historical linguistics aims at organizing this set of inventoried languages into family trees according to their genetic relatedness. Figure 8.1 represents the language family tree for the 11 languages included in the first LId corpus collected by the Center for Spoken Language Understanding (CSLU) at the Oregon Graduate Institute (OGI) [MUT 93]. The Indo-European branch is the family with the largest number of speakers. Spoken languages, although sharing common features which are

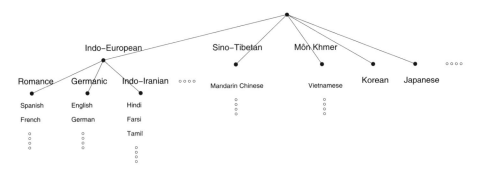

Figure 8.1. *Language family tree for the 11 languages in the first LId telephone speech corpus collected by CSLU-OGI in the early 1990s. Japanese and Korean are isolated in specific families*

exploited to achieve genetic classifications, may vary significantly in their surface forms. This qualitative observation raises the question of how to best quantify these perceived differences. The sound structure of a language can be described at different levels in terms of phonetics [LAD 96, IPA 99], phonemics [IPA 99] and prosody [HIR 99]. Whereas *phonetics* aims at an objective description of sound acoustics with a focus on linguistically-relevant aspects, *phonemics* focuses on the functional and linguistically distinctive roles of abstract sound representations in a language system. *Prosody* deals with fundamental frequency, energy and duration, tones, rhythmic patterns, intonation and prominence. When studied as mere physical events, sounds are referred to as phones, whereas phonemic sounds are termed phonemes. The IPA [IPA 99] (International Phonetic Alphabet), which aims at providing a universally agreed system of notation for the spoken sounds of the world's languages, lists more than 100 distinct elementary phonetic symbol codes (without considering diacritics and diacritical combinations). IPA symbols (or derived alphabets such as SAMPA [WEL 97], Arpabet [SHO 80], Worldbet [HIE 94]) are very popular for the acoustic-phonetic modeling of speech.

A comprehensive description of phonetic alphabet symbols used by linguists and speech scientists to record the sounds of the world's languages can be found in [PUL 96]. Phonemic inventories as well as corresponding phonetic realizations certainly reflect language-specific cues. Even hesitation segments carry some language-specific information [VAS 05]. An interesting question concerns the efficient use of rare, but salient segment information [HOM 99], e.g. clicks in African languages. Language-specific inventories generally range from 20 to 60 symbols. For example German has twice as many phonemic symbols as Spanish. The inventory size may significantly increase when tonal or gemination information is included for the concerned languages (e.g. Mandarin for tones, Italian for geminates). Consonant and vowel distributions, as well as co-occurrences of consonants and vowels are highly language-specific, and phonotactic constraints are known to be very important to identify languages and language styles [HOU 77, DEL 65, MAR 13]. Even simple acoustic measures, such as long-term spectra, exhibit differences among languages which might be related to supralaryngeal settings. Syllabic skeletons vary between languages and language-specific prosodic contours are among the earliest acquisitions in an infant's babbling [LEV 91, HAL 91]. Research into language typology aims at extracting language universals and at organizing language differences [GRE 78, RUH 05, VAL 99].

Depending on the languages under consideration, differences in acoustic surface forms spread in variable proportions over the different levels of information. In general, the largest discrepancies can be observed at the lexical level. Less abstract levels can feature differences that are more or less easily detectable depending on the examined language characteristics. The list of languages selected for the classification task influences the identification and detection capacities of humans and automatic devices. To illustrate typical differences between the acoustic signal of spoken languages, Figure 8.2 compares short speech excerpts from a Romance and a Germanic language (French vs Luxembourgish). Language-specific differences may be observed at different levels of the information representation. While the respective phonemic inventories share many symbols, there are important differences, especially in vowel sets: French has nasal vowels that are not part of the Luxembourgish native inventory, whereas the latter (like English) makes extensive use of diphthongs, which do not belong in the French inventory. Diphthongs entail a change of color in the corresponding vocalic segments, whereas the formant frequencies of French vowels tend to remain very stable. Differences in the distributions of phones and phone sequences are also illustrated in Figure 8.2. The example shows that French has a large proportion of CV syllables, whereas complex syllable structures are common in Luxembourgish and other Germanic languages, like English or German. Even if we are not familiar with both languages, human listeners can easily detect differences in their sound structures. However, distinguishing between two close languages of the same family, such as Italian and Spanish, can be much more complex, even for listeners acquainted with both languages [BOU 04].

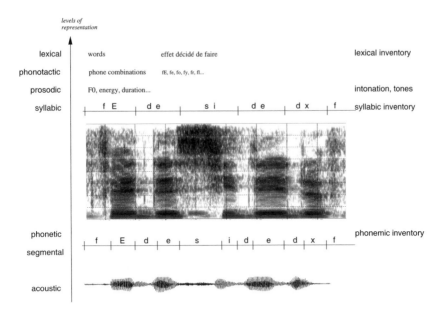

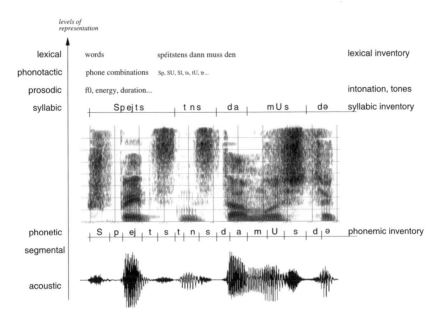

Figure 8.2. *Speech encodes language-specific information at different levels of representation. The first example shows an excerpt of French (Romance language) read speech (...*en effet décidé de faire...* (eng. ...*actually decided to do...*)). The second example displays a Luxembourgish (Germanic language) speech sample (*spéitstens da(nn) muss den... (eng.* by then at the latest it needs...*)) extracted from radio news*

8.3. Language identification by humans

In this section, we present a short overview of human language identification competence. The underlying questions we address are how do humans identify a language, and how important in this are the different levels of information (acoustic, phonetic, phonemic, phonotactic, prosodic, lexical, etc.)? Different identification strategies are probably used depending on the degree of knowledge and of exposure to the languages to be identified by the human listener. The "distance" between languages (in terms of phonetic realizations, phonemic inventories, of syllabic structures, of the use of tonal information, etc.) is certainly another important parameter.

Whereas newborn babies are considered universal phoneticians, they rapidly specialize towards their native language in the first months of their lives, and by the end of the first year, native language characteristics (vowels, consonants, syllables, prosody) are clearly established [DEB 91] in both perception and production. As far as adult capacities are concerned, perceptual experiments have been conducted by many researchers to evaluate human performances in identifying languages [RAM 99, MAD 02]. Different experimental setups can be envisioned: language discrimination corresponding to ABX classifications or language identification corresponding to a necessary choice among a closed set of K languages. Original or modified speech stimuli can be used depending on whether the question of interest is to assess the intrinsic difficulty of discrimination between languages (respectively classifying languages) or, in the case of modified speech, to measure the relative importance of some type of information encoded in the acoustic signal (e.g. prosodic information by filtering out the phonetic content) [RAM 99, MOR 99]. After investigating the relative importance of different information levels, interesting results have been produced by J. Navrátil [NAV 01] on five languages (English, German, French, Mandarin and Japanese). Three series of perceptual tests were conducted: 1) original stimuli, 2) randomly concatenated syllables extracted from the original stimuli and 3) filtered speech with a flattened spectral shape preserving only the $f0$ contour (see Table 8.2). In this experiment,

Test	English	German	French	Mandarin	Japanese	Average
original (3s)	100.0	98.7	98.7	88.7	81.7	93.6
shuffled syll (6s)	98.7	79.7	79.1	57.7	54.6	73.9
f0 contour (6s)	34.3	34.3	69.4	65.9	45.3	49.4

Table 8.2. *Language classification accuracy (%) using different speech stimuli from telephone conversations: original (duration 3s), shuffled syll (duration 6s) corresponds to a random concatenation of unmodified syllables, f0 contour (6s) corresponds to filtered speech keeping f0 information unmodified (after Navrátil 2001)*

with a relatively small number of five languages mainly stemming from different families, identification results were very high (the average of 93.6% on 3 second stimuli increases up to 96% for 6 second excerpts). Results were highest with languages for which listeners had the largest background knowledge. The second condition ensured that segmental information was unmodified and most of the phonotactic information was preserved, but the meaning and most of the prosodic information was lost. This loss of information was responsible for a significant decrease in identification accuracy (22.1% absolute). Whereas the loss was almost negligible for English (the language most familiar to all listeners), the performance decrease was particularly high for languages from the Asian area, which were least familiar to most listeners. These results suggest that the information of language identity may be highly redundant for well known languages, as a partial loss had no major impact. This redundancy also makes it possible to perform reasonably well on less known languages. However, in degraded conditions, subtraction of part of the language identity-related information may entail dramatic losses of performance. This was particularly true in the last condition, where all information, except the $f0$ contour, was filtered out. Here, English was not better identified than German, and both were significantly worse than French and Mandarin. The prosodic contours of both of these languages may prove very informative for LId given the particular perceptual test configuration.

Similar perceptual experiments with French native listeners using unmodified short stimuli (2s) of broadcast news speech from eight languages (English, German, French, Italian, Spanish, Portuguese, Arabic, Mandarin) gave identification results of about 85% [ADD 03], which must be compared to the 93% obtained in the previous experiment with five languages. The stimuli here were significantly shorter and the number of languages higher. Whereas results on Mandarin were almost perfect, the results for the three non-French Romance languages provided a low average accuracy level of only 78%, even though French listeners were more familiar with the Romance languages (without necessarily practicing them) than with Mandarin. These results highlight that perceptual identification results must be examined with respect to the chosen language set: this is an important parameter when interpreting perceptual (and possibly also automatic) identification performances.

The experimental results described in this section show that the information associated with language identity is not exclusively encoded at one information level, but across several different levels, each providing partly redundant information. This hypothesis suggests that different approaches, models and architectures could be used for automatic LId, depending on their reliance on these different levels.

8.4. Language identification by machines

Automatic language identification is the process of using a computer system to identify the language of a spoken speech sample. Despite this very simple formulation, the process may become complex, as the identity of a spoken language may be blurred by code mixing, code switching, dialects [BAR 99], regional or foreign [BOU 06, FLE 03] accents, or other variations of different languages. Language differences are only part of the observed differences between speakers, uttered messages and environmental conditions. In the following sections, we briefly describe the major LId tasks, their corresponding performance measures, and give some pointers regarding evaluation.

8.4.1. *LId tasks*

LId can be considered using different task setups for varying academic and applicative interest, corresponding to different classification tasks [NIS 03]. Traditionally the LId problem has been addressed as a *closed set identification* task, involving identifying a speech input as corresponding to one language among a collection of K *a priori* given languages (Figure 8.3, left). A collection of K language-dependent models is thus required. Whereas this condition is certainly of scientific interest, the closed set assumption is too restrictive for most real-life applications. Another language recognition setup is the *open set detection/verification* task (Figure 8.3, right). The system's task is to decide whether or not the signal corresponds to the target language L. Speech input may come from an open set of languages, and may not necessarily belong to one of the modeled target languages. This task corresponds to selective language filtering. A language detection system can take any language as input and ideally outputs NO for all languages except for the target language L. In addition to a model for the target L, there is generally a complementary model $\bar{L}$, also known as the universal background model (UBM). The detection task matches some of the real application needs better (e.g. language filtering in multilingual audio streams). The most general task, which combines the positive features of the two preceding ones, corresponds to an *open-set identification* or multi-target detection [SIN 04]. This involves either rejecting a speech input, if it stems from an unknown language, or otherwise, identifying it as one of the multiple known languages (multi-target). Open-set identification can be implemented as a closed set identification (producing L^*) followed by a language detection system (producing YES or NO for L^*) or as K detection systems in parallel [SIN 04][1].

8.4.2. *Performance measures*

Depending on the language recognition task, different evaluation measures may apply. For the closed set identification task, the errors correspond to substitutions, and

---

1. Throughout this chapter, if not specified otherwise, LId stands for the general problem of language recognition, either identification or detection.

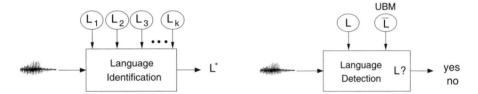

Figure 8.3. *Schematic representation of a language recognition system, which may be implemented either as language identification (left) or as language detection (right)*

LId error rates can simply be measured as average substitution rates. For detection tasks, the situation is slightly more complicated: a speech signal from a target language can be rejected whilst a stimulus for an unknown language can be falsely identified as the target. Hence, performance measures for language detection are generally given as equal error rates (EER), equalizing the contribution of the two types of errors: false alarm and miss rates. This measure corresponds to a specific operating point in the Receiver Operating Characteristics (ROC) curve or equivalently, the Detection Error Trade-Off (DET) curve [MAR 97]. ROC/DET curves give the full set of operating points, i.e. the rejection rate as a function of the acceptance rate. Examples of LId DET curves are given in subsequent sections (Figures 8.9 and 8.13)[2]. Comparing the problems of automatic language to those of automatic speaker recognition, it is worth noting that distinct identities can be uniquely associated with each speaker (at least in ideal conditions), whereas spoken languages may correspond to more or less clearly defined classes, given first the difficulty of defining exact language contours and next the potential bilingual and more generally polyglot capacities of speakers. Independent of the recognition task and from the chosen approach, the decision accuracy largely depends on the stimulus length.

8.4.3. *Evaluation*

Whilst numerous ARPA/NIST evaluation campaigns were run in automatic speech transcription and speaker recognition, only a small number of language recognition evaluations have been organized by NIST (National Institute of Standards and Technology). The first evaluations took place in 1996, followed more recently by other campaigns in 2003 and 2005 [MAR 03, NIS 03, MAR 06b]. NIST is currently scheduling LId evaluations every other year. The evaluation task corresponds to the open set detection/verification task. Figure 8.4 summarizes the objectives of LRE evaluation together with multilingual speech collection efforts in the US. The primary

---

2. Many methodological and technical aspects of automatic language recognition are very similar to those of automatic speaker recognition. For details the interested reader is referred to Chapter 9 in this book.

speech data for these evaluations was the multilanguage CALLFRIEND corpus. This is composed of telephone conversations by native speakers from 12 languages collected in North America. The open test corpus also included conversations in Russian, which is not among the 12 target languages. Significant progress in detection performance results was observed between 1996 and 2003. Since then progress has been achieved without significant evaluation focus for seven years; G. Doddington joked that the *right action to encourage progress is to ignore the problem*. Naturally, speaker recognition evaluations have taken place in the mean time, and the progress achieved has proved to be very beneficial for LId.

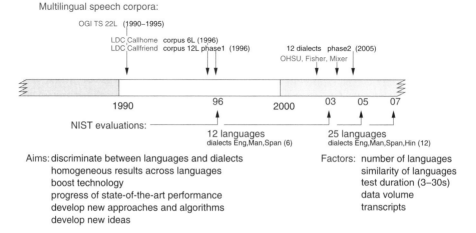

Figure 8.4. *Chronological representation of multilingual data collection efforts in the US and of language recognition evaluations (NIST LRE). The aims of the language recognition evaluation campaigns are recalled and major factors impacting LId performance are listed*

Most successful approaches to automatic language identification rely on a statistical modeling framework that requires appropriate observation sets for model estimation. Multilingual language resource collections are thus vital to LId research.

8.5. LId resources

Until the early 1990s, research in LId suffered from a lack of common, public-domain multilingual speech corpora to evaluate and compare systems, as well as their underlying models and approaches. In the following section, we describe the main efforts employed for collecting multilingual speech corpora to support research in the domain of automatic LId. Although Europe and Member States are actively involved in multilingual language resource collection efforts [MAR 05], coordinated collections dedicated to LId research come mainly from the USA.

OGI_TS 11L and 22L: efforts to collect large-scale multilingual speech corpora began in the early 1990s. The first major multilingual speech corpus for LId was designed and collected by Y. Muthusamy and colleagues [MUT 92] of the former CSLU-OGI institute, which is now known as the OHSU-OGI School of Science and Engineering. The public availability of the OGI Multilanguage Telephone Speech Corpus (OGI_TS) stimulated interest in the field and boosted the development of LId specific approaches. The first OGI_TS corpus contained two hours of speech per language for 11 languages (see Figure 8.1). The speech was collected via domestic telephone calls from native speakers established for years in the USA. For each speaker, read and spontaneous speech was recorded; spontaneous speech took the form of monologue answers to a selection of predefined every day-life questions. Phonetic transcriptions were manually produced for subsets of the recorded speech in 6 languages. Language-dependent acoustic-phonetic models can hence be trained for these languages and many researchers developed LId systems based on acoustic-phonetic decoding, as well as on phonotactic approaches [ZIS 96a, KWA 95]. In the following years OGI extended the multilingual database to twenty-two languages [LAN 95], adding Italian, Portuguese, Swedish, Polish, Czech, Hungarian, Russian, Arabic, Swahili, Cantonese and Malaysian.

The data distributed for the first LId evaluations organized by the National Institute of Standards and Technology (NIST) in 1996 included the OGI_TS corpus. It was also distributed by the Linguistic Data Consortium (LDC) [LDC 07]. The impact of the OGI corpus was most significant for LId research during the 1990s. Later corpora tried to overcome several shortcomings. In particular, later multilingual speech collections aimed at gathering larger amounts of data with native speakers, either through international calls connecting US residents with relatives living in their home countries, or by recording calls locally. However, the OGI corpus may become of renewed interest for research on accented native speech, where the focus goes to the impact of long-time L2 practice on L1.

CallHome 6L: released in the later 1990s by the LDC, the CALLHOME corpus was originally collected to support conversational speech recognition in multiple languages: American English, Arabic (Egyptian), Mandarin, Latin American Spanish, German and Japanese. It has also been used for different NIST language recognition evaluations. CALLHOME consists of 200 international telephone calls per language, each call lasting about half an hour. Along with these 100 hours of speech per language, transcripts have been produced for subsets of the data together with corresponding pronunciation dictionaries, part-of-speech information as well as frequency counts. As CALLHOME conversations connect familiar speakers with their home country, English L2 effects and code switching should be limited. However for each language, differences in telephone networks might leave a signature in the acoustic signal, biasing LId by partial telephone network identification.

CallFriend 12L: the CALLFRIEND collections have specifically been designed and collected for LId research and NIST language recognition evaluations (LRE). They comprise two parts corresponding to different recording phases with different objectives. Whilst the first phase collection [CAL 96] conducted around 1995 was designed for standard LId tasks, the more recent collection phase initiated in 2004 aims at collecting data for more subtle problems, such as dialect and regional variety identification. CALLFRIEND (phase 1) includes 12 languages with 100 half-hour continental telephone conversations per language. Beyond the six CALLHOME languages (American English, Egyptian-Arabic, Mandarin, Latin American Spanish, German and Japanese), CALLFRIEND also includes Canadian French, Farsi, Hindi, Korean, Tamil and Vietnamese. The second phase collection consists of shorter telephone calls (10 minutes per call) recorded as before in the USA and includes more languages (Georgian, Italian, Pundjabi, Russian, Aceh, Amharic, Bengali, Burmese, Chechen, Guarani, Khmer, Lao, Tagalog, Thai, Tigrigna, Urdu, Uzbek) and so-called "dialects" for American English (separated by continental origins: US, Britain, India, Australia), for Arabic, Hindustani and Chinese dialects of Mandarin, Shanghai Wu and Min. This second phase collection, named LVDID (Language, Variety and Dialect Identification) [CAL2] data are to be released by LDC after their use in the corresponding NIST LRE evaluations.

SpeechDat: SPEECHDAT corresponds to a series of multilingual speech data collection projects funded by the European Commission for the past decade. The aim of SPEECHDAT is to enable the development of multilingual teleservices and speech interfaces. Countries involved include Belgium, Denmark, Finland, France, Germany, Greece, Italy, the Netherlands, Norway, Poland, Portugal, Spain, Sweden, UK, etc. Depending on the collections, focus is put either on acoustic conditions: fixed or mobile telephones, in-vehicle recording SPEECHDAT-M, SPEECHDAT-CAR [SPDM], or on geographical and/or linguistic criteria (SPEECHDAT-E for Eastern European languages, ORIENTEL for Mediterranean and Middle East, LILA for languages of the Asian Pacific area, SALA for Latin America). The ORIENTEL project focused on the development of language resources for speech-based telephony applications across the area between Morocco and the Gulf States. Within the SPEECHDAT project series, the projects are launched by industrial consortia including European and international groups such as Siemens, Nokia, IBM, Sony, Nuance, Scansoft, Microsoft, etc. The speech databases are specified concerning content (application words and phrases, phonetic rich words and sentences), coverage concerning dialectal regions, speaker age and gender and recording devices. Information on SPEECHDAT projects and corpora can be found on the following WEB site: http://www.speechdat.org/ and in the proceedings of the LREC (Linguistic Resources and Evaluation Conference) conferences, which are held every second year since the first conference in Granada, Spain in 1998. They have occasionally been used for LId experiments [CAS 98, MAT 05].

GlobalPhone: the GLOBALPHONE project was launched in the mid-1990s at Karlsruhe University by A. Waibel and T. Schultz [SCH 02]. The goal was to produce similar resources in a large number of languages to enable multilingual speech technologies: speech-to-text, speaker and language recognition. Within this ambitious project, text and speech from the world's major languages were collected, with a focus on European and Eastern European languages. GLOBALPHONE languages comprise Croatian, Czech, Polish, Russian, Spanish, French, German, Swedish, Turkish, Arabic, Mandarin, Wu (Shanghai), Tamil, Thai, Japanese, Korean and Brazilian Portuguese. Many different writing systems were experienced and had to be transliterated into a common machine readable format. For each language, news texts were collected via the WEB and recruited speakers read a text composed of about 100 sentences. 100 adult speakers were recorded in each language using close-talking microphones, which resulted in a total volume of 300 hours. Although the GLOBALPHONE corpus seems less appropriate for the development of telephone-speech based LId applications, it may contribute to studying language specificities using controlled read speech, which is generally more carefully articulated than spontaneous conversations.

Fisher: the FISHER corpus collection started in 2002 for the DARPA EARS program [EAR 05], aiming at rich speech transcription and metadata extraction (MDE) in English, Mandarin, Spanish and Arabic. FISHER was designed to collect a large number of short calls from different regions, with speakers dealing with several of a large set of predefined subjects in order to broaden the vocabulary covered by the corpus. The collection was coordinated by a robot operator using a dozen telephone lines in parallel. For the English language, the recorded Fisher subjects indicated their region of origin using, if appropriate, the major dialectal regions [LAB 04] (North, Midland, South, West) or indicating their accent or country of origin. More than 15,000 calls totaling close to 3,000 hours of speech were thus collected from English speakers and organized according to regional origins. Similarly for Mandarin Chinese, Spanish and Levantine colloquial Arabic, about 300 hours of speech were collected for different dialects and thousands of speakers. Beyond their use for text-to-speech and MDE research, these corpora are available for LId research.

Mixer: the MIXER collection was originally designed for speaker recognition research, with a focus on forensic needs. A new feature here is to collect bilingual and multilingual speakers to measure speaker recognition performance independently of the language being spoken. Channel-independence, which is classically seen as independence towards telephone and recording conditions, includes here the language channel. The MIXER collection started in 2003 and totaled about 5000 speakers in English, but also Arabic, Mandarin, Russian and Spanish by 2005. Each speaker was encouraged to accomplish a large number of calls, either in English or in one of his/her other performing languages.

Corpus	Date	#L	Vol. (h)	Speech type	Record
human talks to automated telephone service					
OGI-11*	1992	11	k*10	read & free monologues	domestic tel.
OGI-22*	1995	22	k*10	read & free monologues	domestic tel.
LIMSI-Ideal	1995	4	k*100	read & free monologues	dom./internat.
human reads prompted texts					
GlobalPhone◇	1997	30	k*100	read	close-talk
human-human telephone conversations					
CallHome*	1996	6	k*100	spont. dialogs	internat. tel.
Callfriend*	1996	12	k*100	spont. dialogs	domestic tel.
Fisher*	2002	4	k*1000	spont. dialogs (regional accents)	domestic tel.
Mixer*	2004	3	k*100	spont. dialogs (bilingual speakers)	domestic tel.

Table 8.3. *Examples of (*publically available via LDC, ◇publically available via ELDA) multilingual speech corpora. For each corpus, the number of languages is given in the #L. column. Dates approximately correspond to the start of data collections and total volumes are indicated by orders of magnitude*

Table 8.3 shows examples of multilingual speech corpora which have been collected since the 1990s for language and multilingual speech recognition. These are, or will become, publicly available via the LDC and ELDA, except the LIMSI-IDEAL corpus which is owned by the French company *France-Télécom*. The LIMSI-IDEAL corpus [LAM 98] has been collected according to similar criteria than the OGI corpus. The latter encompasses a relatively high number of languages with all the callers settled in the US, whereas IDEAL comprises only four European languages (French, English, Spanish and German) with a great number of native speakers both calling either from France or from their native countries. Major dialectal regions have been distinguished in the four countries and the speakers have been balanced according to their gender and regions.

Ever-growing efforts to collect huge amounts of speech corpora in multiple languages highlight the increasing importance of multilingual speech technologies, among which language, dialect and accent identification represent key issues. An ISCA special interest group on speaker and language characterization was launched in 1998 [BON 01]. The Odyssey workshop focuses on research in speaker and language recognition and takes place every other year. Furthermore, papers on spoken language, dialect and accent recognition can be regularly found in major speech processing conferences (Eurospeech/Interspeech, ICSLP/Interspeech, ICASSP).

8.6. LId formulation

The problem of language identification can be approached with the help of mathematical formulation, which makes it possible to decompose the problem into simpler or more focused sub-problems. Let X denote the acoustic evidence (speech) on the basis of which the LId decision is to be taken. Without loss of generality, we can make the assumption that X is a sequence of symbols from a finite alphabet $\mathcal{X}$. With a statistical approach [JEL 98] the LId problem can be stated as follows:

$$L^* = \arg\max_{L \in \mathcal{L}} P(L \mid X) \qquad [8.1]$$

L^* being the identified language, X the symbol sequence of the speech sample, $\mathcal{L}$ the set of potential languages, and $P(L \mid X)$ the probability of language L given X. Applying the well known Bayes' formula of probability theory to equation [8.1], the LId problem can be reformulated as follows:

$$L^* = \arg\max_{L} P(X \mid L)\, P(L) \qquad [8.2]$$

where $P(X|L)$ stands for the probability that X is observed, when language L is spoken and $P(L)$ is the a priori probability of language L. With an equiprobability assumption for the different languages, the formula can be simplified as:

$$L^* = \arg\max_{L} P(X \mid L) \qquad [8.3]$$

Written as such, the formula is not yet of great help, as language L is considered as a whole. It merely evokes an acoustic approach to LId, relying on some language-specific acoustic models. However, depending on multilingual resources available for language-specific model training (see Figure 8.2), various decompositions can be proposed. Different LId approaches then correspond to different ways of decomposing $P(X|L)$ and all through the LId literature, proposed approaches and available language-specific resources are tightly coupled.

Figure 8.5 gives a tentative source-channel model of the LId problem, adapted after the corresponding model of speech recognition proposed by F. Jelinek [JEL 98]. Even though the involved cognitive processes are certainly more intricate and complex than depicted in Figure 8.5, it can be instructive to examine the LId problem within this extended source-channel model. A "thought" is to be encoded in one of various different linguistic surface forms, with variable word choices for a given language and across languages. The left (generation and encoding) side illustrates the different language-dependent formulation choices for a given language-independent message. Using Saussure's terminology [DES 15], the language-independent *signifié* is associated with a language-dependent *signifiant*. For example, given a *signifié* of three apples, if the message is to be implemented by a French *signifiant*, the generated

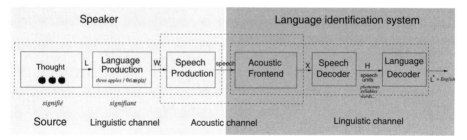

Figure 8.5. *Multi-lingual source-channel model of language*
recognition and multi-lingual speech recognition

word string would be most probably W_{fr}=*trois pommes* with the corresponding phonemic sequence/tʀwa#pɔm/. However in English the *signifiant* would most likely become W_{en}=*three apples* with an underlying phonemic representation/θri#æpəlz/. The right side of Figure 8.5 represents the decoding process with an acoustic front-end converting the acoustic waveform into an acoustic feature stream X which is then processed by a "speech decoder" (acoustic decoder) to produce a hypothesis stream H. The H stream enters the "language decoder", which outputs the most likely language identity and, if desired, the most likely corresponding hypothesis: some unit sequence, which may be of linguistic type (such as phoneme, syllable or word sequences) or of acoustic type (such as acoustic segment or Gaussian label sequences). Various LId implementations have been proposed for the decoding block corresponding to variants of the mathematical formulation of the LId process developed hereafter.

Following the source-channel model and related work on automatic speech and language recognition [GAU 04], $P(X/L)$ can be more generally rewritten as:

$$L^* = \arg\max_{L} \sum_{H} P(X \mid H, L, \Lambda) P(H \mid L) \qquad [8.4]$$

where $P(X \mid H, L, \Lambda)$ corresponds to the speech decoder and $P(H \mid L)$ to the language decoder (in Figure 8.5). Λ is the acoustic model and H a stream of symbols (e.g. phoneme, syllable or word sequences).

The LId problem has thus been formally divided into two blocks. The acoustic decoder can be seen as a speech tokenizer, generating flows of weighted symbols H (either language-dependent or language-independent). The language decoder associates additional weights to the H symbol flow for each language L. The language decision may rely on the language decoder only, the acoustic decoder then merely plays the role of a speech tokenizer considering acoustic and symbolic information simultaneously.

In general, the first term reflecting the acoustic decoder is implemented by approximating the summation over the whole hypothesis space by the best hypothesis, resulting in the one-best H_L^* sequence. This approximation corresponds to:

$$\sum_H P(X \mid H, L, \Lambda) \approx \max_H P(X \mid H, L, \Lambda)$$

which then results in:

$$L^* \approx \arg\max_L \max_H P(X \mid H, L, \Lambda)P(H \mid L) \qquad [8.5]$$

The acoustic decoder may nonetheless contribute to the language-dependent scores, either by weighting via the achieved acoustic likelihoods [GAU 04] or by decoding with language-specific acoustic phone models [LAM 94]. Numerous research contributions aim at minimizing the dependence of the acoustic model (Λ) on the language: it is not required that for each language a specific acoustic model decodes the acoustic input stream X. In the following equation, the speech decoder becomes language-independent, the language-dependent information is captured by the language decoder:

$$L^* \approx \arg\max_L \max_H P(X \mid H, \Lambda)P(H \mid L) \qquad [8.6]$$

Let $H^* = \arg\max_H P(X \mid H, \Lambda)$ be the one-best hypothesis of a language-independent speech decoder, then the LId problem may be rewritten as:

$$L^* \approx \arg\max_L P(H^* \mid L) \qquad [8.7]$$

This formulation of the LId problem globally corresponds to the well known phonotactic approach (see below). Language-independent acoustic models may entail lower phone (or other relevant units) recognition accuracy. It is known that the decoded H stream accuracy correlates with LId accuracy [LAM 95], and the trade-off between language-independent vs language-specific acoustic models is a challenging research issue. In order to take into account the hidden nature of the speech symbols and to increase the amount of information contained in H, limited by the approximation of the sum by the max operator in equation [8.5], Gauvain *et al.* [GAU 04] have recently proposed taking into account hypothesis lattices instead of single one-best hypotheses.

$$L^* = \arg\max_L \sum_{H \in \text{lattice}} P(X \mid H, L, \Lambda)P(H \mid L) \qquad [8.8]$$

The arcs of the lattice are labeled by the decoded speech symbols and may be weighted by the corresponding acoustic likelihoods. A phonotactic lattice-based

approach can then be written as

$$L^* = \arg\max_L \sum_{H \in \text{lattice}} P(H \mid L) \qquad [8.9]$$

The benefits of this approach [GAU 04, SHE 06, ZHU 06] have been demonstrated in terms of LId accuracy and of computational efficiency.

8.7. LId modeling

Beyond language-dependent models, successful LId systems also include effective acoustic feature extraction and language decision modules. In state-of-the art systems [CAM 06, SHE 06, GAU 04, MAT 06], the $\arg\max_L$ operator is implemented, not just as a simple max picking, but with the help of specialized linear or non-linear classifiers. Figure 8.6 shows a block diagram of the major components of an LId system, where the "LId Scoring" block includes both speech and language decoding blocks of Figure 8.5. The acoustic front-end aims at extracting appropriate feature vectors. The decision module (back-end classifier), if appropriately optimized, contributes to significant LId accuracy gains [ZIS 97].

Figure 8.6. *Block diagram of the main modules of an LId system: acoustic front-end, language scoring and a back-end classifier*

If only multilingual audio data are available, language recognition has to rely on mere acoustic properties. Such approaches have successfully been explored using Gaussian mixture models (GMMs) [ZIS 93, TOR 02a, TOR 04, WON 00] and support vector machines (SVMs) [CAM 04, CAM 02]. Additional resources may be required for higher level modeling: phonetic or phonemic labeling and segmentation, transcriptions, pronunciation dictionaries, morphological information, prosodic annotations, etc. Different types of language-specific information can then be modeled ranging from mere acoustics to more sophisticated linguistic levels, including phonemics, information on vowel systems [PAR 99, PEL 00], phonotactics, morphology and prosody. As seen in the previous section on resources, multilingual audio data may be accompanied by orthographic transcriptions or phonetic segmentations (e.g. OGI-TS corpus). These make it possible to train language-specific acoustic phone models or to estimate phonotactic constraints for each language. Standard speech recognition techniques can then be applied to the LId problem. Introducing the highly informative lexical level may result in complex systems, potentially equivalent to multilingual transcription systems. Although this approach certainly guarantees the most reliable identification results, it is at the expense of high development and operating costs.

Acoustic-phonemic and phonotactic modeling approaches raise the question of the phone sets to be used in a multilingual context. For automatic LId using phonotactic constraints, either multiple language-dependent phone sets are used by recognizers in parallel, or a single global phone set [HAZ 97, COR 97], or even a combination of both [ZIS 96a, MAT 99] have been implemented. Defining a single global phone set is an interdisciplinary research issue of its own, which ranges from phonetic and phonemic domains to multilingual speech recognition [IPA 99, ADD 03, ANT 04, SCH 02]. LId system development costs are closely linked to the number of mandatory language-specific resources, hence the success of purely acoustic modeling techniques, such as GMMs, SVMs and unsupervised phonotactic approaches [ZIS 96a, LUN 96].

8.7.1. *Acoustic front-end*

Automatic LId systems generally make use of the same acoustic features as used for speech and speaker recognition, namely MFCC (mel-frequency cepstral coefficients) [DAV 80, CHI 77] or PLP (perceptual linear prediction) features [HER 90, HÖN 05]. Feature vectors (typically 10-15 parameters) are computed at a fixed rate (e.g. 10 ms). First (and second) order derivatives are generally computed from the MFCC or PLP vector flow in order to better take into account the dynamic properties of speech sounds. The dimension of the resulting feature vectors then typically ranges from 20 to 30 parameters and they capture dynamic information of a 50-80 ms time span. Some attempts have been made to develop specific acoustic front-ends for language recognition [DUT 00]. An improved feature set called *shifted delta cepstra* (SDC, see Figure 8.7) [BIE 94] was made

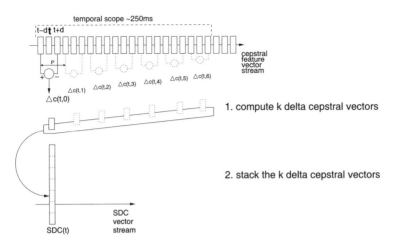

Figure 8.7. *Computation of a SDC (shifted delta cepstral) vector obtained by stacking k=7 consecutive delta cepstral vectors (deltas use +/- d=1 vectors around t), shifted by P=3*

popular by Torres-Carrasquillo *et al.* [TOR 02b] within a Gaussian mixture model approach. SDC vectors are an extension of delta-cepstral coefficients: they consist of stacked delta vectors of typically 7 consecutive deltas. The temporal scope of such vectors are around 250 ms, including information of at least one syllabic unit. Such long-span features allow for an implicit linguistic unit modeling and they happen to be not only language-specific, but also corpus and topic specific, the most frequent syllables being best represented. These features are hence very effective if there is no mismatch between training and test data. Other recent studies to improve feature vectors, based on split temporal context and neural nets are described in [MAT 05].

8.7.2. *Acoustic language-specific modeling*

Purely acoustic approaches to LId only require language-specific audio data for each of the considered languages. The advantage is that no language specific knowledge (generally linguistically informed transcriptions) is needed. Hence LId system development and further extensions to additional languages become straightforward. Early acoustic-based approaches included filterbanks and LPC based approaches [LEO 74, CIM 82] as well as vector quantization-based approaches [FOI 86, SUG 91]. Nowadays, GMMs and SVMs are the most commonly used approaches for acoustic modeling.

Gaussian mixture models

GMMs are the most popular acoustic approach to LId, especially for language detection. A GMM model is estimated for each language and the only prior knowledge to train the language-specific models consists in the language identity of the audio corpus. Figure 8.8 gives a schematic representation of a Gaussian mixture model. The Gaussians, estimated from the acoustic feature vectors produced by the front-end, attempt to model the entire acoustic space. Equation 8.10 gives the acoustic likelihood $P(X \mid L)$ computation of the Gaussian mixture model G based on N Gaussians.

$$P(X \mid L) = \prod_{t=1}^{T} P(x_t \mid L) = \prod_{t=1}^{T} \left(\Sigma_{n=1}^{N} w_n G_n(x_t \mid L) \right) \qquad [8.10]$$

N typically ranges from 64 to 1024. Figure 8.8 illustrates a GMM with $N = 5$ Gaussians. As a general rule, GMMs do not tend to capture temporal dependencies very well, hence the introduction of SDC acoustic features. State of the art performance was achieved using Shifted Delta Cepstra (SDC) feature vectors [TOR 02b]. These excellent language recognition performances [TOR 04, TOR 02b, WON 02] have established GMMs as a major LId approach in addition to the successful PPRLM (parallel phone recognition followed by language modeling) [ZIS 96a].

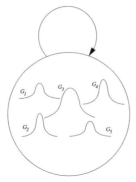

Figure 8.8. *Schematic representation of a Gaussian mixture model with 5 Gaussians representing the acoustic space of a spoken language*

GMM models, generally employed to measure the acoustic adequacy between the input X and the language L, can also be used to act as speech tokenizers [TOR 02a], i.e. converting the acoustic vector stream into a discrete label stream. This is done by replacing each vector x_t of X by the label of the Gaussian which produces the highest contribution in equation 8.10. This then makes it possible to apply PPRLM approaches to the LId problem, replacing the linguistic tokenizing units (phones, syllables) by acoustic vector units.

Support vector machines

Support vector machines (SVM) [VAP 97], a popular tool for discriminative classification [CRI 00, COL 01], have recently been introduced in the area of language recognition [CAM 04]. The assumption here is that two languages are separable by a hyperplane, provided an appropriate acoustic feature space is used for the spoken language representation.

A support vector machine (SVM) is a two-class classifier constructed from sums of kernel functions. The SVM training process aims at modeling the boundary between the two classes as opposed to the generative Gaussian mixture models which represent language-dependent probability densities. SVMs are hence particularly adapted to language detection and verification problems (yes/no decision). Extensions to multi-class classification can be implemented as several two-class SVMs in parallel.

Figure 8.9 compares SVM and GMM results as DET curves [MAR 97] achieved by Campell *et al.* [CAM 04] using NIST 2003 evaluation data (Callfriend corpus, 12 languages). LId improvements correspond to a DET curve shift towards the origin (left, bottom). The curves indicate that SVMs perform slightly worse than GMMs,

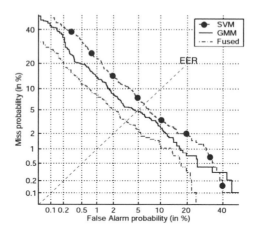

Figure 8.9. *DET curves of LId results achieved on NIST LRE 2003 test data using GMM, SVM and fusion of both (after W. M. Campbell et al. MIT-LL [CAM 04])*

but SVMs carry complementary information, as the fusion of both methods improves the overall performances. Furthermore, as SVMs have been introduced for LId only recently, their potential might not be fully exploited yet and additional improvements could be expected in the future. Beyond SVMs' acoustic classification capacities, SVMs have also served as back-end classifiers of an LId system [WHI 06].

8.7.3. *Parallel phone recognition*

In contrast to acoustic modeling approaches, LId systems based on parallel phone recognition (PPR) require an important amount of language-specific knowledge and resources, namely phone inventories and correspondingly labeled acoustic training data. However, these resources generally exist for the world's major languages, and in particular those for which automatic speech recognition systems have been developed.

In the early 1990s, language-specific phone decoders in parallel were a popular approach to automatic LId [YAN 96b, MUT 93, LAM 94, ZIS 94]. The assumption here is that different languages make use of different sound inventories and, even though part of these sounds are shared among languages, their spectral realizations as well as their occurrences might differ significantly from one language to another. These differences may be taken into account by language-specific sets of phone-based hidden Markov models (HMM) and researchers have taken advantage of the progress accomplished in speech recognition to investigate HMMs for LId [UED 90, NAK 92, LAM 94, ZIS 94]. Figure 8.10 representing a phone-based HMM model illustrates the additional modeling capacities as compared to the acoustic GMM approach: improved temporal modeling (three left-to-right GMM

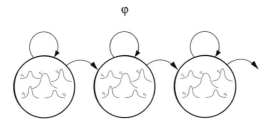

φ

Figure 8.10. *Schematic representation of a phone-based HMM model representing the acoustic realizations of a given phoneme φ. Each state corresponds to a GMM (5 Gaussians per state)*

states); different GMMs for different states and sounds (φ). The number of Gaussians per state remains limited (typically between 16 and 32) as compared to the acoustic GMM approach described before. A language can then be viewed as a source of phonemes, modeled by a fully connected Markov chain. Its higher level structures are approximated by phonotactic constraints. Figure 8.11 then shows a schematic representation of such a parallel phone recognition approach, termed PPR in Zissman's terminology [ZIS 96a]. PPR systems correspond to implementations of equation [8.5] [LAM 94, ZIS 94]. For each language L_i a phone inventory

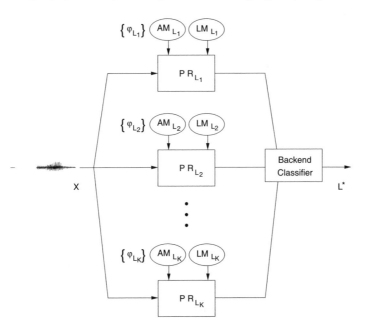

Figure 8.11. *Schematic representation of a LId system based on language-dependent phone recognizers (PR) in parallel: the PPR (parallel phone recognition) approach*

($\{\varphi_{L_i}\}$) is defined and a set of acoustic phone models (AM_{L_i}) is trained from appropriately labeled acoustic training data. Phonotactic language models (LM_{L_i}), generally bigrams or trigrams representing phone co-occurrence specificities, are also estimated from such labeled training data. The likelihood scores resulting from PPR incorporate both acoustic and phonotactic information, the latter contributing to improved automatic phonemic transcriptions. The scores are then used by the back-end classifier to determine the identity of the language being spoken. The PPR approach can be implemented using many variants: the phone sets corresponding to the basic modeling units can be changed. The sets can be extended to include longer units such as syllables [NAG 06] or words [MAT 98]. Articulatory features may be used instead of linguistic units [KIR 02]. Acoustic HMM models may or not depend on gender, phonemic context or channel conditions and the complexity of language model constraints may be differently tuned to fit the amount of training data available. The PPR approach is expensive both in terms of implementation and of computation resources in operating mode. However, it yields excellent identification results, in particular for shorter durations, as acoustic, phonemic and phonotactic information levels jointly contribute to the decision. A serious bottleneck of the approach is the extension to minority languages and spoken varieties for which the required language-specific knowledge about phonemic systems and pronunciations as well as the segmented and labeled audio data are lacking.

To address these shortcomings, a smart modification to the PPR approach [ZIS 94] exploits the fact that phonotactic constraints for one language can be estimated based on *automatically decoded phone streams* using the acoustic models of a different language, even if the quality of the produced phone label streams certainly differ from a labeling achieved by a human phonetician. The advantage is then, that phone recognizers need not be developed for all the target languages. This idea has led to the most successful parallel phone recognition followed by the language modeling (PPRLM) approach, which will be developed in the section on *phonotactic modeling* below.

8.7.4. *Phonotactic modeling*

Similarly to the PPR approach, the phonotactic approach views a language as a source of phonemes (or some other structuring units), modeled by a fully connected Markov chain. The phonotactic approach addresses the LId problem using the formulation of equation [8.7]. In contrast to the PPR approach, the acoustic decoding component no longer influences the language score directly. The acoustic decoder merely serves as a speech tokenizer, producing a discrete symbol sequence H^*. The first automatic LId systems using phonotactic constraints date back to the early 1980s [LI 80]. They exploit the findings of House and Neuburg [HOU 77], who experimented LId with hand-labeled broad phonemic classes corresponding to phonemic transcriptions of texts.

Figure 8.12 gives a simple overview of language identification using a phonotactic approach. Both the language-dependent phonotactic models (LM_{L_i}) estimated from automatically decoded language-specific training data, and the test sequence X to be identified depend on the accuracy of the acoustic model (AM) and phone recognizer (PR) [LAM 95, MAT 05]. The implementation

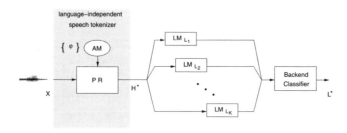

Figure 8.12. *Automatic language identification based on a phonotactic approach (PRLM: phone recognition followed by language modeling)*

of the speech tokenizer raises many challenging research issues, in particular concerning the choice of a "language-independent" symbol inventory $\{\varphi\}$. Such inventories may be limited to broad phonetic classes which are shared among languages [HOU 77, MUT 93], IPA-like inventories [COR 97, ZHU 05, NAV 01] or the union of several language-dependent phone sets. The size of the inventory may then range from ten to several hundred. As a general rule, results increase with the size of the phone inventory $\{\varphi\}$, provided that the amount of available training data is appropriate. Instead of using phone-like units, longer units representing syllables or sub-words have been experimented [ANT 04, ZHU 06, TUC 94, MAR 06a], with the objective of achieving higher decoding accuracy and hence better language identification. However, longer units are generally performing less well. A potential explanation is that the training data remain insufficient, as the number of units drastically increases (thousands) as compared to phone-like units (tens). A set of units may also be automatically determined from a multilingual speech corpus using unsupervised clustering techniques. Whatever the approach and the resulting set of acoustic models, the speech tokenizer tends to best represent the languages included in the training data.

The phonotactic constraints are approximated by n-gram language models (LMs). Standard smoothing techniques have been shown to work well for LId [SIN 03]. The orders of the LMs are generally limited to bigrams and trigrams. Important performance gains are observed for trigrams (30-40% relative) compared to the lower order model. Fourgrams or even higher orders prove to remain globally inefficient in test conditions. Figure 8.13 (left) illustrates performance differences for bigrams and trigrams achieved by the MIT Lincoln Lab [SHE 06] in the NIST 2005 evaluation.

An important improvement to the phonotactic approach consists of replacing the one-best hypothesis H^* of the speech tokenizer by a hypothesis lattice (see equation 8.9) [GAU 04], both for LM training and for the operational phase. This results in more reliable LMs, and more information can be extracted from the unknown test segment for automatic identification. Figure 8.13 (right) shows the achieved gains with a lattice-based approach and PPRLM systems.

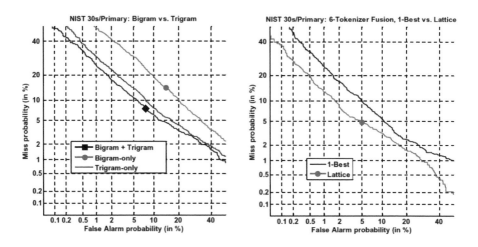

Figure 8.13. *DET curves illustrating improvements to the phonotactic approach. Left: phonotactic constraints implemented as trigrams vs bigrams. Right: standard 1-best versus lattice-based PPRLM approach (Figures after W. Shen et al. MIT-LL [SHE 06])*

Parallel phone recognition followed by language modeling

Instead of progressively enhancing models, methods and techniques to optimize the performance of a given PRLM system, LId results can be improved by running several different PRLM systems in parallel. This corresponds to the most popular approach, PPRLM (parallel phone recognition followed by language modeling) [ZIS 94, YAN 96a, ZIS 96a], represented in Figure 8.14. The same decision rule formulation for PRLM applies (equation [8.7]). However, for multiple recognizers, care must be taken to apply the decision rule across LM scores from different recognizers with bias normalization.

The PPRLM approach is the most efficient *stand-alone* method [SHE 06]. During the LRE05 evaluation, the best MIT-LL subsystem was based on PPRLM with a 4.9% EER (see Figure 8.13 right), whereas the best system (combination of GMM, SVM and PPRLM) was below 3%.

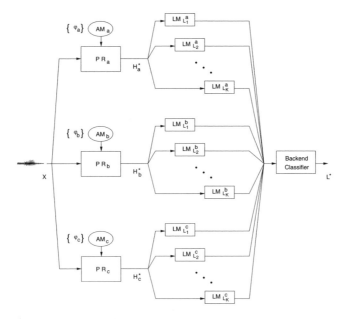

Figure 8.14. *Automatic language identification with a PPRLM approach with 3 phone recognizers (a,b,c) in parallel*

In contrast to the PPR approach, language-specific resources are required only to train the speech tokenizers. An arbitrary large number of languages can potentially be identified, exclusively relying on audio data to train the language-specific models. To avoid using prior knowledge corresponding to linguistic unit sequences, Torres-Carrasquillo and colleagues [TOR 02a, TOR 02b] proposed to use H^* tokens obtained via GMMs, resulting in a "phonotactic" approach requiring exclusively language-specific audio data.

Prosodic modeling

Beyond different sound inventories and sound combinations, different languages can also be characterized by different intonation or rhythm patterns, for which the measurable acoustic correlates are: fundamental frequency $f0$ variations, energy and duration [RAM 99, FAR 01, ROU 05]. The importance of prosodic information in recognizing speech or in discriminating between languages has long been acknowledged. In section 8.3, we saw that humans were able to identify languages, although with limited accuracy, based on prosodic information [NAV 01, RAM 99]. However, this information is often ignored in LId systems. Some early studies including prosodic information to automatic LId examined pitch variation, duration and syllabic rate with marginal success [MUT 93, THY 96, HAZ 93, HAZ 97].

NIST evaluations indicated that prosodic information was largely ignored by most competing systems. However, renewed interest in prosodic cues for LId can be found in recent studies on prosody [ADA 03, OBU 05] for LId. In [ADA 03], the authors make use of temporal trajectories of prosodic cues, i.e. fundamental frequency and short-term energy to segment and label the speech signal into a small set of discrete units. A prosodic labeling system is derived which distinguishes 10 different classes, depending on segment duration (2 classes) combined with either classes of falling and rising $f0$ and energy (4 combinations) for voiced segments and an unvoiced class. The number of modeled classes thus remains much lower than it does in more traditional phone-based or GMM approaches. The approach is evaluated using the NIST Language Identification task, and has achieved promising results. Even though they were significantly worse than more standard approaches, this does demonstrate in principle at least that the prosody dynamics can capture language-dependent information.

System combination and further LId extensions

System combination, exploited for years within the PPRLM approach emerges as a standard practice for present LId systems. The fusion of heterogenous systems allows them to cope efficiently with individual system limitations. Whereas the best NIST LRE 1996 systems made use of PPRLM approaches, in 2003 and 2005 the trend was to combine different approaches in parallel [MAR 03, MAR 06b]. In 2003, the best performance ($< 3\%$ EER on 30 second segments) was achieved by fusing the scores of phonotactic PPRLM, acoustic GMM (Gaussian mixture models) and discriminative SVM (support vector machines) based systems, with a decoding time of roughly 15 times real-time. Therefore, gains were mainly due to an optimal combination of different known approaches, rather than very significant progress for one specific approach. Heterogenous system combinations recall the hypothesis of multiple partly redundant information levels with respect to language identification previously formulated for human perception.

As for prosodic modeling, many different attempts have been made to extend phone-based approaches to include syllable-level knowledge to LId and dialect identification [LI 94, BER 98, BER 99, MAR 06a, ANT 04, ZHU 06]. Adopted approaches range from syllabic spectral features [LI 94], to syllabotactic modeling together with multilingual syllabic inventories [ANT 04, ZHU 06], as well as syllable-like phone triplets [MAR 06a]. Whereas these approaches are individually less efficient than optimized phone-based PPRLM systems, their fusion with existing systems generally increases performance gains. Including lexical information into LId systems [HIE 96, MEN 96, SCH 96, MAT 98] has also been proposed. However, such approaches are extremely resource-greedy and are difficult to extend to any additional spoken language.

8.7.5. *Back-end optimization*

LId results can be significantly improved by an optimized back-end decision module. The simple maximum picking used in the earlier PPRLM systems can be replaced by linear or non-linear combinations of different scores, generally normalized log-likelihood (LLR) scores, stemming from multiple, often heterogenous LId scoring modules [YAN 95b, YAN 96b, ZIS 97, GAU 04]. Score normalization allows reduction of the bias created by different LId scoring modules. The impact

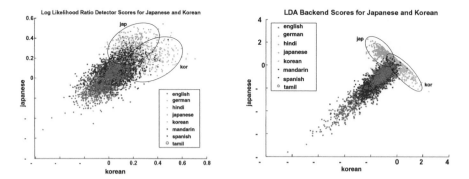

Figure 8.15. *Figures after W. Shen et al. MIT-LL [SHE 06]*

of back-end optimization on LId results range from very low to 40% relative improvement. For example, a comparative analysis of the different MIT-LL systems developed for the NIST LRE 2005 evaluations [SHE 06] revealed important gaps in accuracy for given back-ends depending on LId system configurations (standard vs lattice-based PPRLM). The authors provide a thorough analysis of different back-end classifiers, and Figure 8.15 illustrates these by plotting the correlation between Japanese and Korean language detector scores for the target and impostor language test samples. In the left plot, LLR scores are used, whereas the right plot is achieved by applying a Linear Discriminant Analysis (LDA)-based Gaussian back-end classification. The high cross-class correlation observed in the left plot appears to be significantly reduced in the right plot, thus allowing for improved detection capacities. The difference between both back-ends remains low for the standard PPRLM approach, but it proved to become very efficient for the lattice-based PPRLM system.

8.8. Discussion

Most successful approaches to automatic LId rely on a statistical modeling framework, and require appropriate observation sets in order to estimate models accurately. The progress in LId research over the last 20 years has been strongly linked to the production and the public availability of multilingual speech corpora. OGI have provided an important pioneering contribution with their multilingual telephone speech corpus, collected and distributed in the early 1990s.

A number of different approaches have been implemented for LId, and these use various degrees of language-specific information, ranging from purely acoustic, such as GMMs [ZIS 93, TOR 02a, TOR 02b, WON 02] and SVM [CAM 04], to lexically-informed systems [ITA 93, MAT 98, SCH 96]. Phonotactic approaches have achieved a very interesting trade-off between LId accuracy, implementation ease and cost: competitive results together with very low classification times; the extension to additional languages does not require any language-specific resources, besides audio, as unsupervised phonemic labeling can be carried out.

Many of the approaches applied to spoken LId are derived either from speech recognition or from speaker recognition systems. Whereas the best 1996 LId systems made use of PPRLM approaches (banks of parallel phone tokenizers followed by phonotactic language modeling), in 2003 the trend was to combine different approaches in parallel, fusing the scores of phonotactic PPRLM, acoustic GMM (Gaussian mixture models) and discriminative SVM-based systems. Thus, gains were mainly due to an optimal combination of different known approaches, rather than to a new approach or to an important advance of one single approach. An important exception might be the lattice-based approach [GAU 04], which demonstrated significant improvements to PPRLM, achieving similar results as the best MIT-LL system combination, with a significantly lower decoding time.

Independently from the adopted approach, decision accuracy largely depends on the stimulus length (in general a maximum duration of several tens of seconds). Results are much better for 30-second excerpts than 3-second examples, and phonotactic approaches tend to need longer speech samples than approaches that include acoustic-based scores. As human identification performance behaves asymptotically for much shorter durations of speech utterances, an important amount of information is still lacking in current approaches. This gap on short speech excerpts may be progressively bridged by improving existing methods, by using a larger variety of acoustic parameters or by combining more and more basic LId system variants in parallel, thus producing richer input to optimized back-ends, and last but not least, by using larger audio corpora for model estimation. The gap between humans and machines is clearly related to under-represented or missing information levels, such as prosody, morphology and voice quality or to a lack of discriminating capacity between fine acoustic-phonetic differences. The comparison of machine and human performances is informative to gain insight into the achieved modeling accuracy and to guide future research. Perception experiments suggest that humans exploit multiple partly redundant information levels to achieve language identification or at least a broad classification. Heterogenous system combinations can then be considered as an implementation of this human capacity, rather than as a second best implementation of the complex LId problem.

Today, automatic LId systems are capable of processing a large number of languages (typically several tens), certainly more than even an over-averagely gifted human being. Automatic LId poses the question of how to define a language, and how to distinguish between variations within and across languages. Present research efforts increasingly focus on subtler identification problems, such as dialect and accent identification. NIST LRE evaluations started addressing dialect identification in 2005 and appropriate audio data are still being collected (Callfriend2-LVDID [CAL2]). The labeling of these speech data in terms of language, dialect and accent may be problematic, and may interact with measured identification errors. Detailed analyses of the 2005 NIST evaluation [MAR 06b] revealed problems due to over-simplified labels for Chinese dialects and others due to Indian-accented English often misclassified as Hindi. Accented speech collections include a non-negligible part of code mixing and code switching [NIE 06], representing challenges to LId research in the future. Areas of investigation will increasingly address dialects, regional varieties, and foreign accents, the challenge for the latter being the identification of both the spoken language (L2) and of speaker accent, relative to his/her native language (L1). These issues will undoubtedly benefit from accurate acoustic-phonetic models to improve automatic transcription of accented speech [BAR 07], and more generally rich transcription into a multilingual framework.

Further research directions include language identification of audio-visual and multimodal speech [BEN 00]. Beyond the information delivered by human faces, language-specific information might be found in visemes and in accompanying gestures, potentially contributing to a multimodal LId component. The results of LId research, beyond multilingual indexing, public speech-driven services and intelligence applications, may cross-fertilize other research domains, such as foreign language acquisition/training, corpus-based language typology and characterization as well as dialectal studies.

8.9. References

[ADA 03] ADAMI A. and HERMANSKY H., "Segmentation of speech for speaker and language recognition", *Proc. of Eurospeech-03*, Geneva, September 2003.

[ADD 03] ADDA-DECKER M., ANTOINE F., BOULA DE MAREÜIL P., VASILESCU I., LAMEL L., LIÉNARD J.S., VAISSIÈRE J. and GEOFFROIS E., "Phonetic knowledge, phonotactics and perceptual validation for automatic language identification", *Proc. of ICPhS-03*, Barcelona, 2003.

[ANT 04] ANTOINE F., ZHU D., BOULA DE MAREÜIL P. and ADDA-DECKER M., "Identification des langues par unités phonémiques et syllabiques", *Proc. of JEP-04*, Fès, Morocco, 2004.

[BAR 99] BARKAT M., *et al.*, "Prosody as a distinctive feature for the discrimination of arabic dialects", *Proc. of Eurospeech-99*, Budapest, Hungary, September 1999.

[BAR 07] BARTKOVA K. and JOUVET D., "On using units trained on foreign data for improved multiple accent speech recognition", *Speech Communication*, vol. 49, no. 10-11, pp. 836–846, 2007.

[BEN 00] BENOÎT C., MARTIN J.C., PELACHAUD C., SCHOMAKER L. and SUHM B., "Audio-visual and multimodal speech-based systems", in GIBBON D., MERTINS I. and MOORE R. (eds.), *Handbook of Multimodal and Spoken Dialogue Systems: Resources, Terminology and Product Evaluation*, pp. 102–203, Boston Kluwer Academic Publishers, 2000.

[BER 98] BERKLING K., ZISSMAN M., VONWILLER J. and CLEIRIGH C., "Improving accent identification through knowledge of english syllable structure", *Proc. of ICSLP-98*, Sydney, 1998.

[BER 99] BERKLING K., REYNOLDS D. and ZISSMAN M., "Evaluation of confidence measures for language identification", *Proc. of Eurospeech-99*, Budapest, Hungary, September 1999.

[BIE 94] BIELEFELD B., "Language identification using shifted delta cepstrum", *XIVth Annual Speech Research Symposium*, Baltimore, 1994.

[BON 01] BONASTRE J.F., MAGRIN-CHAGNOLLEAU I., EULER S., PELLEGRINO F., ANDRÉ-OBRECHT R., MASON J. and BIMBOT F., "Speaker and Language characterization (SpLC): A Special Interest Group (SIG) of ISCA", *Proc. of Eurospeech-01*, pp. 1145–1148, Aalborg, Danemark, 2001.

[BOU 04] BOULA DE MAREÜIL P., MAROTTA G. and ADDA-DECKER M., "Contribution of prosody to the perception of Spanish/Italian accents", *Proc. of Speech Prosody*, Nara, Japan, March 2004.

[BOU 06] BOULA DE MAREÜIL P. and VIERU-DIMULESCU B., "The contribution of prosody to the perception of foreign accent", *Phonetica, Int. Journal of Phonetic Science*, vol. 63, pp. 247–267, no. 4, December 2006.

[CAM 02] CAMPBELL W., "Generalized linear discriminant sequence kernels for speaker recognition", *Proc. of ICASSP-02*, pp. 161–164, Orlando, May 2002.

[CAM 04] CAMPBELL W., SINGER E., TORRES-CARRASQUILLO P. and REYNOLDS D., "Language recognition with support vector machines", *Proc. Odyssey-04*, Toledo, June 2004.

[CAM 06] CAMPBELL W., *et al.*, "Advanced language recognition using cepstra and phonotactics: MIT-LL system performance on the NIST 2005 language recognition evaluation", *Proc. of Odyssey-06*, Puerto Rico, June 2006.

[CAS 98] CASEIRO D. and TRANCOSO I., "Spoken language identification using the speechdat corpus", *Proc. of ICSLP-98*, Sydney, Australia, December 1998.

[CHI 77] CHILDERS D.G., *et al.*, "The cepstrum: a guide to processing", *Proc. of the IEEE*, vol. 65, no. 10, pp. 1428–1443, October 1977.

[CHO 77] CHOMSKY N. and HALLE M., *The Sound Pattern of English*, MIT Press, Cambridge, Massachusetts, 1977.

[CIM 82] CIMARUSTI C. and IVES R., "Development of an automatic identification system of spoken languages: phase I", *Proc. of ICASSP-82*, pp. 1661–1663, May 1982.

[CRI 00] CRISTIANINI N. and SHAWE-TAYLOR J., *Support Vector Machines*, Cambridge University Press, Cambridge, 2000.

[COL 01] COLLOBERT R. and BENGIO S., "SVMTorch: Support vector machines for large-scale regression problems", *Journal of Machine Learning Research*, vol. 1, pp. 143–160, 2001.

[COM 90] COMRIE B., *The World's Major Languages*, Oxford University Press, Oxford UK, 1990.

[COM 01] COMPERNOLLE (VAN) D., "Speech recognition by goats, wolves, sheep and ... non-natives", *Speech Communication*, vol. 35., pp. 71–79, 2001.

[COR 97] CORREDOR-ARDOY C., GAUVAIN J.L., ADDA-DECKER M. and LAMEL L., "Language identification with language-independent acoustic models". *Proc. Eurospeech-97*, Rhodes, Greece, September 1997.

[CRY 93] CRYSTAL D., *The Cambridge Factfinder*, Cambridge University Press, Cambridge UK, 1993.

[CRY 97] CRYSTAL D., *The Cambridge Encyclopedia of Language*, Cambridge University Press, Cambridge UK, 1997.

[DAL 96] DALSGAARD P., ANDERSEN O., HESSELAGER H. and PETEK B., "Language-Identification using Language-Dependent Phonemes and Language-Independent Speech Units", *Proc. of ICSLP-96*, Philadelphia, USA, October 1996.

[DAN 96] DANIELS P. and BRIGHT W., *The World's Writing Systems*, Oxford University Press, Oxford 1996.

[DAV 80] DAVIS S. and MERMELSTEIN P., "Comparison of parametric representations for monosyllabic word recognition in continuously spoken sentences", *IEEE Transactions on Acoustics, Speech and Signal Processing*, vol. ASSP-28, no. 4, pp. 357–366, August 1980.

[DEB 91] DE BOYSSON-BARDIES B. and VIHMAN M., "Adaptation to language: evidence from babbling and first words from four languages", *Language* vol. 67, no. 2, pp. 297–319, 1991.

[DEL 65] DELATTRE P., *Comparing the Phonetic Features of English, Spanish, German and French*, Julius Gross Verlag, Heidelberg, 1965.

[DES 15] DE SAUSSURE F., *Cours de linguistique générale*, Payot, Paris, 1915.

[DUT 00] DUTAT M., MAGRIN-CHAGNOLLEAU I. and BIMBOT F., "Language recognition using time-frequency principal component analysis and acoustic modeling", *Proc. of ICSLP-00*, Beijing, China, October 2000.

[EAR 05] EARS: Effective affordable reusable speech-to-text, http://projects.ldc.upenn.edu/EARS/.

[ELD] ELDA: European Language Resources Distribution Agency, http://www.elda.org.

[FAR 01] FARINAS J. and PELLEGRINO F., "Automatic rhythm modeling for Language Identification", *Proc. of Eurospeech-03*, September 2001.

[FLE 03] FLEGE J.E., *et al.*, "Interaction between the native and second language phonetic subsystems", *Speech Communication*, vol. 40, no. 4, June 2003.

[FOI 86] FOIL J.T., "Language identification using noisy speech", *Proc. of ICASSP-86*, pp. 861–864, 1986.

[GAU 04] GAUVAIN J.L., MESSAOUDI A. and SCHWENK H., "Language recognition using phone lattices", *Proc. of ICSLP 04*, Jeju Island, South Korea, 2004.

[GEN 07] GENDROT C. and ADDA-DECKER M., "Impact of duration and vowel inventory size on formant values of oral vowels: an automated formant analysis from eight languages", *Proc. of ICPhS-07*, Saarbrücken, August 2007.

[GEO 04] GEOFFROIS E., "Identification automatique des langues: techniques, resources et évaluations", *Proc. of MIDL-04*, November 2004.

[GOR 05] GORDON R.G. (ed.), *Ethnologue: Languages of the World*, 15th edition, Dallas, Tex. SIL International, 2005. Online version: http://www.ethnologue.com/.

[GRE 78] GREENBERG J., FERGUSON C. and MORAVCSIK E. (eds.), *Universals of Human Languages: Method and Theory*, Stanford University Press, 1978.

[HAL 91] HALLÉ P.A., DE BOYSSON-BARDIES B. and VIHMAN M., "Beginnings of prosodic organization: intonation and duration patterns of disyllables produced by Japanese and French infants", *Language and Speech* vol. 34, pp. 299–318, 1991.

[HAZ 93] HAZEN T. and ZUE V., "Automatic language identification using a segment-based approach", *Proc. of Eurospeech-03*, Berlin, Germany, September 1993.

[HAZ 97] HAZEN T. and ZUE V., "Segment-based automatic language identification", *Journal of the Acoustical Society of America*, (JASA), vol. 101, no. 4, pp. 2323–2331, April 1997.

[HER 90] HERMANSKY H., "Perceptual lineair predictive (PLP) analysis of speech", *Journal of Acoustical Society of America*, (JASA), vol. 87, no. 4, pp. 1738–1752, April 1990.

[HIE 94] HIERONYMOUS J., "ASCII phonetic symbols for the world's languages: Worldbet", *Technical Report AT&T Bell Labs*, 1994.

[HIE 96] HIERONYMOUS J. and KADAMBE S., "Spoken language identification using large vocabulary speech recognition", *Proc. of ICSLP-96*, Philadelphia, USA, 1996.

[HIR 99] HIRST D. and DI CRISTO A. (eds.), *Intonation Systems: A Survey of Twenty Languages*, Cambridge University Press, 1999.

[HOM 99] HOMBERT J.M. and MADDIESON I., "The use of 'rare' segments for language identification", *Proc. of Eurospeech-99*, Budapest, Hungary, September 1999.

[HOM 06] HOMBERT J.M. (ed.) *Aux origines des langues et du langage – Towards Origins of Languages and Language*, Fayard, 2005.

[HÖN 05] HÖNIG F., *et al.*, "Revising Perceptual Linear Prediction (PLP)", *Proc. of Eurospeech-05*, September 2005.

[HOU 77] HOUSE A.S. and NEUBURG E.P., "Toward automatic identification of the language of an utterance. I. Preliminary methodological considerations", *Journal of the Acoustical Society of America*, (JASA), vol. 62, no. 3, pp. 708–713, September 1977.

[IPA 99] IPA association, *Handbook of the International Phonetic Association: A Guide to the Use of the International Phonetic Alphabet*, Cambridge University Press, 1999.

[ITA 93] ITAHASHI S., *et al.*, "Language Identification with phonological and lexical models", *Proc. of Eurospeech-95*, Madrid, Spain, September, 1995.

[JEL 98] JELINEK F., *Statistical Methods for Speech Recognition*, ISBN 0-262-10066-5, The MIT Press, 1998.

[KIR 02] KIRCHHOFF K., PARANDEKAR S. and BILMES J., "Mixed-memory Markov models for automatic language identification", *Proc. of ICASSP-02*, Orlando, 2002.

[KIR 06] KIRCHHOFF K., "Language characteristics", in SCHULTZ T. and KIRCHHOFF K. (eds.), *Multilingual Speech Processing*, Elsevier, 2006.

[KUM 96] KUMPF K. and KING R.W., "Automatic accent classification of foreign accented Australian English speech", *Proc. of ICSLP-96*, Philadelphia, October 1996.

[KWA 95] KWAN H.K. and HIROSE K., "Recognized phoneme-based n-gram modeling in automatic language identification", *Proc. of Eurospeech-95*, Madrid, Spain, September 1995.

[LAB 04] LABOV W., "Phonological atlas of North America", 2004. http://www.ling.upenn .edu/phono_atlas/home.html.

[LAD 96] LADEFOGED P. and MADDIESON I., *The Sounds of the World's Languages*, Blackwell, 1996.

[LAM 94] LAMEL L. and GAUVAIN J.L., "Language identification using phone-based acoustic likelihoods", *Proc. of ICASSP-94*, Adelaide, April 1994.

[LAM 95] LAMEL L. and GAUVAIN J.L., "A phone-based approach to non-linguistic speech feature identification", *Computer Speech and Language*, January 1995.

[LAM 98] LAMEL L., ADDA G., ADDA-DECKER M., CORREDOR-ARDOY C., GANGOLF J.J. and GAUVAIN J.L., "A multilingual corpus for language identification", *Proc. of LREC-98*, Granada, May 1998.

[LAN 95] LANDER T., *et al.*, "The OGI 22 language telephone speech corpus", *Proc. of Eurospeech-95*, Madrid, Spain, September 1995.

[LAV] Laval University, "Les grandes familles linguistiques du monde", http://www.tlfq .ulaval.ca/AXL/monde/familles.htm.

[LDC 07] LDC, http://www.ldc.upenn.edu/Catalog/.

[CAL 96] LDC-Callfriend, "CallFriend Corpus", Linguistic Data Consortium, 1996.

[CAL2] LDC-Callfriend2 LVDID: Language, Variety and Dialect Identification, http://www.ldc.upenn.edu/CallFriend2.

[LEO 74] LEONARD R. and DODDINGTON G., "Automatic language identification", *Technical report* RADC-TR-74-200, Air Force Rome Air Development Center, August 1974.

[LEV 91] LEVITT A. and WANG Q., "Evidence for language-specific rhythmic influences in the reduplicative babbling of French-and-English-learning infants", *Language and Speech*, vol. 34, pp. 235–249, 1991.

[LI 80] LI K. and EDWARDS T., "Statistical models for automatic language identification", *Proc. of ICASSP-80*, pp. 884–887, Denver, Colorado, 1980.

[LI 94] LI K. and EDWARDS T., "Automatic language identification using syllabic spectral features", *Proc. of ICASSP-94*, Adelaide, Australia, April 1994.

[LUN 96] LUND M., MA K. and GISH H., "Statistical language identification based on untranscribed training", *Proc. of ICASSP-94*, Adelaide, Australia, April 1994.

[MAD 02] MADDIESON I. and VASILESCU I., "Factors in human language identification", *Proc. of ICSLP-02*, Denver, 2002.

[MAR 98] MARIANI J. and PAROUBEK P., "Human language technologies evaluation in the European framework", *Proc. of the DARPA Broadcast News Workshop*, Herndon, VA, pp. 237–242, 1998.

[MAR 05] MARIANI J., "Developing language technologies with the support of language resources and evaluation programs", *Journal of Language Resources and Evaluation*, Springer Netherlands, 2005.

[MAR 13] MARKOV A.A., "An example of statistical investigationin the text of *Eugen Onyegin* – illustrating the coupling of tests in chains", *Proc. Acad. of St. Petersburg*, vol. 7, pp. 153–162, 1913.

[MAR 97] MARTIN A., *et al.*, "The det curve in assessment of detection task performance", *Proc. of EUROSPEECH-97*, Rhodes, Greece, September 1997.

[MAR 03] MARTIN A. and PRZYBOCKI M., "NIST 2003 language recognition evaluation", *Proc. of Eurospeech-03*, pp. 1341–1344, Geneva, September 2003.

[MAR 06a] MARTIN T., BAKER B., WONG E. and SRIDHARAN S., "A syllable-scale framework for language identification", *Computer Speech & Language*, vol. 20, pp. 276–302, 2006.

[MAR 06b] MARTIN A. and LE A., "The current state of language recognition: NIST 2005 evaluation results", *Proc. of Odyssey-06*, Puerto Rico, June 2006.

[MAT 98] MATROUF D., ADDA-DECKER M., LAMEL L. and GAUVAIN J.L., "Language identification incorporating lexical information", *Proc. of ICSLP-98*, Sydney, Australia, December 1998.

[MAT 99] MATROUF D., ADDA-DECKER M., GAUVAIN J.L. and LAMEL L., "Comparing different model configurations for language identification using a phonotactic approach", *Proc. of Eurospeech-99*, Budapest, Hungary, September 1999.

[MAT 05] MATEJKA P., *et al.*, "Phonotactic language identification using high quality phoneme recognition", *Proc. of the 9th European Conference on Speech Communication and Technology (Eurospeech)*, Lisbon, Portugal, September 2005.

[MAT 06] MATEJKA P., *et al.*, "Brno University of technology system for Nist 2005 language recognition evaluation", *Proc. of Odyssey-06*, June 2006.

[MEN 96] MENDOZA S., GILLICK L., ITO Y., LOWE S. and NEWMAN M., "Automatic language identification using large vocabulary continuous speech recognition", *Proc. of ICASSP-96*, Atlanta, Georgia, April 1996.

[MON 06] MONTERO-ASENJO A., *et al.*, "Exploring PPRLM performance for NIST 2005 language recognition evaluation", *Proc. of Odyssey-06*, Puerto Rico, June 2006.

[MOR 99] MORI K., TOBA N., HARADA T., ARAI T., KOMATSU M., AOYAGI M. and MURAHARA Y., "Human language identification with reduced spectral information", *Proc. of Eurospeech-99*, Budapest, Hungary, September 1999.

[MUT 92] MUTHUSAMY Y., COLE R. and OSHIKA B., "The OGI multi-language telephone speech corpus", *Proc. of ICSLP-92*, Alberta, October 1992.

[MUT 93] MUTHUSAMY Y., "A segmental approach to automatic language identification, a segment based automatic language identification system", PhD Thesis, Oregon Graduate Institute of Science and Technology, July 1993.

[NAG 06] NAGARAJAN T. and MURTHY H.A., "Language identification NAG 0620 using acoustic log-likelihoods of syllable-like units", *Speech Communication*, vol. 48, pp. 913–926, 2006.

[NAK 92] NAKAGAWA S., UEDA Y. and SEINO T., "Speaker-independent, text-independent language identification by HMM", *Proc. of ICSLP-92*, pp. 1011–1014, Alberta, October 1992.

[NAV 01] NAVRÁTIL J., "Spoken language recognition: a step towards multilinguality in speech processing", in IEEE *Transactions on Speech and Audio Processing*, vol.9, no. 6, pp. 678–685, 2001.

[NAV 06] NAVRÁTIL J., "Automatic language identification", in SCHULTZ T. and KIRCHHOFF K. (eds.), *Multilingual Speech Processing*, Elsevier, 2006.

[NIE 06] NIESLER T. and WILLETT D., "Language identification and multilingual speech recognition using discriminatively trained acoustic models", *Proc. of MULTILING-06*, Stellenbosch, South Africa, April 2006.

[NIS 03] NIST, *The 2003 NIST language recognition evaluation plan*, http://www.nist.gov/speech/tests/lang/index.htm.

[OBU 05] OBUCHI Y., *et al.*, "Language identification using phonetic and prosodic HMMs with feature normalization", *Proc. of ICASSP-05*, Philadelphia, March 2005.

[OGI 07] OGI-22L, http://cslu.cse.ogi.edu/corpora/22lang/.

[PAR 99] PARLANGEAU N., PELLEGRINO F. and ANDRÉ-OBRECHT R., "Investigating automatic language discrimination via vowel system and consonantal system modeling", *Proc. of ICPhS-99*, San Francisco, 1999.

[PEL 00] PELLEGRINO F. and ANDRÉ-OBRECHT R., "Automatic language identification: an alternative approach to phonetic modeling", *Signal Processing*, Elsevier Science, North Holland, vol. 80, pp. 1231–1244, July 2000.

[PUL 96] PULLUM G.K. and LADUSAW W.A., *Phonetic Symbol Guide*, University of Chicago Press, 2nd edition, 1996.

[RAM 99] RAMUS F. and MEHLER J., "Language identification with suprasegmental cues: A study based on speech resynthesis", *Journal of Acoustical Society of America, (JASA)*, vol. 105, 1999.

[ROU 05] ROUAS J.L., "Modeling long and short-term prosody for language identification", *Proc. of Eurospeech-05*, Lisbon, Portugal, September 2005.

[RUH 05] RUHLEN M., "Taxonomy, typology, and historical linguistics", in JAMES W.M. and WILLIAM S.-Y. WANG (eds.), *Language Acquisition, Change and Emergence: Essays in Evolutionary Linguistics*, City University of Hong Kong Press, 2005.

[SCH 96] SCHULTZ T., ROGINA I. and WAIBEL A., "LVCSR-based language identification", *Proc. of ICASSP-96*, Atlanta, Georgia, May 1996.

[SCH 02] SCHULTZ T., "Globalphone: a multilingual text and speech database developed at Karlsruhe University", *Proc. of ICSLP-02*, Denver 2002.

[SCH 06] SCHULTZ T., *et al.*, *Multilingual Speech Processing*, Elsevier, 2006.

[SCH 97] SCHWARTZ J.L., BOË L.J., VALLÉE N. and ABRY C., "Major trends in vowel system inventories", *Journal of Phonetics*, vol. 25, no. 3, pp. 233–253, 1997.

[SHE 06] SHEN W., CAMPBELL W., GLEASON T., REYNOLDS D. and SINGER E., "Experiments with lattice-based PPRLM language identification", *Proc. of Odyssey-06*, Puerto Rico, June 2006.

[SHO 80] SHOUP J., "Phonological aspects of speech recognition", in LEA W. (ed.), *Trends in Speech Recognition*, Prentice-Hall, Englewood Cliffs, NJ, pp. 125–138, 1980.

[SIN 03] SINGER E., TORRES-CARRASQUILLO P., GLEASON T., CAMPBELL W. and REYNOLDS D., "Acoustic, phonetic and discriminative approaches to automatic language recognition", *Proc. of EuroSpeech-03*, Geneva, September 2003.

[SIN 04] SINGER E. and REYNOLDS D., "Analysis of multitarget detection for speaker and language recognition", *Proc. of Odyssey-04*, Toledo, 2004.

[SPDM] Speechdat-(M) mobile phone: EU-project LRE-63314, http://www.phonetik .uni-muenchen.de/SpeechDat.html.

[SUG 91] SUGIYAMA M., "Automatic language recognition using acoustic features", *Proc. of ICASSP-91*, pp. 813–816, Toronto, Ontario, May 1991.

[TEI 97] TEIXEIRA C., TRANCOSO I. and SERRALHEIRO A., "Recognition of non-native accents", *Proc. of Eurospeech-97*, Rhodes, Greece, September 1997.

[THY 96] THYMÉ-GOBBEL A.E. and HUTCHINS S.E., "On using prosodic cues in automatic language identification", *Proc. of ICSLP-96*, Philadelphia, October 1996.

[TOR 02a] TORRES-CARRASQUILLO P., REYNOLDS D. and DELLER J., "Language identification using Gaussian mixture model tokenization", *Proc. of ICASSP-02*, Orlando, 2002.

[TOR 02b] TORRES-CARRASQUILLO P., SINGER E., KOHLER M., GREENE R., REYNOLDS D. and DELLER J., "Approaches to language identification using Gaussian mixture models and shifted delta cepstral features", *Proc. of ICSLP-02*, Denver, 2002.

[TOR 04] TORRES-CARRASQUILLO P., GLEASON T. and REYNOLDS D., "Dialect identification using Gaussian mixture models", *Proc. of Odyssey-04*, Toledo, 2004.

[TUC 94] TUCKER R., CAREY M. and PARRIS E., "Automatic language identification using sub-word models", *Proc. of ICASSP-94*, Adelaide, Australia, April 1994.

[UED 90] UEDA Y., and NAKAGAWA S., "Prediction for phoneme/syllable/word category and identification of language using HMM", *Proc. of ICSLP-90*, pp. 1209–1212, Kobe, 1990.

[VAL 99] VALLÉE N., SCHWARTZ J.L. and ESCUDIER P., "Phase spaces of vowel systems: a typology in the light of the Dispersion-Focalisation Theory", *Proc. of ICPhS-99*, San Francisco, 1999.

[VAP 97] VAPNIK V.N., Support vector learning machines – Tutorial at NIPS'97, Denver (CO), December 1997.

[VAS 05] VASILESCU I., CANDEA M. and ADDA-DECKER M., "Perceptual salience of language-specific acoustic differences in autonomous fillers across eight languages", *Proc. of Eurospeech-05*, Lisbon, Portugal, September 2005.

[WAN 99] WANNEROY R., BILINSKI E., BARRAS C., ADDA-DECKER M. and GEOFFROIS E., "Acoustic-phonetic modeling of non-native speech for language identification", *Proc. of MIST-99*, Leusden, The Netherlands, September 1999.

[WEL 97] WELLS J., "SAMPA computer readable phonetic alphabet", in GIBBON D., MOORE R. and WINSKY R. (eds.), *Handbook of Standards and Resources for Spoken Language Systems*, Mouton de Gruyter, New York - Berlin, 1997.

[WHI 06] WHITE C., SHAFRAN I. and GAUVAIN J.L., "Discriminative classifiers for language recognition", *Proc. of ICASSP-06*, Toulouse, France, May 2006.

[WIT 99] WITT S. and YOUNG S., "Off-line acoustic modeling of non-native accents", *Proc. of Eurospeech-99*, pp. 1367–1370, Budapest, Hungary, September 1999.

[WON 00] WONG E., PELECANOS J., MYERS S. and SRIDHARAN S., "Language identification using efficient Gaussian mixture model analysis", *Proc. of Australian Int. Conference on Speech Science and Technology*, 2000.

[WON 02] WONG E. and SRIDHARAN S., "Methods to improve Gaussian mixture model based language identification system", *Proc. of ICSLP-02*, Denver, 2002.

[YAN 95a] YAN Y. and BARNARD E., "An approach to automatic language identification based on language-dependent phoneme recognition", *Proc. of ICASSP-95*, pp. 3511–3514, Madrid, Spain, September 1995.

[YAN 95b] YAN Y. and BARNARD E., "A comparison of neural net and linear classifier as the pattern recognizer in automatic language identification", *Proc. of ICNNSP-95*, Nanjing, 1995.

[YAN 96a] YAN Y. and BARNARD E., "Experiments with conversational telephone speech for language identification", *Proc. of ICASSP-96*, Atlanta, Georgia, May 1996.

[YAN 96b] YAN Y., BARNARD E. and COLE R., "Development of an approach to language identification based on phone recognition", *Computer Speech & Language*, vol. 10, pp. 37–54, 1996.

[ZHU 05] ZHU D., *et al.*, "Different size multilingual phone inventories and context-dependent acoustic models for language identification", *Proc. of Eurospeech-05*, Lisbon, Portugal, September 2005.

[ZHU 06] ZHU D. and ADDA-DECKER M., "Language identification using lattice-based phonotactic and syllabotactic approaches", *Proc. of Odyssey-06*, June 2006.

[ZIP 49] ZIPF G.K., *Human Behaviour and the Principle of Least Effort*, Addison-Wesley Publishing, Reading, MA, 1949.

[ZIS 93] ZISSMAN M., "Automatic language identification using Gaussian mixture and hidden Markov models", *Proc. of ICASSP-93*, pp. 399–402, Minneapolis, April 1993.

[ZIS 94] ZISSMAN M. and SINGER E., "Automatic language identification of telephone speech messages using phoneme recognition and N-gram modeling", *Proc. of ICASSP-94*, vol. 1, pp. 305–308, Adelaide, Australia, April 1994.

[ZIS 96a] ZISSMAN M., "Comparison of four approaches to automatic language identification of telephone speech", *IEEE Transactions on Speech and Audio Processing*, vol. 4, no. 1, pp. 31–44, January 1996.

[ZIS 96b] ZISSMAN M., GLEASON T., REKART D. and LOSIEWICZ B., "Automatic dialect identification of extemporaneous conversational, Latin American Spanish speech", *Proc. of ICASSP-96*, Atlanta, Georgia, 1996.

[ZIS 97] ZISSMAN M., "Predicting, diagnosing and improving automatic language identification performance", *Proc. of Eurospeech-97*, pp. 51–54, Rhodes, Greece, September 1997.

[ZIS 99] ZISSMAN M. and BERKLING K., "Automatic language identification", *Speech Communication*, vol. 35, no. 1-2, pp. 115–124, August 2001.

Chapter 9

Automatic Speaker Recognition

9.1. Introduction

A spoken message does not merely convey the meaning of the message produced by the person who utters it. It also conveys information *about* the speaker, and more particularly, his or her identity. Each of us is indeed able to recognize familiar people by their voice.

In general, the main objective of spoken communication is centered on the spoken message. Variability due to the speaker (i.e. the transmitter of the message), is compensated for by the interlocutor (i.e. the receiver).

In speech recognition, a key goal is to neutralize inter-speaker variability. Conversely, in *speaker recognition*, the focus is put on the characteristics of the transmitter. Inter-speaker variability factors become essential to *characterize* the voice of a particular speaker. At the same time, *intra-speaker* variability becomes predominant, and *acoustic factors* such as noise, microphone or channel variations play a limiting role.

9.1.1. *Voice variability and characterization*

Voice variability between people originates primarily from the existence of *morphological* differences in speech production organs, in particular the length of the vocal folds, the *shape* of the oral and nasal cavities and of the various

Chapter written by Frédéric BIMBOT.

articulators. These elements evolve during childhood, and they stabilize in adulthood, during which they hardly change, even though specific pathologies (such as colds) can modify their characteristics more or less permanently.

Another dimension of inter-speaker variability comes from *physiological* differences which impact motor control, such as muscle tonicity, motor accuracy, movement coordination and, more generally speaking, all phenomena which influence the *functioning* of the articulators. This second set of factors is also subject to (usually unintentional) fluctuations, depending on the speaker's fitness, his or her mood, stress level, emotional state and on the possible consumption by the speaker of various types of substances (for instance, alcohol or drugs).

Finally, a third series of factors influence a speaker's voice: the socio-cultural context and, above all, the educational context, which induce unconscious imitation and identification in the way people speak. Let us mention for instance the importance of family, school, professional environment, geographic origin (regional accents), age class (with their specific registers), articulation, style and lexical habits within a community. All these factors (which can be viewed as codes) are also prone to being spontaneously modified consciously by the speaker, either to hide some characteristics or to mimic others.

Thus, differences between voices can be explained by a set of morphological, physiological and socio-cultural factors that represent the background for speaker recognition. However, these factors are in themselves interlinked with intra-speaker variability, in particular the speaker's health, emotional state, state of mind and intentions[1], making voice characteristics non-deterministic, difficult to reproduce and easy to modify.

Moreover, any technological process that aims at identifying particular characteristics of a person requires a recording phase during which the speech signal is almost inevitably altered by microphone and channel distortions and other factors related to the recording environment.

In that respect, the term *voice print*, which is sometimes used (including by the media) is totally inadequate, as it conveys the idea that voice has characteristics which are as reliable as fingerprints or DNA, whereas this is absolutely not the case. Strictly speaking, voice characterization can not be considered as a biometric identification technique[2], as it is not solely based on biological factors but also on *behavioral* processes.

---

1 As well as other factors which are more difficult to designate.
2 Biometric: *based on the measurements of an organ* (Petit Larousse, 1999).

To avoid confusion, it is more appropriate to use the term *voice signature*, which better accounts for the various sources of variability, whether they are intentional or not, which impact the realization and the measurement of the signal (as is also the case for a handwritten signature).

9.1.2. *Speaker recognition*

Conventionally, the term *speaker recognition* refers to any decision-making process that consists of using characteristics of a speech signal to determine the speaker's identity for a given utterance. If, moreover, this process is achieved without any human intervention (including pre-processing, algorithmic steps or interpretation of the results), the task is termed *automatic speaker recognition*.

In spite of the limiting factors mentioned in the previous section, many applicative contexts exist for which the possibility of identifying automatically a speaker's identity is of utmost interest, even if the task is achieved with some error margin. Moreover, the massive rise in computational resources and their simultaneous drop in price makes deploying such technologies at a commercial scale possible.

Like other branches of automatic speech processing, speaker recognition results from the convergence of several disciplines, including signal processing, pattern recognition, statistics, probabilities, decision theory and also phonetics, linguistics and ergonomics.

In this chapter, we will mostly focus on technological aspects of speaker recognition. Section 9.2 presents a typological overview of automatic speaker recognition systems and how they operate. Section 9.3 deals with theoretical and algorithmic aspects by presenting state-of-the-art approaches in the context of the probabilistic framework. Section 9.4 is dedicated to assessment methodology for speaker recognition system. Section 9.5 addresses applicative aspects and underlines specific constraints resorting to a given applicative context.

The following text contains inevitably a number of simplifications, shortcuts or omissions so as to provide a reasonably global overview of the domain. A list of pointers to recent publications is proposed at the end of the chapter for readers who would like to know more on the domain, after having gone through (and hopefully enjoyed) this chapter.

9.2. Typology and operation of speaker recognition systems

Before detailing the fundamentals of the most common speaker recognition techniques, let us describe some aspects related to the typology and the operation of automatic speaker recognition systems and introduce some terminology.

9.2.1. *Speaker recognition tasks*

One way to categorize speaker recognition systems is to classify them depending on the task they achieve:

– *closed-set speaker identification* involves identifying the speaker among a (closed) set of L reference speakers;

– *speaker verification* checks whether the current speaker is indeed the expected or claimed speaker;

– *open-set speaker verification* is similar to closed-set speaker verification but an additional hypothesis is considered, namely the possibility that the current speaker may be none of the L reference speakers[3].

Additional tasks can also be considered (even though some of them have no known solution as yet), which aim at identifying some characteristics of a speaker, without trying specifically to recognize his/her identity; for instance, gender recognition, age bracket recognition, voice pathology detection, drunkenness, etc. All these problems can be gathered under the general term of *speaker classification*.

Finally, let us mention a number of other tasks which relates to speaker characterization:

– *speaker segmentation* (or speaker change detection), which involves localizing automatically the various speakers which intervene in an audio recording (without necessarily identifying them);

– *speaker detection* which aims at detecting the presence of a given speaker in an audio recording (without necessarily localizing the speaker);

– *speaker tracking* which involves localizing a particular speaker in an audio recording (i.e. detection and segmentation);

– *speaker counting*, which aims at finding the number of distinct speakers in an audio recording.

---

3 This task can be viewed as a closed-set speaker identification step followed by a verification step, for which the claimed identity is that found in the identification step.

– *speaker diarization*, i.e. the task of making an inventory of the various speakers taking part in an audio recording and tracking them on that recording (without any prior knowledge on the speakers);

– *speaker selection*, i.e. the task of finding, in a speaker database, the speaker(s) which is/are closest to a given speaker, even if that speaker does not belong to the database.

The rest of this chapter will focus on the tasks of closed-set identification and of speaker verification, as they are central to the domain. However, the term "speaker recognition" will still be used to refer to these two tasks from a more global point of view.

9.2.2. *Operation*

From the user viewpoint, the operation of a speaker recognition system involves two phases: a *training* phase and an *operational* phase (also called *test* phase, when referring to laboratory experiments).

The training phase involves collecting one or several spoken training *utterances* from a speaker, during one or several recording *sessions*. These sessions are preferably spread over time and consist, when possible, of various recording conditions (different microphones, different environments, etc.) so as to correspond at best to the various conditions in which the system is designed to operate. The system analyzes and models the speaker's voice characteristics and stores the model parameters and the corresponding identity in a database of reference voice templates.

In operation mode, the process differs between a speaker identification system and a speaker verification system. In identification mode, the *client* speaker produces a test utterance Y. The system returns an identity index $\hat{X}_\ell$ that corresponds to the recognized speaker. In verification mode, together with the spoken test utterance Y, the client speaker also claims a specific identity X. The system returns a binary decision in terms of acceptance or rejection (respectively denoted $\hat{X}$ and $\tilde{\hat{X}}$) indicating whether the system has actually identified the client speaker as the genuine claimed speaker or considers him/her as an *impostor*.

Figure 9.1 schematically depicts these two scenarios (closed-set identification and verification). It also illustrates the fact that open-set speaker identification can be obtained by cascading these two scenarios.

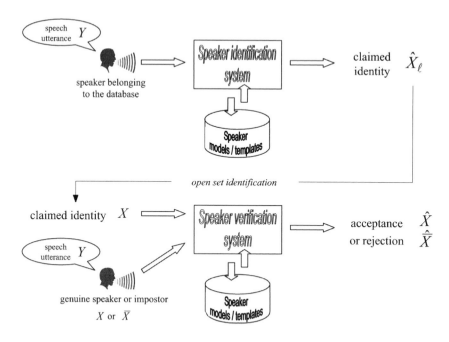

Figure 9.1. *Diagrammatic representation of a speaker identification system (top) and a speaker verification system (bottom)*

From a practical viewpoint, most applications have to be able to operate with a small number of training sessions (typically one or two) so as not to discourage new users with training procedures running over too long a lapse of time. This ergonomic constraint has a significant impact on the system performance, but it can be partly overcome using incremental training procedures for refining speaker models during operation mode.

9.2.3. *Text-dependence*

When considering speaker recognition systems from the type of input speech which they accept, the tradition is to distinguish between *text-dependent* systems and *text-independent* systems, depending on the existence of constraints on the linguistic content of the speech utterance pronounced by the speaker to be recognized.

In practice, this binary distinction is too simplistic, as there is, in practice, a variety of text-dependence levels corresponding to different application profiles. In fact, five types of system can be distinguished with regard to the type of input speech:

 – *common password* systems, for which all users share the same key-word or key-sentence;

 – *individual password* systems, for which each user chooses the linguistic content of his/her key-word or key-sentence;

 – *text-prompted* systems, which prompt, either in oral or in written form, the text to be pronounced by the user;

 – *specific phonetic event* systems, for which some flexibility is left to the user in what he says, as long as the utterance contains a particular type of phoneme or linguistic event;

 – *text-free* systems, for which no particular constraint is put on the content of the test utterance.

Password-based systems are theoretically more efficient than text-free ones, since the linguistic constraint neutralizes one of the sources of variability. Nevertheless, password-based systems (as well as those that are not text-prompted) are more vulnerable to pre-recorded speech, whereas prompted systems (coupled with a *speech verification* system[4]) provide a better protection against this type of imposture. Note, however, that some applicative contexts require non-constrained text, for instance when verification takes place on spontaneous speech.

More generally, the choice of a given text-dependence level for a particular application results from a combination of ergonomic considerations and security requirements.

9.2.4. *Types of errors*

For closed-set identification, a system error results in a *recognition error*, i.e. speaker X_ℓ is confused with another speaker $X_{\ell'}$, with $\ell \neq \ell'$. Using the terminology derived from decision theory, this can be equivalently considered as both a missed detection of speaker X_ℓ and a false alarm on speaker $X_{\ell'}$. Performance is evaluated in terms of recognition error rate (or probability), which is calculated as the average missed detection probability $P(\hat{\bar{X}}|X)$ (or as the average false alarm probability $P(\hat{X}|\bar{X})$ which is equal to the previous quantity, in the case

---

4 I.e. a speech recognition system which is used in order to verify that the speaker has indeed produced the expected text.

of closed-set identification). Note that, for speaker identification, the intrinsic complexity of the task (and therefore, the system error rate) increases with the size L of the reference speaker population.

For speaker verification, the two types of errors mentioned above can occur independently. Either the client speaker is rejected whereas it was indeed the genuine claimed speaker: this is called *false rejection* and corresponds to a missed detection (of the correct speaker); or the client speaker is accepted even though he/she is an impostor; in that case, the error is called *false acceptance* and it corresponds to a false alarm (on the correct speaker). Depending on the way the system is tuned, the false rejection and false acceptance rates (denoted respectively as $P(\hat{\overline{X}}|X)$ and $P(\hat{X}|\overline{X})$ in the rest of the chapter) vary in opposite directions and the global performance of the system is measured by assigning some costs to both types of errors. Note that, as opposed to speaker identification, speaker verification performance is not affected by the size of the reference population: the task to accomplish consists only of a binary choice between two hypotheses.

9.2.5. *Influencing factors*

Factors that influence the performance level of speaker recognition systems are very diverse. Here is a brief list of the most obvious ones:

- the recording environment (especially the presence of noise);
- the quality of the recording equipment (for instance, the type of microphone);
- the type of transmission channel, including the signal's bandwidth, the type of audio compression applied to the signal, the error rate on the transmission channel, etc.;
- the time interval between the training utterance(s) and the test utterance (inter-session variability, voice drift over long periods of time);
- the speaker condition: health (colds, loss of voice), emotional state (stress, nervousness), psychological condition (excitation, depression) and other individual factors;
- the application of intentional voice modifications (imposture, masking);
- the degree of cooperation and/or familiarity of the speaker with the situation (benefit vs. drawback to be recognized, frequent vs. occasional use);
- the quantity and the statistical representativeness of the training speech data, including the number of training sessions, the variety of recording conditions, the linguistic coverage of the utterances;

–   the quantity of recognition data (utterance duration) and their linguistic coverage.

Excellent levels of performance can be achieved in artificially ideal conditions (quiet environment, high quality microphone, contemporaneous training and test, cooperative speaker). However, they degrade drastically in real use conditions, which require specific approaches to improve system robustness.

9.3. Fundamentals

We will now describe the main aspects of state-of-the-art techniques for speaker recognition, which are widely based on probabilistic modeling.

9.3.1. *General structure of speaker recognition systems*

A speaker recognition system can be decomposed into four modules:

–   *acoustic analysis*, the function of which is to transform the raw (mono-dimensional) signal into a sequence of acoustic feature vectors;

–   *speaker modeling*, which is used during the training phase (and sometimes during the test phase) to extract a template or a model of the acoustic feature vector distribution, from one or several spoken utterances;

–   *scoring*, to measure the similarity between a test utterance and one or several reference speaker model;

–   *decision*, assigning an identity (in the case of identification) or opting between acceptance and rejection (in the case of verification), on the basis of one or several scores stemming from the previous module.

Figure 9.2 schematically depicts the way these modules interoperate.

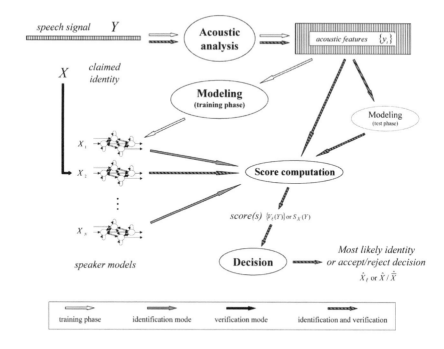

Figure 9.2. *Modular analysis of a speaker recognition system*

The input consists of a speech signal and, in the case of speaker verification, a claimed identity. Both in training and test modes, the signal is submitted to an acoustic analysis step. Speaker models are generated during the training phase. In test mode, the speech utterance (or a model of this utterance) is compared to one or several speaker models in the scoring module. The score(s) thus generated is/are used to select the most probable speaker's identity (speaker identification) or to decide between acceptance or rejection (speaker verification).

9.3.2. *Acoustic analysis*

In speech recognition, the most common acoustic analysis scheme involves slicing the speech signal into fixed-length blocks (called *frames*) of typical duration equal to 25 ms, shifted every 10 ms (i.e. with some overlap) and to extract for each frame a set of *cepstral coefficients*. These coefficients correspond merely to the first coefficients of the Fourier decomposition of the short-term power spectral density of the signal, in logarithmic scale. They characterize the envelope of the short-term spectrum, without accounting for finer details (such as the exact location of the harmonics). Variants can be introduced, such as non-linear frequency scale transformations (Mel scale) or alternate spectral estimation techniques (such as linear prediction).

Given that morphological properties of the vocal tract and articulatory habits of the speaker have a direct influence on the spectral envelope of spoken sounds, it is reasonable to consider cepstral coefficients for speaker recognition. However, additional and/or alternative acoustic representations also appear potentially relevant. For instance, features for describing prosodic information (melody, accentuation, rhythm) are first choice candidates. Not only are the corresponding acoustic features (fundamental frequency, energy, duration) clearly speaker-dependent, but they are also remarkably robust to many channel distortions. Note however, that these features are also easier to modify intentionally by the speaker.

Nevertheless, the manner to efficiently use prosodic information for modeling speaker characteristics remains an open problem, all the more as modeling and classification techniques are not well suited to handling heterogenous, discontinuous and non-uniform information, as is the case for prosodic phenomena.

Similarly, it would sound rather judicious to seek acoustic features specific to speaker recognition, by using for instance discriminant analysis techniques. However, no systematic advantage of such an approach has been proven to date.

As a consequence, most state-of-the-art speaker recognition techniques use rather conventional acoustic representations, namely cepstral coefficients and their time derivatives (*delta*-cepstrum), including (or not) energy coefficients (usually, in delta form only). In practice, the feature vector dimension is in the range of 20 to 30 (including deltas).

Therefore, the same type of acoustic representation is used for speaker recognition systems and speech recognition systems (including speaker-independent systems), etc. What looks like an apparent paradox may in fact not be one: the acoustic analysis step can be understood as a simple change of representation, the goal of which is to give a more compact description of the signal, while preserving as much information as possible. The task-specific process is moved to the modeling and scoring modules, which are designed to detect the speaker-specific characteristic information. Cepstrum coefficients are indeed a good representation for preserving information stemming from the speech signal. However, it is reasonable to imagine that improved performance will ultimately be gained from proper incorporation of prosodic information in speaker recognition systems (and speech recognition systems too).

9.3.3. *Probabilistic modeling*

The core of almost all conventional approaches to speaker recognition is a probabilistic model. Conceptually, the training phase involves describing the characteristic of a speaker in terms of a *speaker model*, the parameters of which are learned using more or less elaborate statistical methods. The model describes an approximation of the multidimensional distribution of the speaker's acoustic feature vectors. However, especially for verification tasks, the training phase can also require the learning of contrastive characteristics for the speaker, also called the *non-speaker model*.

In this chapter, we will simplify notations and intentionally will make no difference between a random variable (X) and a realization of that variable ($X = x$). Symbols P and p will be used to denote probabilities. We will denote as $\overline{X}$ the complementary event of X, and $\hat{X}$ the identity (or the decision) provided by the automatic system. Notation $\mathcal{P}$ will denote a probability density function (or distribution) and $\hat{\mathcal{P}}$, an estimation of that density (also called *likelihood* function).

9.3.3.1. *Speaker modeling*

Given a speech utterance Y, composed of a sequence of r-dimensional acoustic feature vectors $\{y_t\}_{1 \le t \le n}$, the model of speaker X aims at describing the probability density function $\mathcal{P}(Y|X)$ for any utterance produced by that speaker. Most state-of-the-art techniques use more or less degenerated variants of *hidden Markov models* (HMMs) with state densities modeled by *Gaussian mixture models* (GMMs), i.e. a linear combination of r-dimensional Gaussian laws (sometimes reduced to a single Gaussian).

Let us recall that, in its most general form, HMM (X) is an automaton composed of a set of q states $\{Q_i\}_{1 \le i \le q}$, each of which being characterized by a probability law $\mathcal{P}(y|X, Q_i)$ and connected to other states by branches to which are attached transition probabilities a_{ij}^x (see Figure 9.3).

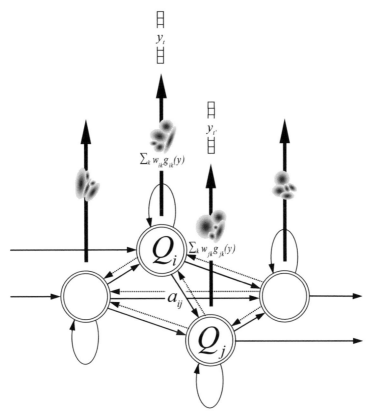

Figure 9.3. *Schematic representation of an HMM with multidimensional GMMs as emission laws*

A random process driven by an HMM X follows a state sequence Q_{i_t} at successive instants t, and at each instant it emits an observation y_t according to the probability law $P(y|X, Q_{i_t})$. For a complete sequence of observations $Y = \{y_t\}_{1 \leq t \leq n}$ and the corresponding sequence of states $Q = \{Q_{i_t}\}_{1 \leq t \leq n}$, the probability density function for HMM X is written:

$$P_{HMM}(Y, Q|X) = p(Q|X)P_{HMM}(Y|X, Q) = \prod_{t=1}^{n} a^{X}_{i_{t-1}i_t} P_{GMM}(y_t|X, Q_{i_t}) \quad [9.1]$$

Each state i is characterized by a distinct GMM emission probability law:

$$P_{GMM}(y|X, Q_i) = \sum_{k=1}^{m} w^{X}_{ik} G_{ik}(y|X) \quad [9.2]$$

where m designates the number of Gaussians in the mixture and where w_{ik}^X are the scalar weights of the Gaussian functions $G_{ik}(y|X)$, themselves defined by a r-dimensional mean vector μ_{ik}^X and a $r \times r$ covariance matrix Σ_{ik}^X. In practice, the covariance matrices are generally constrained to be diagonal, i.e. with only r free parameters.

The HMM formalism is extremely effective. The model states represent the various acoustic events constituting the utterance. The emission probability laws describe the localization and the dispersion of acoustic realizations for these events. The HMM topology, i.e. its connectivity, makes it possible to handle different types of linguistic constraints. *Left-right HMM*, i.e. a model topology with a partial time-order between states, is used for speech utterances with a predetermined structure and possible pronunciation variants (password-based and text-prompted systems).

Ergodic HMM, i.e. a fully interconnected model, or ergodic network of left/right HMM (typically, a large vocabulary continuous speech recognition system with a bigram model) is used when the linguistic structure of the utterance is unconstrained, which is the case for text-independent systems.

Furthermore, many conventional speaker recognition approaches can be expressed in the HMM formalism:

–   *long-term Gaussian statistics* (mean and covariance) are equivalent to a 1-state HMM with a 1-gaussian density function;

–   *dynamic time warping* (DTW) prototypes can be interpreted (provided some restrictions are set on the local constraints), as a left-right HMM with one state per frame and a unit-covariance mono-Gaussian emission probability;

–   *vector quantization* (VQ) is formally equivalent to an ergodic HMM with mono-Gaussian emission probabilities and common covariance matrices;

–   GMM, which correspond to a 1-state HMM with a GMM emission probability function, but which can also be viewed as a HMM with multiple mono-Gaussian states and transition probabilities between states equal to the weights of the Gaussians composing the mixture.

In non-degenerated forms of these configurations, the number of HMM states q and the number of Gaussians m are pre-determined parameters, which are set experimentally on the basis of the phonetic complexity of the utterance to be modeled.

To the above list, it must also be added that a number of alternative approaches exist, which rely on neural networks, auto-regressive vector models or support vector machines. However, it is beyond the scope of this chapter to detail them, but the reader is invited to refer to more technical literature for additional information.

Probabilistic modeling of a speech utterance or a set of speech utterances for a given speaker is based on statistical estimation techniques. These techniques rely on variants of the EM (*expectation-maximization*) algorithm, which aims at optimising a modeling criterion from the training data. In practice, the amount of training data is limited and the *Maximum Likelihood* (ML) criterion leads to models that over fit the training data and have limited generalization abilities. This is why it is preferable to use more elaborate criteria, which can induce constraints on model parameter values: for instance, fixing *a priori* some parameter values, regularizing the ML criterion, resorting to Bayesian learning or using MAP (*maximum a posteriori*) criteria.

9.3.3.2. *Non-speaker modeling*

In speaker verification, the acceptance/rejection decision is usually made by comparing the test utterance to two models: to the claimed speaker model X, but also to a non-speaker model, denoted as $\overline{X}$, which corresponds (at least virtually) to the same utterance as it would be pronounced by any other speaker. This non-speaker model is trained by statistical methods, and two variants are distinguished, depending on the way it is defined:

– *the cohort approach*, for which the non-speaker model $\overline{X}$ is obtained from data (or models) corresponding to a set of other speakers Γ_X selected on the basis of their complementarity with the current speakers (for instance, speakers with similar characteristics); or

– *the world-model approach*, consisting of a single speaker-independent model Ω, common to all speakers X.

Intermediate approaches exist, for instance class-dependent world models, i.e. specific to some *a priori* knowledge (claimed speaker gender, type of microphone, transmission channel properties, etc.).

In any case, the non-speaker model aims at approximating the probability density function $\mathcal{P}(Y|\overline{X})$ of utterances Y under the hypothesis that an impostor has produced them.

9.3.4. *Identification and verification scores*

In this section, Y denotes specifically a *test* utterance produced by a person using a speaker recognition system.

It is worthwhile recalling that, in the case of speaker identification (closed-set case), the system's task is to retrieve the identity of the test speaker, i.e. that to which utterance Y corresponds *best*. In the case of speaker verification, utterance Y

is accompanied by a claimed identity (X) and the function of the system is to decide whether the match between the two is *good enough*.

9.3.4.1. *Likelihood and likelihood ratio*

In the framework of probabilistic approaches, speaker recognition systems use similarity measures between the utterance Y and a speaker model based on the *likelihood* of the utterance Y with respect to model X, i.e. an *approximation* of the probability $\mathcal{P}(Y|X)$ that the observed utterance has been produced by the current speaker.

In the HMM formalism, this likelihood is obtained by summing individual likelihoods resulting from each possible state sequence Q in the model (*Baum-Welch* approach):

$$\hat{P}(Y|X) = \sum_Q \mathcal{P}_{HMM}(Y,Q|X) = \sum_Q p(Q|X)\mathcal{P}_{HMM}(Y|X,Q) \qquad [9.3]$$

where $p(Q|X)$ denotes the probability of the state sequence Q for model X. The quantity $\hat{P}(Y|X)$ is calculated with the *forward-backward* algorithm. However, a frequent approximation (the *Viterbi* approximation) involves evaluating the likelihood only on the state sequence Q^* corresponding to the maximum individual likelihood:

$$\hat{P}(Y|X) \approx \mathcal{P}_{HMM}(Y|X,Q^*) \qquad [9.4]$$

where:

$$Q^* = \arg\max_Q \mathcal{P}_{HMM}(Q|X,Y) = \arg\max_Q \mathcal{P}_{HMM}(Y,Q|X) \qquad [9.5]$$

The state sequence Q^* and the corresponding likelihood are obtained simultaneously by means of the *Viterbi* algorithm.

Let us insist on the fact that the likelihood provided by a probabilistic model is always an approximated value of the real probability density function (denoted $\mathcal{P}$ in this text) that remains inaccessible in practice, as opposed to the model likelihood $\hat{P}$.

In speaker identification, the quantity measuring the similarity between the test utterance Y and a reference speaker X_ℓ is generally based on the model likelihood, via its logarithm, as:

$$V_\ell(Y) = -\log \hat{P}(Y|X_\ell) \qquad [9.6]$$

Thus defined, the metric V_ℓ is low when the test utterance matches better the reference speaker model.

In speaker verification, Bayesian decision theory relies on a score based on the *log likelihood ratio* (LLR):

$$S_X(Y) = \log \frac{\hat{P}(Y|X)}{\hat{P}(Y|\overline{X})} \qquad [9.7]$$

This score is used to make the accept/reject decision on the claimed identity X. When the LLR is high, the test utterance can be considered as specific to the claimed speaker.

It must be noted that the above metrics give a dissymmetric role to the speaker model and to the test data. Some variants explicitly model the test utterance Y and use a similarity measure between models, which can be defined as symmetric.

9.3.5. *Score compensation and decision*

Decision modules used in speaker recognition are usually based on conventional Bayesian decision theory: nearest neighbor for speaker identification, comparison to a threshold for verification. However, some processes of score compensation or normalization are usually necessary, for reasons detailed below.

9.3.5.1. *Decision in speaker identification*

Conventionally, decision in a speaker identification system relies on the *nearest neighbor* rule, which derives from the Bayes formula.

In order to minimize the recognition error rate on the entire population of reference speakers $\{X_\ell\}_{1\leq\ell\leq L}$, the optimal decision rule involves choosing the speaker $\hat{X}_\ell$ for which the *a posteriori* probability is maximum, given the test utterance (namely, the *maximum a posteriori* rule):

$$\hat{X}_\ell = \arg\max_{X_\ell} P(X_\ell|Y) \qquad [9.8]$$

By applying the Bayes formula, the previous rule is rewritten:

$$\hat{X}_\ell = \arg\max_{X_\ell} \frac{\mathcal{P}(Y|X_\ell)p(X_\ell)}{\mathcal{P}(Y)} = \arg\max_{X_\ell} \mathcal{P}(Y|X_\ell)p(X_\ell) \qquad [9.9]$$

where $p(X_\ell)$ denotes the *a priori* probability of speaker X_ℓ and where the suppression of the denominator in the arg max results from the fact that it is a constant quantity for all reference speakers. If, moreover, all speaker priors are assumed equal and if the probability density function is replaced by the model likelihood, the *maximum likelihood* decision rule is obtained:

$$\hat{X}_\ell = \arg\max_{X_\ell} \mathcal{P}(Y|X_\ell) \approx \arg\max_{X_\ell} \hat{P}(Y|X_\ell) = \arg\min_\ell V_\ell(Y) \qquad [9.10]$$

Thus, speaker identification in the *maximum a posteriori* sense on a set of equally probable speakers can be approximated by the nearest neighbor rule with metric V, as defined in the previous section.

9.3.5.2. *Decision in speaker verification*

As opposed to (closed-set) speaker identification, for which the system has to be optimized with respect to a single quantity (the recognition error rate), the optimization of a speaker verification system results from a compromise between two types of errors: the false rejection rate and the false acceptance rate.

This compromise is classically formulated via a *detection cost function* (DCF), which assigns a weight to both types of errors. Taking into account the *a priori* impostor probability, the cost function writes:

$$C = p_X C(\hat{\bar{X}}|X)P(\hat{\bar{X}}|X) + p_{\bar{X}}C(\hat{X}|\bar{X})P(\hat{X}|\bar{X}) \qquad [9.11]$$

In this equation, $C(\hat{\bar{X}}|X)$ and $C(\hat{X}|\bar{X})$ respectively denote the costs of a false rejection and of a false acceptance, $P(\hat{\bar{X}}|X)$ and $P(\hat{X}|\bar{X})$ the corresponding probabilities and p_X and $p_{\bar{X}}$ the *a priori* probabilities that a user be the actual claimed speaker or an impostor $(p_{\bar{X}} = 1 - p_X)$. Setting:

$$\beta = \frac{p_{\bar{X}}}{p_X} \qquad \text{and} \qquad R = \frac{C(\hat{X}|\bar{X})}{C(\hat{\bar{X}}|X)} \qquad [9.12]$$

yields:

$$C \propto P(\hat{\bar{X}}|X) + \beta R P(\hat{X}|\bar{X}) \qquad [9.13]$$

The pair (β, R) defines the *operating conditions* of the system. The product βR appears as the relative weight of false acceptance w.r.t. false rejection, in the performance definition.

In the context of decision theory, the optimal rule that minimizes the DCF involves comparing the ratio between the speaker and non-speaker *a posteriori* probabilities (given the test utterance), to the value of R:

$$\frac{P(X|Y)}{P(\overline{X}|Y)} \underset{\hat{\overline{X}}}{\overset{\hat{X}}{\underset{<}{>}}} R \qquad\qquad [9.14]$$

Using once again the Bayes formula, this decision rule can be rewritten:

$$\log \frac{P(Y|X)}{P(Y|\overline{X})} \underset{\hat{\overline{X}}}{\overset{\hat{X}}{\underset{<}{>}}} \log \beta R \qquad\qquad [9.15]$$

Therefore, the optimal decision rule for minimizing the total cost function involves comparing the logarithm of the ratio of the speaker and non-speaker probability density functions for the test utterance to a threshold $T = \log \beta R$ that only depends on the operating conditions. In practice, the probability density functions are approximated by their corresponding model likelihood:

$$S_X(Y) \underset{\hat{\overline{X}}}{\overset{\hat{X}}{\underset{<}{>}}} \log \beta R \qquad\qquad [9.16]$$

9.3.5.3. *Score compensation*

The decision rule of equation [9.16] applies to the log likelihood ratio and not to the actual ratio of the true probability density functions. Therefore, it is not optimal for the speaker verification task, in the sense that it does not minimize the DCF. The optimal decision function on the likelihood ratio can be written in a rather general manner as:

$$S_X(Y) \underset{\hat{\overline{X}}}{\overset{\hat{X}}{\underset{<}{>}}} \Theta(\beta, R, X, Y) \qquad\qquad [9.17]$$

where the function Θ depends not only on the operating conditions, but also on the claimed speaker and the speech utterance content.

The problem of decision *threshold setting* can be understood as a particular case of the more general problem which involves the *adjustment* (or *compensation*, or *normalization*) of the score S_X in order to correct for the inaccuracies of the speaker and the non-speaker models, given that these inaccuracies are generally speaker-dependent but also variable with the linguistic content of the speech utterance.

The threshold setting problem can therefore be reformulated by considering that, for a given operating condition (β, R), the *raw* LLR $S_X(Y)$ must be submitted to a transformation $\mathcal{F}_{X,Y}$, so that the "optimal" decision rule can be applied to the *adjusted* LLR: $S'_X(Y) = \mathcal{F}_{X,Y}[S_X(Y)]$:

$$S'_X(Y) \underset{\hat{\bar{X}}}{\overset{\hat{X}}{\underset{<}{\gtrless}}} \log \beta R \qquad [9.18]$$

A very common score adjustment procedure involves normalizing the raw score as a function of the utterance duration. In fact, in most cases, the LLR for the entire speech utterance Y can be written as the sum of terms corresponding to each speech frame y_t:

$$S_X(Y) = \sum_{t=1}^{n} s_X(y_t) \qquad [9.19]$$

where $s_X(y_t)$ corresponds to the contribution of each frame y_t to the overall score for the entire utterance. Under the (rather inexact) hypothesis that the frame-based scores are statistically independent, S_X can be normalized as follows:

$$\bar{S}_X^t(Y) = \frac{1}{n} S_X(Y) = \frac{1}{n} \sum_{t=1}^{n} s_X(y_t) \qquad [9.20]$$

so that its expected value does not depend (or depends less) on the utterance length. Other, more elaborated normalizing functions depending monotonically but non-linearly on the utterance length can also be envisaged.

A second approach, also rather common, involves applying a constant, speaker-independent translation to the score $\bar{S}_X^t$, as a function of the operating conditions:

$$S'_X(Y) = \overline{S}^t_X(Y) - \Delta(\beta, R) \qquad [9.21]$$

where Δ is obtained, for a given operating condition, by optimizing a speaker-independent (and utterance-independent) decision threshold Θ_{dev} so as to minimize the DCF on a *development* population. This population is composed of a set of speakers chosen to be as representative as possible of the real population of users (however, this population is distinct from the test population). Once Θ_{dev} has been estimated, the score adjustment is obtained by setting:

$$\Delta(\beta, R) = \Theta_{dev}(\beta, R) - \log \beta R \qquad [9.22]$$

This approach is based on the hypothesis that there is a global difference between the speaker models and the non-speaker model, which causes a systematic bias in the LLR.

A more elaborate (and experimentally more efficient) score compensation technique is called *z-normalization* and involves estimating the *impostor score distribution* mean and standard deviation, i.e. the mean M_X and standard deviation D_X of the distribution of $\overline{S}^t_X$ when Y has been produced by an impostor. These values are obtained by simulating impostor tests from a set of representative speakers (distinct from X), also called pseudo-impostors. The score $\overline{S}^t_X(Y)$ is then centered and reduced, i.e. z-normalized (before being submitted to the translation that corresponds to the operating condition):

$$S'_X(Y) = \frac{\overline{S}^t_X(Y) - M_X}{D_X} + \Delta'(\beta, R) \qquad [9.23]$$

where Δ' is (generally) a negative constant similar to Δ and obtained on a development population. The underlying objective of z-normalization is to make the speaker-independent score translation more efficient by normalizing first the impostor score distribution for the various speakers.

Ideally, more efficient speaker-dependent LLR normalizations could be obtained on the basis of a reliable *speaker score distribution*, i.e. the distribution of $\overline{S}^t_X$ when Y is actually produced by the speaker. However, the sole speaker data available are usually those that have been used to train the speaker model, and the distribution of the score yielded by these data is both inaccurate and very biased. Some approaches try to estimate indirectly some parameters of this distribution, but most LLR compensation techniques use only the impostor score distribution.

The score adjustment as a function of the utterance Y has yielded to several scheme, the most popular of which is the *t-normalization* approach, under which the mean and standard deviation of the impostor score for utterance Y is estimated by computing, on the *test* utterance (i.e. during the test phase), the mean $M_X(Y)$ and the standard deviation $D_X(Y)$ of $\overline{S}_X^t(Y)$ for a number of speaker models which are distinct from the claimed speaker. These quantities can then be used similarly as in equation [9.23], but note that this requires a significant computation overhead during the test phase, whereas z-normalization can take place once for all before the system is operational.

9.3.6. *From theory to practice*

The theoretical elements presented above represent a general framework within which a number of additional steps are required to obtain a system robust to real use conditions. Some of them are based on well-defined concepts, other resort to experience, engineering or ergonomics.

It is not feasible to list all these techniques in this chapter. We limit ourselves to a non-exhaustive list, which illustrates several problems and possible solutions.

   –   *robustness to noise, recording conditions and transmission channel*: spectral subtraction, cepstral transformation on a subset of frequency bands, cepstral mean subtraction (CMS), RASTA coefficients, microphone dependent world-model, score compensation according to the type of microphone (h-norm), multi-band approaches, latent factor analysis (LFA);

   –   *robustness to the lack of training data (or sessions)*: predefinition, copy or thresholding of some model parameters (transitions probabilities set equal, diagonal covariance matrices, speaker-independent variances and/or weights), model learning by Bayesian adaptation, introduction of artificial variability during training;

   –   *robustness of score computation*: voice activity detection, removal of low-energy and/or highly distorted frames (in the case of transmission errors), forced and/or synchronous alignment, Gaussian pre-selection from a speaker-independent model, score fusion;

   –   *robust decision*: multi-level decision (acceptance, doubt, rejection), sequential decision with multiple verification steps, decision fusion;

   –   *robustness to voice variability and drift:* incremental adaptation along successive test sessions;

   –   *robustness to intentional imposture*: discriminative training of speaker models with speech utterances of similar speakers, detection of pre-recorded and/or artificial speech (technological imposture);

  – *robustness to the consequences of a false rejection* (in particular, the risk of offending a genuine user): triggering of alternative procedures in the case of a negative verification or a high level of doubt on the user's identity, in particular verbal information verification and/or switching to a human operator;

  – *robustness to the consequences of false acceptance* (risk mitigation): adaptation of the service coverage to the level of confidence, triggering of confirmation procedures for transactions with high stakes;

  Experience proves that taking carefully into account these various factors is essential to the efficient functioning of a speaker recognition system in the context of real world applications.

9.4. Performance evaluation

  In this section, we present an overview of the most common performance measures used to evaluate speaker identification and verification systems. We will refer to the *evaluation set* as the set of speech data used in the evaluation protocol. Classically, the evaluation set decomposes into a training set (used for training the speaker models), a development set (used for tuning the system) and a test set (used for measuring the performance).

9.4.1. *Error rate*

  In general, the performance of a speaker identification system is measured by calculating, on the test set, the mean of the *recognition error rate* for each pair of referenced speakers:

$$\mathcal{T}_{RER} = \hat{P}(\hat{\bar{X}}|X) = \frac{1}{L}\sum_{\ell}\frac{1}{L-1}\sum_{\ell'\neq\ell}\mathcal{T}_{RER}(\hat{X}_{\ell'},X_\ell) = \frac{1}{L(L-1)}\sum_{\ell}\sum_{\ell'\neq\ell}\hat{P}(\hat{X}_{\ell'}|X_\ell)$$

[9.24]

where L denotes the number of speakers in the evaluation set and $\hat{P}(\hat{X}_{\ell'}|X_\ell)$ an estimation of the probability of misrecognising speaker X_ℓ as speaker $X_{\ell'}$ on the test set. As opposed to a direct sum over all test utterances, this calculation as the mean of partial speaker-dependent error rates compensates for a possible unbalanced number of utterances per speaker. Note that the error rate $\hat{P}$ will tend to increase with the size L of the reference speaker population.

The measure of performance for a speaker verification system in a given operating condition calls for two error rates: the false rejection rate (FRR) and the false acceptance rate (FAR). These error rates are obtained respectively as:

$$T_{FR} = \hat{P}(\hat{\bar{X}}|X) = \frac{1}{L}\sum_{\ell} T_{FR}(X_\ell) = \frac{1}{L}\sum_{\ell} \hat{P}(\hat{\bar{X}}_\ell|X_\ell) \qquad [9.25]$$

and

$$T_{FA} = \hat{P}(\hat{X}|\bar{X}) = \frac{1}{L}\sum_{\ell} T_{FA}(X_\ell) = \frac{1}{L}\sum_{\ell} \frac{1}{L'(\ell)} \sum_{\ell'\neq\ell} \hat{P}(\hat{X}_\ell|X_{\ell'}) \qquad [9.26]$$

Here, L denotes the size of the reference speaker population, $L'(\ell)$ the number of distinct impostors tested with the claimed identity X_ℓ and where $\hat{\bar{X}}_\ell$ and $\hat{X}_\ell$ denote respectively the rejection or acceptance decision with respect to this claimed identity. The pair $(\mathcal{T}_{FA}, \mathcal{T}_{FR})$ is called the *operating point*.

For obvious methodological reasons, it is essential that test speakers do not belong to the development population and that impostor speakers belong neither to the non-speaker population (used to train the non-speaker model(s)) nor to the development population (pseudo-impostors used for score adjustment). However, nothing prevents using, in a test protocol, other reference speakers as impostors against a given speaker.

The profile of the impostor population has of course a significant impact on the performance of a system, especially the proportion between male and female impostors. In general, the population of impostors tested against a reference speaker must systematically be either gender-balanced (same proportion of male and female) or composed solely of speakers of the same gender as the claimed speaker (same-gender impostor configuration). Note also that laboratory tests using databases collected outside of any specific applicative context often show a clear lack of realism for impostor tests, which do not account for real, intentional, impostor attempts.

9.4.2. *DET curve and EER*

In the specific case of speaker verification, false rejection and false acceptance rates are dependent on the decision threshold Θ, itself a function of the operating conditions. When the threshold varies, one of the error rates increases and the other one decreases. The two rates can be depicted as a function of each other, as a curve

$\mathcal{T}_{FR} = \Phi(\mathcal{T}_{FA})$, which is called *receiver operating characteristic* (ROC). This curve is located in the first quadrant and is monotonously decreasing.

If the *x*- and *y*-scale is transformed into a *normal deviate* scale, the ROC curve becomes a DET (*Detection Error Tradeoff*) curve. Under this representation, the DET curve of a speaker verification system that would have Gaussian distributions of the speaker and impostor scores (with same variances) would be a straight line with slope -1. The intersection of the DET curve with the *xy* bisecting line is closer to the origin when the two distributions are further apart.

In practice, the score distributions are not really Gaussian, but do not differ drastically from a Gaussian. The DET representation provides a way to compare visually several systems on a wide range of operating conditions. Figure 9.4 gives an example of DET curves.

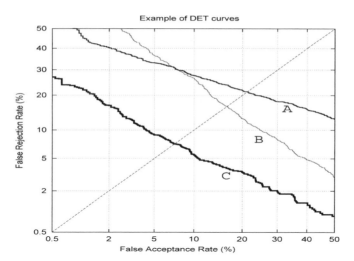

Figure 9.4. *Example of DET curves. System A performs better than system B for operating conditions corresponding to low false acceptance rates, whereas system B performs better for low false rejection rates. System C performs better that A and B in all operating conditions. The EER for systems A, B and C are respectively 22%, 17% and 7%, approximately*

A point of interest on the DET curve is its intersection with the *xy* bisecting line. This point corresponds to operating conditions for which the false acceptance rate is equal to the false rejection rate. It is currently called EER, (*Equal Error Rate*):

$$\mathcal{T}_{EER} = \arg_{\mathcal{T}} \{\Phi(\mathcal{T}) = \mathcal{T}\} \qquad\qquad [9.27]$$

This figure may not correspond to a realistic operating condition. However, it remains a rather popular figure to evaluate the system's separation capability between speakers and impostors.

9.4.3. *Cost function, weighted error rate and HTER*

For a given decision threshold, the performance of a verification system is characterized by a specific operating point $(\mathcal{T}_{FA}, \mathcal{T}_{FR})$. For the corresponding operating conditions (β, R), the system cost function writes: $C \propto \mathcal{T}_{FR} + \beta R \mathcal{T}_{FA}$. By normalizing this cost by the sum of the weights on each type of error, the *weighted error rate* (WER) is obtained:

$$\mathcal{T}_{\beta R} \propto \frac{\mathcal{T}_{FR} + \beta R \mathcal{T}_{FA}}{1 + \beta R} \qquad [9.28]$$

that is, a quantity which can be interpreted as a weighted average between false acceptance and false rejection error rates, with weights proportional to their respective costs. In predetermined operating conditions, $\mathcal{T}_{\beta R}$ makes it possible to compare two systems (or the same system with two tuning procedures) via a single figure, homogeneous to an error rate and minimum when the optimal functioning conditions are reached.

In the particular case when both costs are equal $(R = 1)$ and genuine/impostor attempts are equiprobable $(\beta = 1)$, the WER is equal to the arithmetic mean of the false acceptance and the false rejection rates. This quantity is sometimes termed as *Half Total Error Rate* (HTER). Note that the HTER is, in general, not equal to the EER.

Finally, the WER obtained with a system for which the decision rule has been tuned *a priori* has to be distinguished from the WER obtained *a posteriori* (i.e. by optimizing the cost function on the test set). Of course, the latter is always more advantageous than the former, but it does not correspond to a realistic scenario. Nevertheless, the difference between both scores is useful in evaluating the quality of the threshold setting procedure.

9.4.4. *Distribution of errors*

Beyond performance measures consisting of one or two global figures, or a DET curve, it can be extremely relevant to examine more closely the system errors as a function of the speakers.

In fact, for most systems, it is usually observed that the errors are very unevenly distributed among speakers: a large proportion of false rejection are often concentrated on a small subgroup of so-called *goats* (speakers with very selective models) whereas false acceptances affect an other small subgroup of lambs (i.e. speakers which are very vulnerable to imposture, because their model has a very broad coverage). Once these two sub-populations are put aside (typically a small percentage), performance on the rest of the population (i.e. the *sheep*) turns out to be much more favorable. This is why it can be interesting to re-evaluate the performance of a system after having removed a fraction of α % of speakers for which the individual error rates are the highest:

$$\mathcal{T}_\alpha = \frac{1}{R_\alpha} \sum_{r < R_\alpha} \mathcal{T}(X_r)$$
[9.29]

where r denotes the rank index of speakers according to an increasing individual error rate $\mathcal{T}(X_r)$ and R_α the rounded value of αL. In practice, $\alpha = 0.95$ is a typical value.

This approach must be considered as a way to evaluate the performance of a system on the vast majority of "standard" users, provided an alternative mean of verification can be offered to the small group of "problematic" users. Ideally, the latter should be detected during the training sessions.

9.4.5. *Orders of magnitude*

It is virtually impossible to provide an exhaustive overview of the performance levels reached by speaker recognition technology, given the variety of system profiles, experimental protocols and use conditions.

It is however informative to provide typical orders of magnitudes for some speaker verification system performance, tested in laboratory conditions on data set which are somehow comparable to the requirements of applicative contexts (typically, one or two training sessions and test utterances of a few tens of seconds).

In good recording conditions and for text-constrained approaches, it is possible to reach EERs lower than 0.5% (for instance, on the YOHO database). In text-independent mode, EERs range rather around 2% (for good recording conditions).

For telephone speech, the audio quality degradation causes a significant increase of the error rates, which can reach 1-2% for digit sequences (SESP database) up to 8-10% in text-independent mode (Switchboard database). These figures may rise further when systems are evaluated with an *a priori* tuning of the decision rule.

Current speaker recognition technologies have to be combined with a careful ergonomic design to achieve a reasonable level of performance. Even so, speaker identification and verification are far from being among the more reliable authentication techniques. However, some studies show that voice is a natural verification modality that is generally well accepted by human users. It is also the sole possible modality when a microphone is the only sensor available. For these reasons, speaker recognition remains a relevant authentication modality.

9.5. Applications

Speaker recognition has many possible applications. They cover almost all sectors where user identification and/or authentication is required to secure actions, transactions or interactions, in the real or in the virtual world, in particular for the banking sector and for telecom service providers.

Given that it does not provide an absolute level of security, speaker recognition is interesting in contexts where the goal is to dissuade and to limit fraud rather than to completely prevent it. One advantage of speaker recognition lies in its relatively low deployment cost and its high level of user-acceptance, especially when it is coupled with other voice technologies.

Five application profiles can be listed for speaker recognition: physical access control, protection of remote transactions, audio information indexing, education/entertainment and forensic applications.

9.5.1. *Physical access control*

Under this category, we consider all the applications that require the actual physical presence of the user in front of the system for carrying out the recognition. This context corresponds to situations where a material interaction is intended between the user and the protected place (house, vehicle, room), object (safe), device (cash dispenser), terminal (workstation), goods, and so on. In these cases identification/verification systems commonly in use are keys, smart cards, passwords and PIN (personal identification number).

In the case of physical access, the environment in which the system is operating can be controlled, so that the recording set-up can be calibrated and isolated from unwanted noise. Speaker recognition can take place either locally or remotely (but with good transmission quality).

From the implementation point of view, the user can bear his/her voice characteristics (template/model) on a material object such as a smart card and the authentication then involves verifying the consistency between the template on the card and the actual user's voice. This application profile has some dissuasive effect, as it is possible to trigger an alarm procedure in the case of a doubt on the user's identity (remote transmission of a picture, intervention of a watchman, and so forth).

Note however that for physical access control, it is usually possible to use alternative identification techniques (including biometric techniques such as fingerprint, iris, retina or hand-shape recognition), some of them performing much better than speaker recognition (even though they may be less well-accepted by the users and rather costly to implement).

Today there are very few such applications, none of them being deployed on a large scale. However, voice may find its place among these many modalities, in a multi-modal identity verification framework.

9.5.2. *Securing remote transactions*

In this section, we consider that a remote transaction is any transaction that takes place at a distant location from the user, via a terminal such as a telephone or a personal computer. In this case, the goal is to protect access to services and functionalities reserved to subscribers (telecommunication networks, server, database, web sites), or to authenticate the person who is engaged in a particular operation (bank transfer, e-trade). Today, these services are essentially protected by a password or a PIN code, a stroke on a telephone or a computer keyboard (sometimes combined with the recognition of the terminal itself).

For these applications, the quality of the speech signal is rather uncontrolled, because of the wide range of terminals, microphones, compression schemes, transmission channels, etc., and the signal can be very poor (especially in the case of mobile terminals).

Voice characteristics are usually centralized in a large capacity server. The dissuasive impact of such systems is relatively low, because of a high level of anonymity. However, it is quite easy to transfer a call to a human operator who will handle doubtful authentications.

Despite these limitations, speaker recognition on the telephone is a rather natural authentication modality and a quite simple technology to deploy, as it does not require specific sensors. It can also be combined with speech recognition in an integrated way, both technically and ergonomically.

This is why the banking and telecommunication sectors have shown interest in speaker verification functionalities as a means to secure transactions and services without penalizing the naturalness of the interaction with the users. The profile of the application is usually designed so as to reduce fraud (without trying to suppress it completely), while making life easier for genuine users. Why not imagine, in the near future, a secure telephone with no keyboard!

9.5.3. *Audio information indexing*

A third category of applications for speaker verification technologies covers audio information indexing; for instance, speaker-based audio archive annotation, speaker indexing, speaker change detection, etc. Such needs stem mainly from the cinema and video production industry, from media groups, audio-visual archiving institutes and from journalists. However, more general contexts emerge such as speaker indexing for audio recordings of meetings and conferences.

The specificities of such systems are twofold: firstly, it is quite common to have a large quantity of training speech data, at least for some speakers (for instance, public figures or dominant speakers) and secondly, the system usually does not need to function in real time (and several passes are therefore conceivable). Moreover, the detection of occurrences from a same speaker in a given recording is usually easier because some sources of variability are absent, or less salient (same day, same microphone, same channel, etc.). However, the segmentation in speakers is not known, nor the exact number of speakers and it is quite common to observe several speakers speaking simultaneously.

Audio indexing is an emerging domain and tomorrow it will be feasible to navigate in audio recordings, as we do today in textual contents, to search for program types, topics and, of course, speakers. As well as speech/music classification, speech transcription, keyword and key-sound spotting, speaker detection and recognition will play a relevant role in audio indexing.

9.5.4. *Education and entertainment*

Another application domain, under-developed so far, but which could represent an interesting outlet for speaker recognition technology, is the education and entertainment sector: educational software, toys, video consoles, CD-ROMs or Internet games, etc.

Indeed, educational software and entertainment products show increasing interactivity and user-customization in their design and their functionalities. With technological progress, voice is bound to become a common modality within a few years. In this context, speaker recognition may emerge as a way of recognizing the player in a game, the owner of a toy, or even to integrate voice features as one element of the game (an imitation game, for example).

The imagination of the manufacturers and the reactivity of the users will be the driving force of such a potential development, given that in such contexts, perfect recognition is not vital to the application, which can tolerate some error margins. Nevertheless, only trial and experience will tell whether such a market really exists or not.

9.5.5. *Forensic applications*

By considering now the applications of speaker recognition in the forensic domain, this section addresses a context for which the matters at stake and their consequences are considerably more serious than those of commercial products. This application domain covers not only the legal and judicial applications, but also intelligence activities and home incarceration.

In some situations, a voice recording is a key element, and sometimes the only one available, for proceeding with an investigation, identifying or clearing a suspect and even supporting an accusation or a defense in a court of law.

In this context, the use of speaker recognition must face situations of immense difficulty and diversity. Typically, the anonymous voice has been recorded in poor technical conditions, in unusual speaking situations (fear, stress), and may have been submitted to artificial transformation using technological devices. This voice has to be compared to reference recordings from one or several suspects, which have spoken in very different conditions than that of the anonymous voice. Moreover, the suspect may be more or less cooperative in recording his/her voice.

The public perception is that voice identification is a straightforward task and that there exists a reliable "voiceprint" in much the same way as a fingerprint or a genetic DNA print. This is not true because the voice of an individual has a strong behavioral component and is only remotely based on anatomical properties.

The way speaker recognition technologies are used in legal cases and the weight that is given to them is rather variable in different countries. In France, the procedures and the (semi-automatic or automatic) techniques used by human experts in forensic contexts are very poorly documented in scientific literature. Moreover, they are not submitted to any evaluation protocol that would allow us to define and measure their validity and their limits. This is the reason why representatives of the

French and French-speaking Speech Science community (AFCP[5], formerly GFCP) have asked for a moratorium on speaker recognition for legal purposes in France, until the technologies used by the experts are scientifically validated and evaluated through a well-defined protocol.

9.5.6. *Perspectives*

Rather than functioning as a key that allows or forbids access to a place or a service, speaker recognition appears as a technology that must be integrated into *a set* of authentication tools that contribute jointly to a user's profile definition and tracking. Besides other identification means, voice can help to detect early the fraudulent use of a service by an impostor. In this context, speaker recognition must be approached as an *alarm bell* for triggering, if necessary, a heavier additional verification procedure, if the transaction has important stakes and/or if other elements of the user profile are discordant: for instance, the transfer to an operator who will use other means of identity verification.

In practice, many applicative contexts fulfill these requirements. Moreover, voice recognition has a high level of user acceptance and offers interesting synergies with speech recognition and oral dialogue systems. Therefore, speaker recognition should gain importance over the years, as a means of authentication beside others, in the context of secure transactions, user verification and audio information indexing.

At the same time, it is essential that members of the legal professions and forensic experts become aware of the technological, but also intrinsic limitations of the speaker identification tasks, which make it impossible, given the current knowledge, to state with absolute certainty that two voices belong (or do not belong) to a same person. In parallel, it is absolutely necessary to set up concerted evaluation protocols and procedures for estimating the reliability of any speaker identification technology that could be used in a legal framework.

9.6. Conclusions

Speaker recognition is situated in a very specific position within the topics covered by speech processing. Dealing with the characteristics of the source of the message, rather than with the message itself, it may appear as a distinct autonomous field. In fact, state-of-the-art techniques for speaker recognition rely on rather similar approaches to those used in speech recognition (probabilistic formalism,

---

5 Association Francophone de la Communication Parlée, also a Regional SIG (Special Interest Group) of ISCA (International Speech Communication Association).

HMM models, etc.), thus showing the connection and the complementation between both technologies.

From an application perspective, speaker recognition has reached a level of maturity. Its good user-acceptability and the possible synergies with speech recognition techniques make it possible to integrate it progressively in some commercial applications, especially those with vocal components. Current levels of performance are already good enough to deploy applications that can reduce fraud in the context of remote commercial transactions with an appropriate ergonomic design. In parallel, speaker recognition technologies are emerging in the domain of audio indexing.

Some research and progress on the subject concern the consolidation of existing approaches (robustness to noise, finer speaker models, improved statistical estimation, efficient decision). More prospective work tends to go beyond the state-of-the-art in several directions, by using parametric speaker transformations, adaptive speaker representations, the selection of speaker-specific information, the direct estimation of the likelihood ratio, the use of neural networks, source separation techniques and, recently, the use of support vector machines (SVM) with some success. Some of these tracks will lead to clear improvements in terms of performance and robustness. Some others will facilitate the integration of speaker recognition techniques within a growing number of applications.

9.7. Further reading

Specialized Workshop Proceedings

Proceedings of ESCA. "Workshop on Automatic Speaker Recognition, Identification and Verification", Martigny (Switzerland), 1994.

Proceedings of RLA2C: "Workshop on Speaker Recognition and its Commercial and Forensic Applications", Avignon (France), 1998.

Proceedings of 2001: "A Speaker Odyssey, ISCA/ITRW workshop", Chania (Crete, Greece), 2001.

Proceedings of IEEE Odyssey: "The Speaker and Language Recognition Workshop", Toledo (Spain), 2004.

Proceedings of IEEE Odyssey: "The Speaker and Language Recognition Workshop", San Juan (Puerto Rico, USA), 2006.

Proceedings of Odyssey, "The Speaker and Language Recognition Workshop", Spier (Stellenbosch, South Africa), 2008.

Special issues in international journals

Speech Communication: "Special Issue on Speaker Recognition, Identification and Verification", Vol. 17, no.1-2, Elsevier, August 1995, F. BIMBOT, G. CHOLLET, A. PAOLONI (eds.).

Digital Signal Processing: "Special Issue on NIST 1999 Speaker Recognition Workshop". Vol. 10, no. 1-3, Academic Press, January-April-July, J. CAMPBELL (ed.).

Speech Communication: "Special Issue on Speaker Recognition and its Commercial and Forensic Applications", Vol. 31, no. 2-3, Elsevier, June 2000, R. ANDRÉ-OBRECHT (ed.).

Computer Speech and Language: "Special Issue on Odyssey 2004: the Speaker and Language Recognition Workshop". Vol. 20, Issues no. 2-3, April-June 2006, J. CAMPBELL, J. MASON, J. ORTEGA-GARCIA (eds.).

IEEE Trans. on Audio, Speech and Language Processing: "Special Section on Speaker and Language Recognition"*,* Vol. 15, Issue no. 7, September 2007, K. BERKLING, J.-F. BONASTRE, J. CAMPBELL (eds.).

Recent technical tutorials

J. CAMPBELL: "Speaker Recognition: A Tutorial", *Proc. of the IEEE,* Vol. 85, no. 9, September 1997, pp. 1437-1462.

F. BIMBOT ET AL.: "A tutorial on text-independent speaker verification.*" EURASIP Journal on Applied Signal Processing* 2004(4), pp. 430-451.

S. PARTHASARATY ET AL.: "Speaker Recognition (Part F)"*,* in *Handbook of Speech Processing*, J. BENESTY, M.M. SONDHI, Y. HUANG (eds.), Springer 2008.

NIST website on speaker recognition evaluation

http://www.nist.gov/speech/tests/sre/

Chapter 10

Robust Recognition Methods

10.1. Introduction

Progress made in automatic speech recognition during the past decades has resulted in systems providing high performance levels in well-controlled operating conditions. Such systems are currently used in various domains: office automation (dictation machines), Internet applications, voice command, etc. [RAB 93]. The level of performance is strongly dependent on the complexity and the level of difficulty of the application under consideration. Thus, typical laboratory error rates in speaker-independent mode range from 0.3% (for digit sequences), 5% (for 20,000 word continuous speech recognition), up to 8% for spelled letter, and even above 50% for spontaneous telephone conversations!

In addition, the error rate tends to increase dramatically where there is a mismatch between the training conditions and the actual conditions of use (especially noise levels [MOK 98]). For example, experiments [SIO 95] have shown that, for a system trained with clean speech, the continuous speech recognition error rate varies from 3% with a very low SNR (signal-to-noise ratio) of 36 dB, up to 97% with a very noisy signal with a SNR = 0 dB (note: in these experiments, the noise is additive, white and Gaussian).

Current systems are therefore very sensitive to variability factors, even when they seem minor to the ear and sometimes even completely imperceptible. The current systems clearly lack robustness.

Chapter written by Jean-Paul HATON.

Variability factors in speech can be classified into three categories, depending on their sources, which include:

   – the environment of the speaker: noise correlated to speech (reverberation, reflections) or additive (surrounding noise);

   – the speakers as such, their condition and mode of expression (stress, breathing rhythm, Lombard effect, speech rate);

   – the recording conditions: type of microphone, distance from the microphone, transmission channel (distortion, echo, electronic noise).

The Lombard effect is a response of speakers to a noisy environment that leads them to modify their voice to compensate for the presence of noise. This effect is very dependent on the speaker, the noise level and the context: it is therefore difficult to quantify or model. Its main manifestations are, for instance, an increase in the average fundamental frequency, a frequency shift of the F1 and F2 formants, and an increase in phoneme durations [JUN 93]. Figure 10.1 illustrates this phenomenon when a Gaussian white noise is injected in the speaker's ears via headphones.

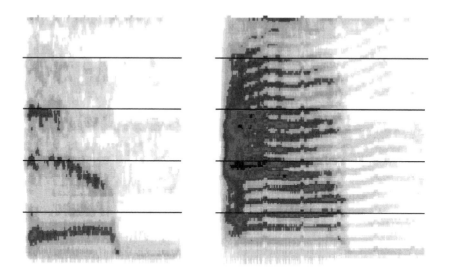

Figure 10.1. *Lombard effect: an illustration. Spectrogram of the English consonant /m/ in the absence (left) and the presence (right) of surrounding noise*

Several methods have been proposed to increase the robustness of automatic speech recognition (ASR) systems, especially in terms of the effects of noise [DEM 98, GON 95, HAT 95, JUN 96, MEL 96]. These techniques can be affiliated to three types of approaches, which will be presented in the forthcoming sections:

– signal preprocessing, so as to reduce the influence of noise;

– speech parameterization with robust description methods;

– system adaptation to new conditions.

10.2. Signal pre-processing methods

Noise filtering of a speech signal can be achieved in several ways. In the next section we review a few of the most popular methods used in the field of ASR.

10.2.1. *Spectral subtraction*

Spectral subtraction involves subtracting from a speech frame (represented in the frequency domain), an estimation of the noise that is affecting the signal, assuming that the speech and noise signals are additive and uncorrelated [BER 79, BOL 79]. This process is repeated on all the speech frames in the utterance. Figure 10.2 illustrates the process. It is very difficult to obtain an accurate estimation of the noise, which requires a reliable speech/non-speech detector so that the noise can be estimated on the non-speech segments. Several improvements have been achieved [DEL 99], in particular the non-linear spectral subtraction technique [LOC 92] and the use of multi-windowing [HE 99]. Today, this technique is well-mastered and efficient for situations involving stationary noise.

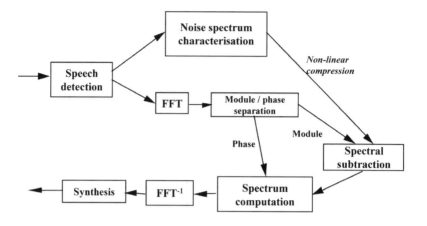

Figure 10.2. *Principles of spectral subtraction*

10.2.2. *Adaptive noise cancellation*

This approach is based on the use of two microphones, one for recording the noisy speech signal and the second for noise only. The principle is to calculate an adaptive filter that models the unwanted noise, and to remove it from the speech signal, as illustrated in Figure 10.3.

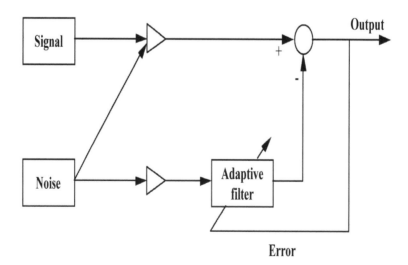

Figure 10.3. *Principle of adaptive noise cancellation*

Several filtering methods have been studied. The interest of this approach is that it does not require any *a priori* knowledge of the noise and can be applied to non-stationary noise. An extension of this approach uses several microphones or microphone arrays and relies on beam-forming techniques developed in the field of sonar [OMO 97]. The latter are of great interest, especially for echo cancellation and de-reverberation [MAR 98, SAR 99].

10.2.3. *Space transformation*

Space transformations involve defining a transform (in a time domain or in a given parametric space) that allows a relatively simple retrieval of the "clean" speech signal from the noisy one [GON 95]. This is generally achieved by first quantifying the acoustic space. Each acoustic class in the clean space is mapped with the corresponding class in the noisy space. The most common methods are:

– CDCN (*codeword dependent cepstral normalization*) [ACE 90] which relies first on a vector quantization of the acoustic space, followed by the learning of a corrective bias to map each noisy sound with its clean counterpart;

– SCDN (*SNR-dependent cepstral normalization*) [ACE 91] which is an extension of CDCN where the corrective vectors are first learnt on a noisy database for several SNR values, and where only the SNR is estimated in the operational phase;

– POF (*probabilistic optimal filtering*) [NEU 94] which builds a linear transformation between noisy and clean acoustic spaces, in a similar way to CDCN, but which uses GMM models to represent the transformations;

– SPLICE, which is a de-noising method based on the Bayesian framework [DEN 04]. It consists of estimating the posterior probability $p(x|y,n)$ of the clean signal x given the noisy signal y and an estimate of the noise n. This *a posteriori* probability is computed from the likelihood $p(y|x,n)$ and the *a priori* probabilities $p(x)$ and $p(n)$.

10.2.4. *Channel equalization*

Channel equalization is particularly relevant for telephone speech. In this case, voice signal distortions are observed which add to the noise in the speaker's environment. Some of these distortions are non-linear, and therefore are difficult to compensate for. Others can be approximated linearly. In the cepstrum domain, some of these effects manifest themselves as a bias that is added to the cepstral parameters. In general, this bias varies from one call to another, but can be assumed constant during a given call [MOK 93]. A popular solution to this problem is cepstral mean subtraction (CMS) [MOK 93], whose principle is to calculate the mean of the cepstral parameters over a complete call and then to subtract from the cepstral vectors, with the underlying (reasonable) assumption that speech has a zero mean cepstrum. The method can be adapted to on-line processing, on a frame-by-frame basis [RAH 96]. CMS can also be achieved by some high-pass filtering on the cepstral parameters, which removes the slowly varying channel distortions. This is the case for the RASTA analysis (see section 10.3.1). This method has also been successfully extended to MFCC (mel-frequency cepstrum coefficient) parameters [HAN 99]. Channel equalization and spectral subtraction methods can be combined in order to compensate simultaneously additive and convolutive effects, in particular in the context of cellular transmission [MOK 95].

10.2.5. *Stochastic models*

Stochastic models can be used as more elaborated models for the task of speech de-noising. For instance, in [EPH 89] and [EPH 92], the models used are hidden Markov models (HMMs). The principle is to estimate the *a priori* model densities and to iterate towards the real distributions using a MAP (*maximum a posteriori*) criterion, optimized with an EM (expectation-maximization) algorithm. The reported results are convincing for additive noise up to a SNR of 10 dB. With very noisy signals, the process converges to local optima that can be very far from the correct solution if an incorrect initialization has been chosen. To remedy this drawback, it was proposed [MAT 98] to carry out the decoding operation in the cepstral space. This yields a good estimate of the probability distributions, and the de-noising can either be performed on the spectrum or the time signal. The method achieves good results, even for low SNR values (typically 5 dB), but it is expensive in terms of its computation load.

10.3. Robust parameters and distance measures

Several works have been conducted with the aim of increasing the noise robustness of voice signal parameterization techniques. These approaches rely on signal processing and statistical estimation techniques, but also on studies based on auditory perception. Two major types of signal representation play a key role in robust speech recognition: spectral representations (either parametric or not) and auditory-based representations.

10.3.1. *Spectral representations*

10.3.1.1. *Non-parametric methods*

Homomorphic analysis [OPP 68], the principles of which are recalled in Figure 10.4, is very widespread in ASR, especially the MFCC coefficients. These coefficients are robust because, on the one hand, they provide a separation between the transfer function of the vocal tract and the pitch characteristics and, on the other hand, they are barely sensitive to the power of the signal. Expanding the cepstrum coefficients with their first and second time-derivative makes the resulting parameter set even more robust to speaker and environment variability. More generally, dynamic parameters improve the robustness of speech recognition [AIK 88, FUR 86]. Derivative coefficients are usually calculated over several successive speech frames (typically 5) as regression coefficients.

Linear Discriminant Analysis can be applied to spectral parameters such as MFCC, so as to increase their robustness. This statistical data analysis method offers a reduction of the representation space and an increased discrimination power of the new parameter set. Thus, several such variations around the IMELDA system [HUN 89, HUN 91] have provided good results for the recognition of isolated words in noisy conditions.

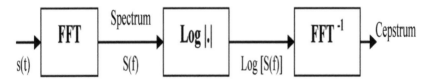

Figure 10.4. *Homomorphic analysis: basic steps*

10.3.1.2. *Parametric methods*

10.3.1.2.1. Short-time modified coherence (SMC)

Short-time modified coherence (SMC) analysis [MAN 88] takes advantage of the fact that the estimation of an LPC speech model in the autocorrelation domain is more robust to noise than in the signal domain. As it is not trivial to obtain an LPC estimate from an autocorrelation sequence, the authors propose to use a spectral shaping operation based on the square root of the power spectrum density. Figure 10.5 summarizes the corresponding sequence of operations. Experience has proved that this approach significantly increases the SNR.

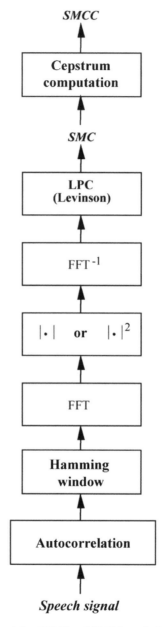

Figure 10.5. *Principle of SMC and SMCC analysis (the latter being obtained from SMC followed by a cepstrum calculation)*

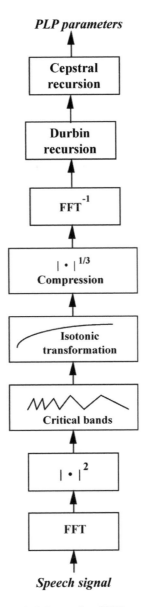

PLP parameters

Figure 10.6. *Principle of PLP analysis*

10.3.1.2.2. Perceptually-based Linear Prediction (PLP) and RASTA-PLP

The introduction of psycho-acoustical knowledge in the estimation of an LPC model has motivated the proposal for PLP analysis [HER 87]. As indicated in Figure

10.6, this knowledge concerns the integration of critical bands, a non-linear pre-emphasis of the speech signal following an isotonic curve and a compression of the resulting spectrum. This method revealed itself as more efficient than a conventional LPC analysis, for noisy speech recognition [JUN 89].

The principle of the RASTA (*RelAtive SpecTrAl*) analysis [HER 91] is to reduce long-term time-varying communication noise by means of a filtering in the log-spectral domain. A combined RASTA-PLP analysis has been shown to yield good results in various operating conditions [HER 94].

Another type of time filtering, in the autocorrelation domain, has also been proposed in association with the MFCC coefficients [YUO 99]. Experiments have shown the interest of the approach for noisy speech recognition, compared to MFCC coefficients and their derivatives.

10.3.2. *Auditory models*

The human auditory system has an exceptional ability to recognize speech in difficult conditions and to adapt rapidly to new situations. Therefore, it makes sense to try to understand how it operates and to draw one's inspiration from it, so as to design robust speech analysis techniques. The four following steps are usually modeled:

– signal conditioning by the outer and middle ears, by a low-pass or band-pass linear filtering;

– signal analysis carried out in the inner ear by the basilar membrane, comparable to a non-linear (mel) scale filter-bank;

– mechanic-to-electric transduction at the output of the cochlea. This non-linear operation is based on automatic gain control and lateral inhibition mechanisms which play an important role in the analysis;

– transformation of the information from the cochlea into coded neuronal pulse trains. This mechanism is far less well known than the previous ones.

Several models have been proposed as analysis procedures for ASR. They often yield better results for noisy recognition than conventional ones do (LPC, MFCC, filter-banks), but their performance degrades in the case of clean speech. Among the most well known models, the following can be mentioned:

– Lyon's model [LYO 82] which is essentially a model of the cochlea which has inspired other representation of the speech signal, such as the *Reduced Auditory Representation* [BEE 90];

– Seneff's model [SEN 84] which consists of three processing steps: a Mel-scale filter bank decomposition, a non-linear compression of the filter bank output and an amplification of the formant frequencies. This model has been used successfully for noisy speech recognition [HUN 87];

– Ghitza's model [GHI 87] which is founded on time-domain information processing in the auditory system. It provides a representation of speech as *Ensemble Interval Histograms* which possess a high non-uniform frequency resolution and a good immunity to noise [SRE 90];

– Gao's model [GAO 92] which focuses on the central auditory system. It includes spatial and time-domain signal processing and implements an efferent mechanism which models the feed-back effect of the auditory system on the inner ear. Experiments in spelled letters recognition have shown the robustness of this analysis scheme for low SNRs.

10.3.3. *Distance measure*

The use of a robust speech parameterization method for speech recognition purposes implies the definition of a distance measure, itself robust, associated to the speech representation and allowing for the numerical comparison between two acoustic patterns. Several distances have been proposed for speech recognition (and, more generally, for pattern recognition), not all of them being metrics in the strict mathematical sense. Here are a few methods that have been shown to be efficient for noisy speech recognition:

– *weighted spectral distances*, based on the weighting of spectral peaks as a function of the difference between the spectrum of the noisy signal and the spectrum of the noise (assumed to be known). A similar principle was used for the weighted Itakura distance for LPC parameters [SON 88];

– *weighted cepstral distances*, designed to use cepstral analysis efficiently in the context of noisy signals. These distances are often written as:

$$d = \sum_i \omega_i \left(C_i(f) - C_i(g) \right)^2 \qquad [10.1]$$

where f and g are the two spectra under comparison and ω the weighting factor. The most common examples are the *Mahalanobis distance* [HUN 89, TOK 86], where ω

is the inverse of the covariance matrix, the *root-power sums*, where $\omega = i$, which makes the weighted distance equivalent to a distance between spectral slopes [HAN 86], and the *band-pass lifter* [JUA 87];

    –   *cepstral projection distances*: the addition of a white noise to a speech signal results in an amplitude reduction and a rotation of the cepstral vectors [MAN 89]. It is possible to compensate for these effects by means of a projection operation on the cepstral vectors. This projection distance has provided good results in multi-speaker word recognition, and in situations when the Lombard effect is significant [JUN 89].

Of the approaches listed above, it is difficult to give clear guidance as to the most efficient: several comparative studies have been carried out, which confirmed their positive impact, but did not elicit a specific optimal solution.

10.4. Adaptation methods

As previously mentioned, the main cause of poor performance in recognition systems result from the difference between training and test conditions for the system. It can prove fruitful to transform the recognition model parameters so it adapts to new conditions. Such an adaptation process has been widely studied in the context of ASR, in order to adapt a system's reference patterns (words, codebooks) from one speaker to another [FUR 89]. In a second phase, the same approach was considered as a way to adapt a system, usually via HMMs, to a new environment. Among the techniques used in the latter case, two categories of model composition and statistical adaptation techniques were used.

10.4.1. *Model composition*

A widely used method in this category that has given rise to many variants is the *parallel model combination* (PMC) [VAR 90, GAL 93]. This method applies to stochastic models (HMM or trajectory models). It involves working in a signal representation space where the noisy speech model is obtained by combining a simple "clean" speech model and a noise model that is assumed stationary. Figure 10.7 summarizes this process.

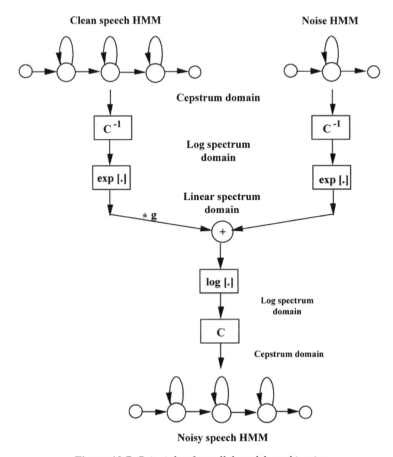

Figure 10.7. *Principle of parallel model combination*

This method is interesting, but raises several issues, in particular the need for a noise model and an estimate of its power. In order to overcome these problems, several techniques have been proposed to apply noise to the models directly in the cepstral domain, for instance by Jacobian adaptation [CER 04].

10.4.2. *Statistical adaptation*

Regression methods can be used to adapt the parameters of a recognition model. The *maximum likelihood linear regression* (MLLR) is certainly among the most popular approaches of this type. Using this approach, the means of the new model are expressed as linear combinations of the means of the original model, plus a bias term [LEG 95]. The same method can also be used to adapt the variances of the model [GAL 96]. A limitation of this approach is the intrinsic non-linearity of the underlying phenomena being modeled. Several improvements have been proposed since [BEL 97]. The coefficients of the MLLR transform can be optimized using the *maximum a posteriori* criterion rather than the *maximum likelihood*. This approach is known as the MAPLR method [CHE 99]. It has also been used to adapt the variances of the Gaussians [CHO 03].

Another method uses a Bayesian MAP estimation [GAU 94, LEE 91]. Here, the shape of the HMM state probability distributions is chosen *a priori* (their real shape being unknown). The parameters are then estimated so as to optimize the MAP criterion, as opposed to the conventional ML criterion. An on-line adaptation scheme, based on the quasi-Bayes method, has also been proposed [HUO 97]. This incremental method is interesting, but generally requires more adaptation data than the MLLR methods, which have more parameters to adapt. A variant, called structural MAP or SMAP [SHI 97, SHI 01], increases the convergence speed of the MAP adaptation by grouping hierarchically the Gaussians within a regression tree.

The current trend is to combine several of these approaches to achieve on line adaptation (see for instance [AFI 97], [CHI 97]).

10.5. Compensation of the Lombard effect

As previously mentioned, a major source of degradation in ASR comes from modifications of the voice characteristics when the speaker is submitted to noise (or, more generally, to stressing factors), termed the *"Lombard effect"*. This effect is difficult to handle, as it is highly variable across speakers and can take various forms. Several solutions have been proposed to compensate the distortions it introduces.

A simple solution, which is valid for small vocabularies, is to achieve training in multiple conditions, i.e. to present to the system the same word uttered in different conditions with several speaking modes [LIP 87]. Another solution involves adapting the vectors from a *codebook* (which have initially been trained with clean speech [ROE 87]) to *"Lombard speech"*.

Lombard compensation can also be achieved in the cepstral domain. A first solution is to assume that the distortions due to the Lombard effect can be compensated by a linear transform applied to the cepstrum. Although this is a very strong hypothesis, it has yielded a 40% reduction of the error rate in speaker-dependent isolated word recognition [CHE 87]. A similar idea has been exploited by [HAN 90], with a compensation transform corresponding to the type of sounds (voiced, unvoiced, transition). Other studies have been carried out with a speech database uttered in stressful conditions, so as to determine specific statistics for compensating the Lombard effect [HAN 94, HAN 95].

As the Lombard effect influences both the spectrum and the time-domain structure of the speech signal, a technique for adapting state durations in an HMM has also been proposed [NIC 92, SIO 93].

10.6. Missing data scheme

In the presence of colored noise, part of the spectral parameters are masked by noise. One solution is to discard the masked parameters during recognition [REN 00].

Two key problems must then be solved:

–   firstly, to identify the masked parameters using various techniques [RAM 00], none of them being fully satisfactory;

–   secondly, to modify the recognition algorithm so as to account for the fact that some parameters are masked [MOR 98, BAR 00].

These techniques have to be refined, but some have already achieved promising results.

10.7. Conclusion

Robustness becomes highly significant for ASR systems when they are used in real conditions: occasional untrained speakers, noisy public places and so forth. Progress in this area still needs to be made, as the current systems see their performance degrading rapidly when the conditions of use, in the wide sense of the term (i.e. speaking style, noise level, transmission conditions) diverge from those in which the system has been trained.

Significant steps have been made in the past decades, but progress is still needed at all levels of the speech recognition process, chiefly:

–   speech acquisition and pre-processing: including the use of microphone arrays, echo cancellation, more robust speech activity detection, and signal-to-noise ratio improvement;

–   speech parameterization: the search for more robust speech parameters remains a research challenge, especially speech representations which incorporate auditory and production-based knowledge;

– recognition model adaptation: the adaptation process to real conditions calls for the development of an improved set of techniques (model combination, spectral transformations, Bayesian estimation) targeted at achieving real on-line system adaptation.

10.8. References

[ACE 90] A. ACERO, Acoustical and Environmental Robustness in Automatic Speech Recognition, PhD Thesis, Carnegie Mellon University, 1990.

[ACE 91] A. ACERO and R. STERN, "Robust speech recognition by normalization of the acoustic space", *Proc. ICASSP-91*, pp. 893–89, Toronto, 1991.

[AFI 97] M. AFIFY, Y. GONG and J.P. HATON, "A Unified Maximum Likelihood Approach to Acoustic Mismatch Compensation: Application to Noisy Lombard Speech Recognition", *Proc. ICASSP-97*, pp. 839-842, Münich, 1997.

[AIK 88] K. AIKAWA and S. FURUI, "Spectral Movement Function and its Application to Speech Recognition", *Proc. ICASSP-88*, pp. 223-226, New York, 1988.

[BAR 00] J. BARKER J. *et al.*, "Soft decisions in missing data techniques for robust automatic speech recognition", *Proc. ICSLP*, pp. 373–376, Beijing 2000.

[BEL 97] J. BELLEGARDA, "Statistical techniques for robust ASR: review and perspectives", *Proc. Eurospeech*, pp. 33-36, Rhodes, 1997.

[BER 79] M. BEROUTI, R. SCHWARTZ and J. MAKHOUL, "Enhancement of speech corrupted by acoustic noise", *Proc. ICASSP-79*, Washington DC, pp. 208-211, 1979.

[BOL 79] S. F. BOLL, "Suppression of acoustic noise in speech using spectral subtraction", *IEEE Trans. ASSP*, 27, pp. 113-120, 1979.

[CER 04] C. CERISARA *et al.*, "Alpha-Jacobian environmental adaptation", *Speech Communication*, 42, pp. 25–41, 2004.

[CHE 87] Y. CHEN, "Cepstral domain stress compensation for robust speech recognition", *ICASSP-87*, pp. 717-720, Dallas, 1987.

[CHE 99] C. CHESTA, O. SIOHAN and C.H. LEE, "Maximum *a posteriori* linear regression for hidden Markov model adaptation", *Proc. EUROSPEECH*, , pp. 211–214, Budapest, 1999.

[CHI 97] J.T. CHIEN, C.H. LEE and H.C. WANG, "Improved Bayesian learning of hidden Markov models for speaker adaptation", *Proc. ICASSP-97*, pp. 1027-1030, Munich, 1997.

[CHO 03] W. CHOU and X. HE, "Maximum a posteriori linear regression variance adaptation for continuous density HMM", *Proc. EUROSPEECH*, pp. 1513–1516, Geneva.

[DEL 99] L. DELPHIN-POULAT, Utilisation de modèles de Markov cachés pour une compensation synchrone à la trame dans un contexte de reconnaissance de la parole, PhD Thesis, University of Rennes 1, 1999.

[DEM 98] R. DE MORI, (Ed.) *Spoken Dialogues with Computers*, Academic Press, New York, 1998.

[DEN 04] L. DENG *et al.*, "Estimating cepstrum of speech under the presence of noise using a joint prior of static and dynamic features", *IEEE Trans. on Speech and Audio Processing*, 12, 218–233, 2004.

[EPH 89] Y. EPHRAIM, D. MALAH and B.H. JUANG, "On the application of hidden Markov models for enhancing noisy speech", *IEEE Trans. ASSP*, 37, no.12, pp. 1846-1856, 1989.

[EPH 92] Y. EPHRAIM, "Statistical model-based speech enhancement systems", *Proc. IEEE*, 80, no.10, pp. 1526-1555, 1992.

[FUR 86] S. FURUI, "Speaker independent isolated word recognition using dynamic features of speech spectrum", *IEEE Trans. Acoustics, Speech and Audio Processing*, 34, pp. 52-59, 1986.

[FUR 89] S. FURUI, "Unsupervised speaker adaptation method based on hierarchical spectral clustering", *Proc. ICASSP-89*, pp. 286-289, Glasgow, 1989.

[GAL 93] M. GALES and S. YOUNG, "HMM recognition in noise using parallel model combination", *Proc. Eurospeech*, pp. 837-840, 1993.

[GAL 96] M. GALES and P.C. WOODLAND, "Mean and variance adaptation within the MLLR framework", *Computer Speech and Language*, 10, no. 4, pp. 249-264, 1996.

[GAO 93] Y. GAO, T. HUANG, and J. P. HATON, "Central auditory model for spectral processing", *Proc. ICASSP-93*, pp. 704-707, Minneapolis, 1993.

[GAU 94] J.L. GAUVAIN and C.H. LEE, "Maximum *a posteriori* estimation for multivariate Gaussian mixture observations of Markov chains", *IEEE Trans. Speech and Audio Processing*, 2, no. 2, pp. 291-298, 1994.

[GHI 95] O. GHITZA, "Robustness against noise: the role of timing-synchrony measurement", *Proc. ICASSP-87*, pp. 2372-2375, Dallas, 1987.

[GON 95] Y. GONG, "Speech recognition in noisy environments: a survey", *Speech Communication*, 16, no.3, pp. 261-291, 1995.

[HAN 86] J. H. HANSEN, and M. A. CLEMENTS, "Spectral slope based distortion measures for all-pole models of speech", *Proc. ICASSP-86*, pp. 757-760, Tokyo, 1986.

[HAN 90] J. H. L. HANSEN, and O. N. BRIA, "Lombard effect compensation for robust automatic speech recognition in noise", *Proc. ICSLP-90*, pp. 1125-1128, Kobe, 1990.

[HAN 94] J.H.L. HANSEN, "Morphological constrained feature enhancement with adaptive cepstral compensation for speech recognition in noise and lombard effect", *IEEE Trans. Speech and Audio Processing*, 2, no. 4, pp. 598-614, 1994.

[HAN 95] J.H.L. HANSEN and M.A. CLEMENTS, "Source generator equalization and enhancement of spectral properties for robust speech recognition in noise and stress", *IEEE Trans. Speech and Audio Processing*, 3, no. 3, pp. 407-415, 1995.

[HAN 99] J. HAN and W. GAO, "Robust telephone speech recognition based on channel compensation", *Pattern Recognition*, 32, pp. 1061-1067, 1999.

[HAT 95] J.-P. HATON, "Automatic recognition of noisy speech" in *Speech Recognition and Coding. New Advances and Trends*, in A. J. RUBIO AYUSO, J. M. LOPEZ SOLER (Eds.), NATO ASI Series, Vol. 147, pp. 1-13, 1995.

[HE 97] C. HE and G. ZWEIG, "Adaptive two-band spectral subtraction with multi-window spectral estimation", *Proc. ICASSP-99*, pp. 793-796, Phoenix, 1997.

[HER 87] H. HERMANSKY, "An efficient speaker-independent automatic speech recognition by simulation of some properties of human auditory perception", *Proc. ICASSP-87*, pp. 1159-1162, Dallas, 1987.

[HER 94] H. HERMANSKY and N. MORGAN, "RASTA processing of speech", *IEEE Trans. Speech and Audio Processing*, 2, no. 4, pp. 578-589, 1994.

[HER 91] H. HERMANSKY, N. MORGAN, A. BAYYA and P. KOHN, "Compensation for the effect of communication channel in auditory-like analysis of speech (RASTA-PLP)", *Proc. Eurospeech-91*, pp. 1367-1370, Genoa, 1991.

[HUN 87] M. J. HUNT and C. LEFEBVRE, "Speech recognition using an auditory model with pitch-synchronous analysis", *Proc. ICASSP-87*, pp. 813-816, Dallas, 1987.

[HUN 89] M. J. HUNT and C. LEFEBVRE, "A comparison of several acoustic representations for speech with degraded and un-degraded speech", *Proc. ICASSP-89*, Glasgow, 1989.

[HUN 91] M. J. HUNT, S. M. RICHARDSON, D. C. BATEMAN and A. PIAU, "An Investigation of PLP and IMELDA Acoustic representations and of their potential for combination", *Proc. ICASSP-91*, pp. 881-884, Toronto, 1991.

[HUO 97] Q. HUO and C-H. LEE, "On-line adaptive learning of the continuous density hidden Markov model based on approximate recursive bayes estimate", *IEEE Trans. Speech and Audio Processing*, 5, no. 2, pp. 161-172, 1997.

[JUA 87] B. H. JUANG, L. R. RABINER and J. G. WILPON, "On the use of bandpass liftering in speech recognition", *IEEE Trans. ASSP*, 35(7), pp. 947-954, 1987.

[JUN 93] J.C. JUNQUA, "The Lombard reflex and its role on human listeners and automatic speech recognizers", *J. Acoust. Soc. Amer.*, 93, pp. 510-524, 1993.

[JUN 89] J.-C. JUNQUA and H. WAKITA, "A comparative study of cepstral lifters and distance measures for all pole models of speech in noise", *Proc. ICASSP-89*, pp. 476-479, Glasgow, 1989.

[JUN 96] J.-C. JUNQUA and J.-P. HATON, *Robustness in Automatic Speech Recognition*, Kluwer Academic Publishers, Dordrecht, 1996.

[LEE 91] C.H. LEE, C.H. LIN and B.H. JUANG, "A study on speaker adaptation of the parameters of continuous density hidden Markov models", *IEEE Trans. Signal Processing*, 39, no. 4, pp. 806-814, 1991.

[LEG 95] C.J. LEGGETTER and P.C. WOODLAND, "Maximum likelihood linear regression for speaker adaptation of continuous density hidden Markov models", *Computer Speech and Language,* 9, no. 2, pp. 171-186, 1995.

[LIP 87] R. P LIPPMANN, E. A. MARTIN, and D. B. PAUL, "Multi-style training for robust isolated-word speech recognition", *Proc. ICASSP-87*, pp. 705-708, Dallas, 1987.

[LOC 92] P. LOCKWOOD *et al.*, "Non-linear spectral subtraction (NSS) and hidden Markov models for robust speech recognition in car noise environments", *Proc. ICASSP-92*, pp. 265-268, San Francisco, 1992.

[LYO 82] R. F. LYON, "A computational model of filtering, detection, and compression in the cochlea", *Proc. ICASSP-* 82, pp. 1282-1285, Paris, 1982.

[MAN 88] D. MANSOUR, and B. H. JUANG, "The short-time modified coherence representation and its application for noisy speech recognition", *Proc. ICASSP-88*, pp. 525-528, New York, 1988.

[MAN 89] D. MANSOUR and B. H. JUANG, "A family of distortion measures based upon projection operation for robust speech recognition", *IEEE Trans. ASSP*, 37(1), pp. 1659-1671, 1989.

[MAR 98] C. MARRO *et al.*, "Analysis of noise reduction and dereverberation techniques based on microphone arrays with postfiltering", *IEEE Trans. Speech and Audio Processing*, 6, no. 3, pp. 240-259, 1998.

[MAT 98] D. MATROUF et J.L. GAUVAIN, "Utilisation des modèles de Markov cachés pour le débruitage", *Actes des Journées d'Etude sur la Parole*, pp. 327-330, Martigny, 1998.

[MEL 96] H. MELONI, rédacteur, "Fondements et perspectives en traitement automatique de la parole", *AUPELF-UREF*, Paris, 1996.

[MOK 93] C. MOKBEL, J. MONNÉ and D. JOUVET, "On-line adaptation of a speech recognizer to variations in telephone line conditions", *Proc. Eurospeech-93*, pp. 1247-1250, Berlin, 1993.

[MOK 95] C. MOKBEL and D. JOUVET, "Recognition of digits over GSM and PSN networks", *Proc. IEEE Workshop on ASR*, pp. 167-168, Snowbird, 1995.

[MOK 98] C. MOKBEL, D. JOUVET, J. MONNÉ and R. DE MORI, "Robust speech recognition", in *Spoken Dialogues with Computers*, DE MORI (Ed)., Academic Press, 1998.

[MOR 98] A. MORRIS, M. COOKE and P. GREEN, "Some solutions to the missing feature problem in data classification, with applications to noise robust ASR", *Proc. ICASSP*, pp. 737–740, Seattle.

[NEU 94] L. NEUMEYER L. and M. WEINTRAUB, "Probabilistic optimum filtering for robust speech recognition", *Proc ICASSP-94*, Vol. 1, pp. 417–420, Adelaide, 1994.

[NIC 92] N. NICOL, S. EULER, M. FALKHAUSEN, H. REININGER, D., WOLF, and J. ZINKE, "Improving the robustness of automatic speech recognizers using state duration information", *Proc. ESCA Workshop on Speech Processing in Adverse Conditions*, pp. 183-186, Cannes-Mandelieu, 1992.

[OMO 97] M. OMOLOGO, "On the future trends of hands-free ASR: variabilities in the environmental conditions and in the acoustic transduction", *Proc. ESCA Workshop Robust Speech Recognition for Unknown Communication Channels*, pp. 67-73, Pont-à-Mousson, 1997.

[OPP 68] A. OPPENHEIM, R. SCHAFER and T. STOCKHAM, "Nonlinear filtering of multiplied and convolved signals", *Proc. IEEE*, 56, pp. 1254-1291, 1968.

[RAB 93] L. RABINER and B.H. HUANG, "Fundamentals of speech recognition", Prentice Hall, Englewood Cliffs, 1993.

[RAH 96] M. RAHIM and B.H. JUANG, "Signal bias removal by maximum likelihood estimation for robust telephone speech recognition", *IEEE Trans. Speech and Audio Processing.*, 4, no. 1, pp. 19-30, 1996.

[RAM 00] B.R. RAMAKRISHNAN, Reconstruction of incomplete spectrograms for robust speech recognition, PhD Thesis, Carnegie Mellon University, Pittsburgh, 2000.

[REN 00] P. RENEVEY, Speech recognition in noisy conditions using missing feature approach, PhD Thesis, Ecole Polytechnique Fédérale de Lausanne, 2000.

[ROE 87] D. B. ROE, "Speech recognition with a noise-adapting codebook", *Proc. ICASSP-87*, pp. 1139-1142, Dallas, 1987.

[SAR 99] H. SARUWATARI, S. KAJITA, K. TAKEDA and F. ITAKURA, "Speech enhancement using nonlinear microphone array with complementary beamforming", *Proc. ICASSP-99*, pp. 69-72, Phoenix, 1999.

[SEN 84] S. SENEFF, "Pitch and spectral estimation of speech based on auditory synchrony model", *Proc. ICASSP-84*, pp. 36.2.1-36.2.4, San Diego, 1984.

[SHI 97] K. SHINODA and C-H. LEE, "Structural MAP speaker adaptation using hierarchical priors", *Proc. IEEE Workshop, on Automatic Speech Recognition and Understanding Processes*, pp. 381-388, Santa Barbara, 1997.

[SHI 01] K. SHINODA and C.H. LEE, "A structural Bayes approach to speaker adaptation", *IEEE Trans. on Speech and Audio* 9, pp. 276–287, 2001.

[SIO 93] O. SIOHAN, Y. GONG and J. P. HATON, "A Bayesian approach to phone duration adaptation for Lombard speech recognition", *Proc. Eurospeech-93*, pp. 1639-1642, Berlin, 1993.

[SIO 95] O. SIOHAN, Reconnaissance automatique de la parole continue en environnement bruité : application à des modèles stochastiques de trajectoires, PhD Thesis. Université Henri Poincaré, Nancy 1, 1995.

[SOO 88] F.K. SOONG and M. M. SONDHI, "A frequency-weighted Itakura spectral distortion measure and its application to speech recognition in noise", *IEEE Trans. ASSP*, 36(1), pp. 41-48, 1988.

[SRE 90] T.V. SREENIVAS, K. SINGH and R. J. NIDERJOHN, "Spectral resolution and noise robustness in auditory modeling", *Proc. ICASSP-90,* pp. 817-820, Albuquerque, 1990.

[TOK 86] Y. TOKHURA, "A weighted cepstral distance measure for speech recognition", *Proc. ICASSP-86*, pp. 761-764, Tokyo, 1986.

[VAR 90] A.P. VARGA and R. MOORE, "Hidden Markov model decomposition of speech and noise", *Proc. ICASSP-90*, pp. 845-848, Albuquerque, 1990.

[YUO 99] K.H. YUO and H.C. WANG, "Robust features for noisy speech recognition based on temporal trajectory filtering of short-time autocorrelation sequences", *Speech Communication,* 28, no.1, pp. 13-24, 1999.

Chapter 11

Multimodal Speech:
Two or Three Senses are
Better than One

11.1. Introduction

The title of this chapter is an (imperfect) plagiarism. In November 1996, during the ICSLP Conference in Philadelphia, a special session was organized by Lynne Bernstein and Christian Benoît. It was entitled: "The Senses of Speech Perception", and was introduced with the following foreword: "For speech perception by humans or machines, three senses are better than one" [BER 96].

Two years later in 1998, a summer school was organized in Grenoble and we asked Christian Benoît to make a presentation on the theme of "Multimodality". However, fate dictated otherwise. So let us quote from Christian, as a furtive tribute to his scientific heritage – other tributes have been, and will doubtless be paid to him on a much wider scale – a few of the phrases he was gifted enough to invent for promoting and developing this domain, to which he has brought so much; in brief, to "make speech visible as much as possible".

"The bimodality of speech is at the center of a wide fundamental research area aiming at better understanding how human beings speak and coordinate the

---

Chapter written by Jean-Luc SCHWARTZ, Pierre ESCUDIER and Pascal TEISSIER.

command of half a dozen articulators, a fraction of which are entirely or partly visible. In this respect, bimodality is at a scientific crossroad, where phoneticians, psychologists and engineers converge. It is also one of the mainstays of speech-impaired rehabilitation and foreign language teaching. Finally, it is the basis for a number of applications in the cartoon industry, in the field of image synthesis and in the design of man-machine interfaces" [BEN 96, p.7].

Thus, although speech is "mostly meant to be heard", it "also deserves to be seen". Hence attempts to measure the parameters needed to describe lip geometry and kinematics, for instance with the LIPTRACK system, an independent, programmable, real-time "Labiometer" using chroma-key [ADJ 97]. These parameters are ideal candidates for the control and animation of talking faces (in the context of Audiovisual text-to-speech synthesis (AVTTS) [LEG 96]) or for use as entry points in Audiovisual automatic speech recognition systems (AVASR) [ADJ 96]. A new theoretical and experimental research framework, encompassing a wide range of potential technologies, stands out, in which human-human and human-machine communication is centered on the relations between the ear, the eye, the sound and the lips. In Grenoble, this reached fruition in the famous "Labiophone", a leading project of the ELESA Federation. This project "gathered several research groups from ICP and LIS, and brought together their knowledge and common tools for the analysis, transmission, recognition and synthesis of talking faces at rates compatible with the transmission of audio and visual speech" (ICP Progress Report, 1998, p. 34). Together with other actual projects, such as the MIAMI project (*Multimodal Integration for Advanced Multimedia Interfaces*), the AMIBE project (*Applications Multimodales pour Interfaces et Bornes Evoluées*) and others which were just imaginary, (including the BASC project (*Bimodal Aspects of Speech Communication*) and the FASCIAL project (*Facial And Speech animation for multiple Cultures And Languages*), the collective aim was to "take up the major challenge of bimodal integration" [BEN 96, p. 55]. A few years later, this included multimodal integration, which led to the initial development of a synthesis system for Cued Speech [VAR 98], which aimed at producing an "augmented telephone" that contributed to "a real telecommunication and rehabilitation tool for auditory impaired people, as opposed to the genial and essential invention of Bell, who have been excluding them from telecommunications during 122 years..." (ICP Progress Report, p. 34).

It is therefore multimodal speech that must be the focus of our studies, models and systems. In this chapter, we will first go back over a number of experimental facts that demonstrate the multimodal nature of speech cannot be considered as marginal, and on the contrary, show it is at the core of the speech communication mechanism. Then, we address the issue of bimodal integration architecture, which is essential for accessing the linguistic code from the joint audio and visual streams produced by the speaker. Finally, we consider how AVASR systems can exploit principles of bimodal fusion to improve their robustness to noise.

11.2. Speech is a multimodal process

Since it is possible to talk and understand speech without any eye contact, it should follow that multimodality is a luxury when dealing with spoken communication, or some kind of unnecessary side interest that is sought out gratuitously by researchers on the fringe of the audio aspects of speech? The contrary is true: we will show (by progressive convergences) that multimodality – and essentially bimodality – is *at the very heart of the speech communication mechanisms.*

11.2.1. *Seeing without hearing*

Seeing acts as the first step as *a relay of audition* to understand when there is no choice as to what concerns the modality: this is of course the case for deaf people using lip reading [FAR 76] (see a comprehensive review on visual perception of speech in [CAT 88]). Lip reading on its own can capture approximately 40 to 60% of the phonemes in a given language, and 10 to 20% of the words, even though individual variability is considerable at this level, the best lip readers (generally, deaf people) can reach scores above 60% (see [BER 98] for a review). In the rare yet existing case of deaf-blind people, a method has been developed to perceive speech by sensing by touch the speaker's vocal tract (see Figure 11.1): this is the Tadoma method, studied in detail by the group of C. Reed at MIT, with phonetic identification scores ranging from 40 to 60%, i.e. comparable to those of lip reading, even though the nature of confusions is very different [REE 82a, REE 82b]. For lip reading, as is the case for the Tadoma method, the principles consist of approaching the motor aspect of speech – the vocal tract movements – not only in terms of its audible consequences (i.e. those on which the focus is generally put), but also with respect to other sensory aspects: vision and touch of the moving articulators. As the speaker's voluntary movements combine to produce three types of stimuli (auditory, visual and somesthesic), which are all perceivable and interpretable to some extent, the picture of multi-sensorial speech is drawn.

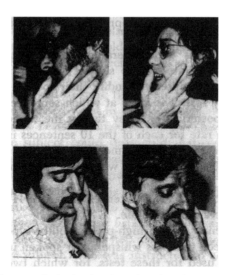

Figure 11.1. *The Tadoma method. The thumb is placed on the lips of the speaker, the index on the cheek and the other fingers on the neck, right under the mandible (from [NOR 77])*

In a second step, a further investigation involved sending the human speech perception and comprehension system other stimuli than those coming directly from the vocal tract. Tactile stimuli, to start with, by recoding speech with vibro-tactile devices for communicating to the skin information that is not visible but essential for phonetic decoding (for example, related to voicing information) [BER 89, EBE 90]: this provides an indirect perception of the vocal tract gestures. Let us also mention visual stimuli such as hand movements corresponding to non-vocal gestures in sign language, or from Cued Speech that will be presented later on, in which the hand comes as an auxiliary source to the vocal gestures and provides missing information on non-visible features, such as voicing, nasality or hidden movements of the tongue. This constitutes what could be called second level multi-modality, i.e. composed of a diversity of sensorial inputs based on multiple motor actions, the convergence of which is ensured by a single speech comprehension system. We will come back later to this convergence, which has been effectively demonstrated by the capacity of normal and handicapped human subjects to take advantage of the variety of perceptual stimulation to improve their performance in capturing the spoken message.

11.2.2. *Seeing for hearing better in noise*

Seeing is also useful *in supporting audition*, in the case of *adverse communication situations*. The paradigm proving this statement has been well-established since the beginning of the 1950s ([ERB 69, MIL 51, NEE 56, SUM 54] and is supported by more recent data in French [BEN 94]). This work shows that the intelligibility of given linguistic material mixed with acoustic noise increases when the subject can see the speaker in action (see Chapter 4, Figure 4.1). In general terms, labial information is estimated to contribute the equivalent of a noise reduction of 11 dB to improving intelligibility [MAC 87]. Three important points need to be underlined, however.

Firstly, the capacity of exploiting visual information in the form of labial reading seems immediately operational (little dependence on training time) even though differences are significant across subjects. Secondly, audition and vision complement each other in their ability to bear phonetic information. Thus, [SUM 87] presents the compared audio and visual intelligibility of consonants in terms of perceptual proximity trees (see Figure 11.2): it appears that weak auditory contrasts in noisy environments, i.e. those on close auditory branches (essentially contrasts on the place of articulation) are the most visible, i.e. on far apart visual branches (mainly when they oppose movements which involve in different ways the front of the vocal tract). Our own work [ROB 98] gave similar results for French vowels (see Figure 11.3).

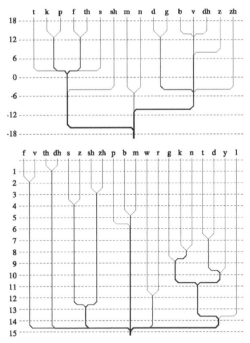

Figure 11.2. *Auditory and visual confusions between consonants (from [SUM 87]). Top: tree representation of the auditory confusions between consonants. The vertical scale indicates the signal-to-noise ratio (in dB) for which the corresponding groups of consonants are mistaken. Bottom: tree representation of the visual confusions between consonants. The results are obtained as a result of hierarchical classification applied to confusions observed from hearing-impaired adults. The vertical scale corresponds to the level of grouping*

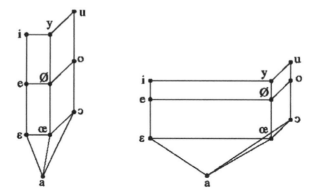

Figure 11.3. *Auditory and visual geometry of French vowels. Qualitative diagram of auditory distances (on the left, based on spectral distances and noisy misrecognitions) and visual distances (on the right, based on geometric parameters and lip reading confusions). From [ROB 95a]*

Finally, in the context of the "noisy audio + lip reading" paradigm, bimodal integration leads to intelligibility scores which are always better with two sensory inputs than with one (hence the title of this chapter). This synergy between audition and vision always leads to gains in intelligibility, at the level of entire sentences, words, phonemes or even phonetic features, as shown by our results in Audiovisual identification of vowels (see Figure 11.4).

Thus, there seems to be some form of "double optimality" in Audiovisual perception, both at the level of available information (*complementarity*) and in the way this information is processed (*synergy*). Moreover, it is reasonable to think that the visual input plays, at a low level, some role in the audio grouping and scene analysis processes as   well, which could lay the foundations of an entirely new Audiovisual [SCH 04] (and more generally multimodal) scene analysis paradigm (see Chapter 5).

11.2.3. *Seeing for better hearing… even in the absence of noise*

Even in the absence of acoustically adverse conditions, lip reading facilitates the subject's ability to understand a message. To demonstrate this, it is of course necessary to design experimental configurations in which auditory comprehension is not straightforward. The claimed result is demonstrated by [REI 87] in the context of a so called *shadowing* experiment, in which subjects must repeat (on the fly) some words pronounced by the experimenter: when the text has a relatively high complexity level (for example, *"Critique of Pure Reason"* by Kant), visual information helps the speaker and the number of words correctly repeated increases. The same trend is observed during comprehension of a foreign language [DAV 98, REI 87]: think how important it is to be able to see one's interlocutor to facilitate comprehension in a foreign country. This is probably the capacity offered by lip-reading: relieving   low-level   processes   (in   particular,   by   aiding   in   syllabic segmentation) facilitates the higher-level comprehension task.

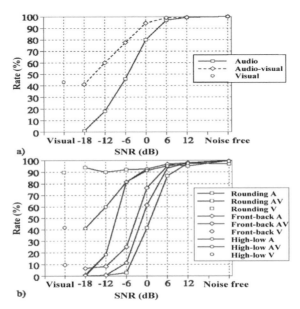

Figure 11.4. *Audiovisual synergy. Auditory, visual and audiovisual identification of (a) the 10 French oral vowels and of (b) each phonetic feature (A for audio, V for visual, and AV for audiovisual): audiovisual performance levels are always higher than auditory-only or visual-only performance (from [ROB 95a])*

11.2.4. *Bimodal integration imposes itself to perception*

The McGurk effect, which we present in this section, is one of the most famous phenomena encountered in speech perception. Even though it has often been presented erroneously in the literature (see [CAT 94], pp. 1-6), this effect can be stated simply: when a subject is presented with a *conflictual* stimulus, in which the sound (of a [ba]) does not match the image (the one of a [ga]), the subject identifies an integrated percept, which matches neither the sound nor the image, but indeed corresponds to a *third kind* category (.a [da] or, for English speaking subjects, a [ða]). This is referred to as a *fusion illusion*. The essential and striking characteristic of this illusion is its immediate and cognitively inaccessible nature. In other words,

384   Spoken Language Processing

it is difficult to hear anything else but a [da], even though, when closing one's eye, the sound heard is clearly a [ba]. The absence of a visual gesture corresponding to [ba] prevents its correct perception. Note that, whereas the [ba]$_A$ + [ga]$_V$ combination gives the illusion of an Audiovisual [da]$_{AV}$, the inverse configuration [ga]$_A$ + [ba]$_V$ procures the impression of a consonant group which includes both plosives, such as [bga]$_{AV}$ or [baga]$_{AV}$: this is referred to as a *combination illusion.*

This effect, initially described by McGurk and MacDonald [McG 76], has been widely studied since, and displays a very rich phenomenology [GRE 96, GRE 98]. A similar effect can be obtained by combining the audition of a [ba]$_A$ with a touch of a [ga]$_T$, in a experimental set up based on the Tadoma method [FOW 91], even with untrained subjects. Here again, the additional modality (touch) somewhat modifies the auditory perception though with less clear fusion blends than in the audiovisual condition. Displaying the written syllable [ga] on a screen instead provides at best very weak effects on the auditory perception of [ba] $_A$. The McGurk effect therefore seems to concern mainly the fusion of sensory information related to the *perception of vocal tract movements.*

The convergence of two or three sensory modalities towards a common level of representation is also evidenced by a whole series of experiments on short-term memory. This is referred to as the *recency effect*: during a task involving memorizing a list of items, it is observed that the last items (i.e. the most recent ones) are always memorized best, as the subjects more often correctly recall them. This effect is observed for auditory lists, but not in the case of written words (visual display on a screen for instance): it is therefore considered that this relates to a pre-phonological auditory encoding which can take various forms depending on the proposed models (echoic memory, pre-categorical acoustic storage, etc.). It appears that the recency effect also exists with items that are presented in lip reading form. Moreover, if an additional item is added at the end of the list (a *suffix*), this suppresses the recency effect, by bringing down the memorization of the last items to the same level as the others. A series of experiments have demonstrated that it is possible to have these auditory and visual modalities interfere with each other in the framework of short-term memory interactions: a suffix read on the lips disturbs the subject's ability to recall an auditory list, and conversely, the recency effect on a list presented in lip reading form is lowered by the audition of an acoustic suffix (see [CAM 82]). However, some nuances indicate that a higher efficiency is obtained from a suffix when it belongs to the same modality as the list [DEG 92]. Here again, these interactions rely on the audition and the vision of vocal gestures, but not on vision in general: a suffix displayed on a screen has no effect on diminishing the recency effect. Thus, convergence is once again in action – here, in the context of phonological memory mechanisms – by fusing sensorial information concerning exclusively vocal gestures.

11.2.5. *Lip reading as taking part to the ontogenesis of speech*

The key role played by lip reading in language acquisition was demonstrated by Mills [MIL 87] in his overview on language acquisition by blind children. It appears that blind children experience difficulties in learning an acoustically weak contrast such as the [m]/[n] opposition. This was shown to be true for German, English or Russian children, who displayed a similar effect for the contrast between [θ]/[f]. Mills also shows that children who could see at the beginning of their language acquisition phase (between 1 and 2 years old) learned with greater ease than blind children how to produce sounds corresponding to a visible articulation, like bilabial plosives [p b m]. Mills cites other works which show blind children aged between 12 and 18 years old exploit their labial dynamics less, while producing normal acoustic contrasts, which suggests the use of some articulatory compensation (see also a recent study with similar effects, in [DUP 05]).

As a confirmation of these results, it must also be noted that production of bilabials predominates in the first stages of language acquisition [VIH 85], in particular for hearing impaired children [STO 88], but not in the case of blind children [MUL 88] (leastways, not to the same extent, according to Mills): lip reading may therefore play a role in this. Finally, the previous observations can be related to the ability for newborn babies to reproduce facial gestures, i.e. to convert visual perception into movement. This was observed as early as 72 hours after birth (and even less than one hour for the youngest cases) in the classic work of Meltzoff and Moore [MEL 83] (see Figure 11.5).

At a different level, a review article by Alegria *et al.* [ALE 92] on phonological representations by deaf-by-birth people, underlines the issues related to language acquisition. In a first step, Alegria and co-authors show the existence of pieces of evidence indicating that, for some deaf people, a phonological system is actually in place: this is demonstrated by the effect of phonological proximity on short term memory in list recalling tasks, the so-called *Stroop effect* (an interference between production and lexical access), in the spelling of known and unknown words. It is important to note that the development of such a phonological system, which seems to play a role in lexical access, varies across deaf subjects (with no known explanation) in a way that is strongly correlated with their ability to produce intelligible speech, even with comparable auditory capacities (which are, in that case, zero or almost zero). So what is the origin of this phonological system and how does it operate in the absence of any auditory input? Algeria *et al.* showed that lip reading plays a crucial role in this process, possibly with additional assistance from gesture information to resolve any ambiguities. This is also true for *cued speech* (developed by Cornett [COR 67]), which attempts to augment communication with a number of gestures for disambiguating labial doubles (see Figure 11.6). Alegria *et al.* showed that young children brought up from their early age in a communication context exploiting lip reading and Cued Speech were particularly capable of developing a phonological system close to normality (see also [LEY 98]).

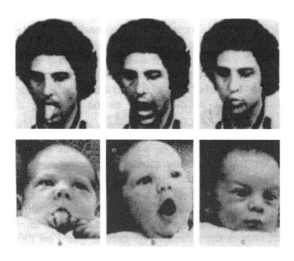

Figure 11.5. *Face imitation by newborns (from [MEL 83])*

11.2.6. ...and to its phylogenesis?

Whilst lip reading is clearly involved in language acquisition mechanisms, does it also take part in the selection and evolution mechanisms that contribute to the evolution of linguistic systems in the languages of the world? This is a quite natural question in the context of a "substance oriented" linguistic science, for which the principles of a sensory-motor economy, including articulatory effort and the distinctiveness (or salience) of the sensory objects, are supposed to rule the structuring of phonological systems [BOE 00]. We may wonder whether phonemes that are more visible on the lips have increased use in the languages of the world.

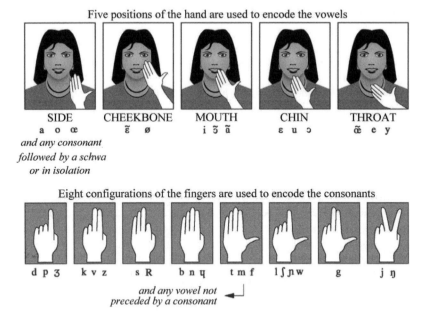

Figure 11.6. *Cued language gestures. Vowels are coded by the position of the hand. Consonants are described by the position of the fingers (see the ALPC website:* http://www.alpc.asso.fr/code01-c.htm)

There is no experimental evidence that can yield a definite answer to this question. On the contrary, bilabials are not favored over dentals or velars within the category of plosives, nor the labiodentals within the fricatives. With vowels, the visible triangle retracted-rounded-open only constitutes a support to the basic acoustic contrast between extreme positions in the F1-F2 triangle, so as to form the canonical triplet [i] (minimum F1, maximum F2, retracted lips) – [u] (minimum F1, minimum F2, rounded lips) – [a] (maximum F1, opened lips).

Conversely, we may wonder whether some fundamental contrasts in the languages of the world, such as the opposition between bilabials and dentals [m]-[n] in 94% of the languages in the UPSID database (see Boë and Vallée, personal communication) would be viable without a visible opposition between them, as this is suggested by the fact that these contrasts are difficult to learn when sight is impaired, or whether the contrast [i]-[y], attested in about 8% of the languages of the world, despite its very weak acoustic distinctiveness [SCH 97] does not benefit from an optimal lip contrast which helps it to survive. This question remains unanswered, and constitutes an exciting challenge in the more general goal of modeling quantitatively the sensory-motor economy of phonological systems.

In conclusion, we have progressively moved our focus from lip reading as an aid for those with impairments, to lip reading as a complementary system, useful to everyone in a wide range of situations from language comprehension to acquisition, and evolution, and, more generally, to a network of sensory afferences which converge towards an integration module for identifying and memorizing phonological units. In other words, we have moved from auditory speech to speech conceived as essentially multi-sensory in nature.

11.3. Architectures for Audiovisual fusion in speech perception

A general theory of multi-sensory speech must address issues including the sensory processing of all channels involved, the relations between substance and code, and hence the dialectics between variability and invariance, the nature of the phonetic code, the mechanisms of memorizing and their relationship with the nature of the sensory modalities This section will be focused on a single issue, crucial in this complex context: the nature of the fusion process of sensory information (here, auditory and visual information) in the decoding process.

The question can be put as follows: we assume that somewhere in the brain (see [CAM 98, CAL 03a, BER 04] for reviews on the cortical localization of the visual and Audiovisual speech perception system), or in an automatic speech recognition system, some module is dedicated to making a decision on the linguistic content of the message. It uses two streams of sensory input, namely the sound and the image of the speaker (see Figure 11.7). In this section, we will not make any hypothesis regarding the specific content of these input streams, nor of the output code. We are only interested in the particular form of the multi-sensory decision mechanism, represented in Figure 11.7 by a question mark.

Many audiovisual integration models for the perception and automatic recognition of speech have been proposed in the literature. However, these models are generally focused on speech processing. By taking a step back from these proposals, we have observed that they can be related to more general considerations, either arising from the field of cognitive psychology and dealing with sensory interactions in general, or from the field of information processing and, particularly, the issue of the fusion of multiple sensors in decision processes [SCH 98].

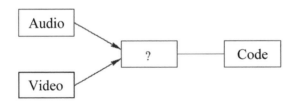

Figure 11.7. *The core of the* Audiovisual *fusion system*

11.3.1. *Three paths for sensory interactions in cognitive psychology*

The issue of sensory integration and inter-modal transfer is a long-standing question in cognitive psychology and has given rise to three main types of models [HAT 93].

Firstly, *language* can constitute an intersection between both sensory paths. This hypothesis, under which inter-modal transfers rely on a symbolic or linguistic mediation has dominated for several years, because of the difficulty in observing such transfers with non-verbal (animal) or pre-verbal (baby) subjects. Since the 1970s, new experimental protocols have lead to observation of such transfers with animals or babies, involving audition, vision, touch, proprioception and prehension, which indicates they do not occur through language. However, linguistic mediation is essential for some transfers (just think how reading a menu can make one's *mouth* water) and facilitates some others, which can be observed with chimpanzees trained for language [SAV 88].

Secondly, a *dominant modality* can also exist in a perceptual task involving several senses, and imposes its own representations. Such a scheme has been proposed for example, in spatial localization, for which vision is probably the dominant modality in human beings, given its higher reliability for this type of tasks [RAD 94, WEL 86]. Thus, auditory information that exploits cues such as the intensity and phase inter-aural differences is treated by various centers like the cochlear nucleus and the olive complex, before becoming organized as a spatial map at the level of the lower colliculus [KNU 78] It is then combined with visual and tactile information in the upper colliculus [STE 93].

Finally, it can be assumed that the various sensory modalities deliver *amodal* attributes, that is to say, transverse attributes, which serve as a basis for later interactions. Time is probably represented and exploited in this way for the perception of rhythm in Audiovisual interactions. Another possibility could be that the sensory modalities point directly to the physical structure of the environment, independent of the transducer, i.e., in an amodal way [GIB 66].

11.3.2. *Three paths for sensor fusion in information processing*

Once again, as a consequence of the fact that "several senses are better than one", the study of sensor fusion for robust decisions is considered as an increasingly important challenge information processing, in particular in the field of robotics. In an overview article, Dasarathy [DAS 94] proposes a classification of multi-sensor decision processes, which can be simplified into the three diagrams, depicted in Figure 11.8. In these diagrams, a set of several streams is used to make a global decision, and it is easy to understand that three distinct architectures are possible, depending on the respective positions of decision and fusion within the process.

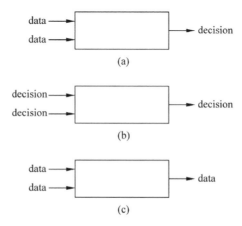

Figure 11.8. *Three architectures for sensor fusion: (a) direct identification of joint data, (b) partial single-sensor decisions followed by a late decision fusion, (c) early data fusion followed by a decision on the fused data (from [DAS 94])*

One option is to make a decision based directly on the set of data (see Figure 11.8a). In this case, there is no actual fusion operation: the global decision-making process operates a global exploitation of the information available from the data. Another option is to make a first level partial or intermediate decision on each data stream, and then to make a global decision from the decision based on single sensors (see Figure 11.8b). This can be viewed as the fusion of elementary decisions. Finally, a third approach is to start with data fusion by compacting individual streams into a global stream, on which the decision process can then be applied (see Figure 11.8c).

11.3.3. *The four basic architectures for audiovisual fusion*

The above analysis leads to four main model categories (Figure 11.9), which offer an efficient interpretation grid that can be proofed against the various proposals available in the literature (see some reviews of these approaches in [ROB 95a], [ADJ 97], [TEI 99a] and in section 11.4).

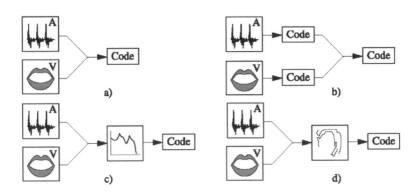

Figure 11.9. *Four basic architectures for audiovisual fusion: (a) direct identification (DI model), (b) separate identification (SI model), (c) recoding in the dominant modality (DR model), (d) recoding in the motor modality (MR model)*

Starting from the previous classification, one can obtain a model that relies on a direct identification of both data streams (DI model, see Figure 11.9a). The main characteristic of such a model is that it does not rely on any intermediate layer corresponding to a common representation of the data. It can be considered, in the taxonomy of psychology (see section 11.3.1) as a model for which language provides the mainspring for the interaction, as there is no contact between the sensory streams before the output of the system, i.e. at the level of the linguistic code.

The second path of section 11.3.2 corresponds to a model termed separate identification (SI model, Figure 11.9b). In this, the sound and the image from the speaker are classified separately and then followed by a so-called late fusion process, posterior to code access in each modality. Classification can either be applied to the whole set of phonetic features or only on particular features assumed specific to each modality. Then the fusion operates on decisions and is generally carried out by a probabilistic process deriving from decision theory. In this second model, it is once again the language that is the convergence point between the sensory paths, here, by means of a phonetic representation, which is common to both modalities.

As to the third path described in section 11.3.2, it is termed "early integration", as it takes place before the code is accessed. However, the form that the common stream takes on after fusion remains unspecified. When this question is approached by investigating what type of implementation may be compatible with that of the brain, we are brought back to the question raised in section 11.3.1: what is the form of a pre-linguistic joint representation of two sensory modalities? This question has two answers. One option is to consider that audition is the dominant modality for speech (nobody would reasonably claim the opposite, as vision only provides an incomplete access to the phonetic code) and that vision has to be recoded under a format which is compatible with auditory representations: this is the model of dominant recoding (DR model, Figure 11.9c). Under this approach, it can be assumed for example that audition and vision converge towards an estimation process of the vocal tract transfer function, in which the source characteristics are restricted to auditory input only and the decision relies on partial information. Another possibility is to enter a theory according to which the common cause between sound and image plays a crucial role – the motor theory [LIB 85] or direct perception theory [FOW 86] – and to assume that there is a common recoding of both sensory input into the motor modality (MR model, Figure 11.9d): the main characteristics of the articulatory gestures are estimated jointly from the auditory and visual modalities and then fed into a classification process in charge of accessing the code.

11.3.4. *Three questions for a taxonomy*

These four models being set, the question of which one should be chosen arises: should one use the most efficient one for the designer of speech recognition systems, or the one most compatible with experimental data for the specialist in human audiovisual perception systems? The choice can be clarified by a series of questions that we have put in diagrammatic form so as to organize these architectures relative to one another (Figure 11.10).

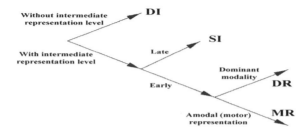

Figure 11.10. *A taxonomy of fusion models*

Firstly, is there a representation common to both modalities? The DI model is distinct from the others, as it is the only one not to include a step in which both the sound and the image can be somehow "compared". In reality, it is perfectly clear that a sound and an image can be effectively compared, in the sense that the dubbing of a film or the asynchrony between the sound and the image can easily be detected (for significant asynchronies of course, i.e. 260 ms when image precedes sounds and 130 ms when sound precedes image [DIX 80]). Experimental results by Summerfield and McGrath [SUM 84] show it is possible to estimate qualitatively a lag between the sound and the image in the case of a conflictual vocalic stimulus, and works by Kuhl and Meltzoff [KUH 82] prove that even 4 months old babies are able to turn their attention towards a mouth which articulates a vowel corresponding to the sound that they hear, rather than towards another producing an inconsistent vowel. Therefore, the DI model is not strictly compatible with experimental data, unless an additional ingredient capable of capturing audiovisual inconsistencies is added.

If there is a common representation, is it an early or a late one, with respect to code access? In other words, does this common representation imply an intermediate classification mechanism? Is it specific to a given language? At this level, a division occurs between the late integration model (SI) and the early ones (DR or MR). The SI model is the most popular one in the field of automatic recognition, as it is easier to configure and benefits both from advances in audio recognition and powerful theories in decision fusion. It is also the basis of one of the most popular perception models, namely the *fuzzy-logical model of perception* (FLMP) by Massaro [MAS 87]. However, it has one major drawback: decisions are made too early if *pre-phonetic audiovisual coherences* are not used. This is not optimal from the information processing point of view: considering the case of two phonetic classes which can be distinguished neither by their average auditory aspect nor by their average visual aspect, but only by their audiovisual covariance, the SI model cannot distinguish between them. This is what happened in the case of a famous experiment in speech perception, in which subjects were presented with the lip opening movement corresponding to a sequence [ba] or [pa] and a low frequency sound synchronized with the voicing phase [BRE 86, GRA 85, ROS 81]. In this case, nothing distinguishes [b] from [p] by audition only or lip-reading only. However, time coordination between sound and image provides a clear cue for the distinction (this could be called an audiovisual *voice onset time* [SER 00]) and subjects were indeed able to discriminate [b] and [p] when presented with the audiovisual stimulus. Other experiments (reviewed in [GRE 98]) show that some pre-phonetic cues belonging to both modalities are fused before classification, which makes the SI model rather improbable.

Finally, if there is a common early representation, what form does it take? If a dominant modality is hypothesized, we obtain the DR model; otherwise, we obtain the MR model. There is actually no definite argument to decide between the two hypotheses. However, the DR model is rather unlikely, as the complementarity between audition and vision for speech perception (see section 11.2 and Figures 11.2 and 11.3) shows that audition is not systematically dominant. The DR model also appears quite inefficient to process speech, particularly in applications which need to be robust to noise [TEI 99b].

For these reasons, the MR model is preferred, in terms of its compatibility with experimental data [ROB 95b]. A number of recent behavioral and neurophysiological studies confirm the role of a perceptuo-motor link in audiovisual speech perception (e.g. [CAL 03b, WAT 03, SAM 05, SKI 07]), however, many variants are possible, so as to circumvent some of the difficulties arising from the various architectures. In any case, the organization that we propose, based on four models and three questions for taxonomy, is essential to shed light on the model design process.

11.3.5. *Control of the fusion process*

The previous discussion was dedicated to the architecture of the decision system consisting of two sensory inputs for identifying the linguistic message (see Figure 11.7). An essential ingredient for this system is the fusion processes itself, particularly the way it depends on the variables available. In a general discussion on sensor fusion for information processing and pattern recognition, Bloch [BLO 96] distinguishes three types of process that differ in their behavior. The simplest one is the constant behavior context-free (CBCF) process, which applies fusion to the inputs with a fixed type of operation (for instance an addition or a multiplication). We may also consider a variable behavior context-free (VBCF) process, in which fusion varies as a function of the input values. This can be a system implementing additive fusion with weighting coefficients that vary as a function of the input level, giving more weight to the one with most intensity. Finally, a context-dependent (CD) process takes external knowledge sources from the system environment into account, which for example makes it possible to weight an input or another on the basis of contextual information.

Various factors thus influence the audiovisual fusion process in speech perception [SCH 98]. They can be classified into two categories, depending on whether they depend or not on the multi-sensor stimulus that is to be identified. First, a certain number of factors exist which are independent of the stimulus. The first one is inter-individual variability. Indeed, not all subjects have the same ability in lip reading (especially their capacity to identify words and phrases), and therefore, the visual modality holds more or less importance for them, in the integration process. Using conflictual Audiovisual data, Cathiard [CAT 94] shows two groups of subjects exist, who differ with respect to the importance they put on the visual modality when dealing with perceptual conflicts. Some characteristics of Audiovisual perception also depend on cultural factors, as work by Sekiyama suggests, by showing that the Japanese community seemingly attributes less weight to the visual modality in conflictual experimental situations [SEK 93]. This topic has yielded a number of experimental works with contradictory conclusions in past years: this could indicate a community who are culturally not used to looking into the people's faces (*"It may be regarded as impolite in Japan to look at someone's face"* [SEK 93, p. 442]), or a simple consequence of variations in the list of linguistic prototypes [KUH 94, MAS 93, though see SCH 06], or could even be the effect of a tone-based language that reinforces concentration on the auditory modality [BUR 98, SEK 96]. Finally, attentional factors seem to be involved in slight modifications to the relative weight put on each modality [MAS 77, TII 04], even though the McGurk effect sets a boundary to these factors, as it is cognitively poorly inaccessible and resists selective orientation of the listener's attention on the auditory component of the conflictual stimulus. A recent line of work also suggests that much more local attentional factors could play a role, by guiding the visual modality in the direction of particular portions of the acoustic input [DRI 96, SCH 04].

At the same time, a number of stimulus-dependent factors may provide entry points to a selection step or to a contextual weighting mechanism, within the fusion process. The first factor is the amplitude of the audio and visual discordance in the case of inconsistent (i.e. asynchronous or conflictual) auditory and visual inputs. Audiovisual integration seems to be a rather robust mechanism, capable of resisting numerous inconsistencies between image and sound: configurational conflicts in the McGurk effect, gender contradictions (male voice + female face and vice versa), reasonable temporal inconsistencies (an intelligibility advantage is brought by visual information in addition to a noisy audio input even for asynchronies up to 1,600 ms [CAT 94]). However, there must be some mechanism for handling interruptions in the case of extreme contradictions [ROB 95a]: for instance, the perception of dubbed movies, for which the visual modality is clearly useless in the identification process. The mechanism that enters in action in this case remains largely unknown.

Moreover, as Welch pointed out [WEL 89, p. 777], "it is a strongly held belief in the area of inter-sensory bias that the weighting of a sensory cue is positively correlated with its precision". This point is crucial for automatic speech recognition and we will come back to it later. In fact, the disturbance of a sensory modality by noise contributes to lower the accuracy of the information supplied by this very modality in the fusion process, and it is crucial to include this phenomenon in models. The issue is how to estimate the relative efficiency of the two sensory inputs and assess the weight that must be attributed to them in a context-dependent fusion process. For Massaro's FLMP model, based on separate identification and multiplicative integration, this equates to the output of each single modality decision process that guides the fusion of decisions towards the most efficient modality, without requiring any external contextual information. Indeed, when one of the modalities provides ambiguous decisions, it plays no role in the final decision. This argument seems open to criticism, because of confusion between two notions: the accuracy (or reliability) of the perceptual measure provided at the input of the classifier, and the ambiguity (or confidence) at the output of the classifier. These two elements are not necessarily identical. Thus, Robert-Ribes [ROB 95a] demonstrated in our laboratory that, in the case of the identification of highly noisy vowels, some subjects make an arbitrary decision that is identical for all the vowels (thus choosing a garbage category), without giving any advantage to that particular category in the case of audiovisual identification. The confusion matrix turns out to be extremely similar to that of visual confusions, and the auditory modality, even though it does not seem ambiguous, is perceived as unreliable and is therefore very weakly weighted in the fusion process. Conversely, experimental data from Lisker and Rossi [LIS 92] on the auditory, visual and audiovisual identification of the rounding feature in vowels, suggest that a modality may be taken into account in the fusion process even when it is ambiguous. In conclusion, it seems preferable to consider a reliability criterion related to the direct estimation of the signal to noise ratio rather than one based on phonetic ambiguity criterion.

11.4. Audiovisual speech recognition systems

Since Petajan's pioneering work [PET 84], the area of audiovisual automatic speech recognition (AVASR) has been very active, and many recognition systems exploit visual information [POT 03]. Several landmarks have played a key role in the progress made in this field: we mention the NATO workshop in Bonas, in 1995, which yielded a publication entitled "*Speech reading by humans and machines*" [STO 96], the specialized AVSP (*audio-visual signal processing*) series of workshops since 1997 and the increasing importance of AVASR in large international conferences such as Interspeech (formerly Eurospeech and ICSLP) and ICASSP. In the French community, it is worth mentioning the AMIBE project (*multimodal applications for advanced interfaces and terminals*) by the Speech pole of the GDR-PRC CHM [MON 94], which has focused research from several French laboratories (LAFORIA, LIUAPV, LIUM, ICP, INRIA, IRIT) on the design of audiovisual speech recognition systems using common corpora and major theoretical developments. It is worth reading [TEI 99a], which proposes a detailed comparative review of AVASR systems. To help us find our way around the rather large inventory of such systems, it is best to consider an objective, a paradigm and three ingredients.

The objective is to use what we have formerly referred to as the "double optimality" of audiovisual perception of speech (section 11.2.2): complementarity of the inputs and synergy of the processes, for finally fulfilling what we have sometimes called the challenge of AVASR – to obtain recognition performances that are higher in bimodal (audiovisual) situations compared to mono-modal performances. We summarize this challenge by the following inequality:

$$AV \geq \sup (A, V) \hspace{5cm} [11.1]$$

This must apply at all levels on which performance is measured: sentence, word, phoneme and features. For this inequality to represent an interesting challenge, the conditions must be difficult, and we proposed [TEI 99b] that the system is tested in the context of an *extrapolation paradigm*. In this, the system is configured (trained) for a given range of degradations (generally acoustic), then tested against a wider range of degradations, to assess its robustness.

Of course, to obtain a satisfactory level of performance, state-of-the-art know-how in speech recognition has to be implemented. Therefore, efficient acoustic parameterization has to be used at the system input (see Chapter 10), as well as adapted decoders (see Chapter 6), based, for instance on hidden Markov models (HMMs). There are also a number of specific ingredients that are required to achieve sensor fusion satisfactorily as well as to address the challenge of audiovisual speech. Our experience in this matter has identified three major ingredients [TEI 99a]: choice of a proper architecture, control of the fusion process (both following the principles presented above) and pre-processing of the data. In the following section, we review the most relevant possibilities and discuss their importance in efficient audiovisual recognition.

11.4.1. *Architectural alternatives*

Among the four possible architectures introduced in section 11.3.3, regardless of their cognitive plausibility (see discussion in section 11.3.4), the most frequently used architectures tend to be based on either the direct identification (DI) model or the separate identification (SI) model (see [ROB 95a]). Two papers ([TEI 98] and [TEI 99b]) describe an implementation of two other architectures and present a comparison to reference architectures).

Basic versions of DI and SI models can be expressed relatively simply. In the former, the decoding process operates entirely on audiovisual feature vectors and the final decision is made on the temporal sequence of these vectors. In SI models, two independent classifiers operate in parallel on auditory and visual sequences, and their individual decisions are fused using various conventional techniques (see

Figure 11.11 and 11.12 for an implementation of these two architectures in an HMM decoder). The first generation of systems used a rule-based approach for fusion [PET 84, PET 87], but most now operate on a probabilistic framework that allows many variants [XU 92] around the multiplicative Bayesian approach, up to the theory of evidence [SHA 76]. The principle of this approach is based on a confusion matrix learned during the training phase, and then used extensively in the decision process (see [TEI 95] for a comparison of several such methods for the recognition of noisy vowels).

Figure 11.11. *The DI architecture in the case of HMM decoders*

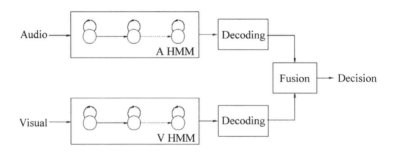

Figure 11.12. *The SI architecture in the case of HMM decoders*

We might anticipate the DI architecture would have an enhanced performance, because it is the only model that can exploit joint time variations between audio and image streams. However, this architecture choice poses two problems. The first is how to weight inputs in the case of context-dependent fusion. We have proposed an original weighting mechanism that does not alter the DI architecture (see next paragraph). However, in a conventional HMM framework, another option involves learning a DI model for which independence between the audio and video streams is assumed at the level of the HMM model states. During the decoding phase, the audiovisual probability for each frame in each state is obtained through a conventional fusion of audio and visual probabilities (on which weighting mechanisms are easily imposed). We call this architecture the "DI-2 streams model" (see Figure 11.13) to indicate that, behind the DI model, two independent streams, fused at the level of the HMM states are being combined [ALI 96, NAK 97, POT 98, ROG 96, ROG 97, SIL 94].

Figure 11.13. *The DI-2 streams architecture, a variant of the DI architecture*

The second problem posed by DI architectures originates in the natural asynchrony between audio and image streams. A key characteristic of bimodal speech is that co-articulation mechanisms show up as a complex set of non-audible yet visible anticipation/perseveration gestures [ABR 97]. Consequently, there is a frequent advance of the image over the audio, e.g. during the labial gesture from an unrounded [i] to a rounded [y], the rounding/protrusion movement begins before the [i] sound is terminated (well before the [y] can be heard) [ABR 95, ABR 96]. This gesture, which is totally inaudible, is nevertheless perfectly visible in lip reading [CAT 91, CAT 95, ESC 96]. As a consequence, in DI architectures, joint segmentation of the audio and image streams into common audiovisual HMM states can create major problems during recognition [MON 94]. Audiovisual asynchrony also poses problems to SI architectures when they are handling the time sequences of decoding units, in connected word and continuous speech recognition.

This crucial issue has led to a series of original proposals (in particular, within the AMIBE project). For sequence recognition with HMM, both [BRO 96, BRO 98] and [JOU 96a, JOU 96b] proposed particular applications of SI architectures that operate a multiplicative fusion at the state level within each model. Using this approach, the "product" of two (audio and visual) HMMs can be defined (see Figure 11.14). This technique enables automatic handling of the asynchrony between the two modalities.

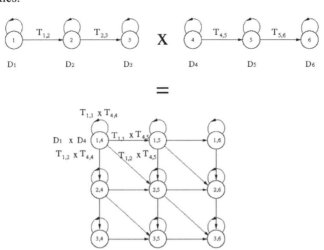

Figure 11.14. *Example of a product HMM (from [JOU 96a])*

Another proposal has been the so-called "master-slave SI architecture" [AND 96, AND 97, JAC 96], which exploits two parallel and correlated HMMs, to dynamically adapt the probability density functions of one HMM as a function of a context modeled by the other. The "master" model is ergodic, and represents three visual configurations (mouth open, mouth half-open, mouth closed): the acoustic HMM functions as the "slave", and its parameters (transition matrix, probability density functions) are probabilistic functions of visual HMM states (see Figure 11.15).

Master HMM (visual observations)

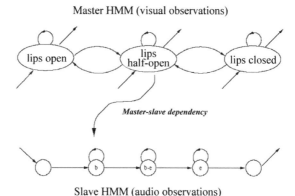

Slave HMM (audio observations)

Figure 11.15. *Relationship between a master HMM (visual modality) and a slave HMM (audio modality) (from [AND 97])*

Finally, we consider a solution provided by the now famous "Multi-Stream Model", which can decode independent streams whilst imposing "forced" meeting points, defined *a priori*, where the two decoders synchronize. This is the framework in which a system proposed by [LUE 98] was implemented (see Figure 11.16).

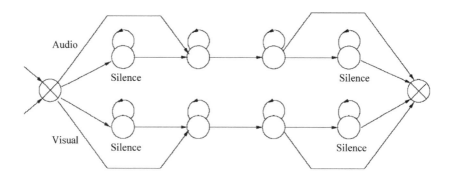

Figure 11.16. *Multi-stream model for* Audiovisual *speech recognition (from [LUE 98])*

In summary, a wide range of variants exists, comprising a *continuum* of solutions ranging from "full dependence" (DI architecture) to "full independence" (SI architecture). The challenge is to find, between these two extremes, how to position the system's cursor at the optimal point to exploit joint variations between audio and image streams, whilst dealing with their asynchrony. Note that recent works on the recognition of French Cued Speech typically encountered the same problem [ATT 04] with the same solutions [ABO 07].

11.4.2. *Taking into account contextual information*

Let us begin with some specific terminology. In this section, we use the word *context* not to convey its classical meaning of temporal context (in relation to coarticulation mechanisms used in automatic speech recognition), but in the same way those involved in sensor fusion use it, to refer to some complementary information provided by the sensors and the context in which the fusion is taking place. In this case, we are essentially using it in the context of auditory noise or, more generally, the respective reliability of the audio and visual sensors. The contextual information enabling selective weighting of a sensory modality during fusion is increasingly being used in AVASR systems, and forms an essential strategy for improving their robustness. In reality, if this problem is not approached carefully, what Movellan and Mineiro call "catastrophic fusion" may happen [MOV 96], that is, the modality disturbed by noise continuously degrades the performance of a reliable modality. Contextual variables must give a true picture of the reliability of each sensor (or at least of the auditory sensor when the noise is only acoustic). This estimation of the reliability must itself be reliable, in order to guarantee a proper control of the fusion process.

The reliability of the auditory sensor can be evaluated from system input data using one of the numerous SNR (signal-to-noise ratio) estimation methods. Note that many studies on audiovisual recognition assume that the SNR is known, so as to test the fusion process only. Whether the SNR is estimated or given *a priori*, an additional step is needed to determine the appropriate context value from the SNR. This can be performed during the training phase [MEI 96] or during the test phase [NAK 97, SIL 96a], though the first option is preferable. The determination is obtained either via some optimization, or by an exhaustive search with an error criterion (minimization of the identification error rate, or maximization of the winning class probability).

Reliability can also be estimated at the output of the classification process. A classifier provides an estimate of how a stimulus conforms to a training set. This approach can be used to determine a context: our reservations on the principle of this approach are discussed in section 11.3.5. Several consistency measures between a test input and the statistical properties of a training set can provide information on the sensor's state, such as the *a priori* probability of the observation given the winning class, the entropy, or any other measure of the dispersion of the *a posteriori* class probability.

In the case of SI architectures, the weighting of each stream is easy, especially in the conventional case of multiplicative fusion. Generally, a geometric weighting is used:

$$P_{AV}(i/\mathbf{x}_{AV}) = \frac{P_A^\alpha(i/\mathbf{x}_A) * P_V^{1-\alpha}(i/\mathbf{x}_V)}{\sum_{j=1}^{C} P_A^\alpha(j/\mathbf{x}_A) * P_V^{1-\alpha}(j/\mathbf{x}_V)} \qquad [11.2]$$

Here i is the class index , C the number of classes, α the context value (deduced from the SNR estimation or from the conformity of the stimulus with the learnt statistics), $\mathbf{x}_A$, $\mathbf{x}_V$ and $\mathbf{x}_{AV}$ the audio, visual and audiovisual inputs respectively and $P_{AV}(i/\mathbf{x}_{AV})$ the *a posteriori* probability that the stimulus $\mathbf{x}_{AV}$ belongs to class i. This equation departs from the Bayesian framework. However, it provides a convenient approximation of the actual variance weighting method in the case of Gaussian classifiers [TEI 99b]. The introduction of the context in the case of DI architectures is more complex, and can only been approached using some simplifications (DI-2 streams models), which reduces it to the previous model. We have proposed, in the case of Gaussian statistics, a more general method using the weighting of a covariance matrix for computing the distance between a stimulus and a prototype [TEI 99b].

Finally, it is worth mentioning an original and promising approach proposed by Movellan and Mineiro [MOV 96] in line with previous work on context competition [CLA 90]. The principle is to introduce in the probabilistic classification algorithm (here, a conventional SI model based on HMM) an explicit context term (namely, the level of noise or the level of sound or image degradation) and to determine by a Bayesian approach, the most probable context given the stimulus and the corresponding phonetic category. Thus, instead of the classical equation:

$$i = \arg\max_j [\log(P(\mathbf{x}_A|j)) + \log(P(\mathbf{x}_V|j)) + \log(P(j))] \qquad [11.3]$$

in which i is the recognized class, $P(\mathbf{x}_A|j)$ (resp. $P(\mathbf{x}_V|j)$) the probability of the audio (resp. visual) observation for class j, and $P(j)$ the *a priori* probability of class j, a more complex equation is used, namely:

$$i = \arg\max_{j}[\max_{\sigma_A} \log(P(\mathbf{x}_A | j, \sigma_A)) + \max_{\sigma_V} \log(P(\mathbf{x}_V | j, \sigma_V)) + \log(P(j))] \qquad [11.4]$$

where σ_A^2 and σ_V^2 are the respective variances of the audio and visual noises, whose statistics are known *a priori* (for example, a Gaussian white noise).

Movellan and Mineiro showed, on a simple 2-class example with Gaussian noise, that estimation of audio and visual noise by context competition adequately tunes the weight of each modality in the fusion process. Thus, this attractive (yet complex) approach both provides an estimation of the context and enables control of the fusion process itself.

11.4.3. *Pre-processing*

In pattern recognition problems, and particularly in speech processing, data pre-processing is essential to a system's performance (see Chapter 10). However, in a sensor fusion based system, this role can become dominant. To illustrate this point, we present in Figure 11.17 the results of a simple experiment in noisy vowel recognition using a SI model (relying on Bayesian fusion of audio and visual Gaussian classifiers). In this experiment, the input presented to the acoustic classifier is either an auditory spectrum over 20 channels, or the reduction of this vector to a decreasing number of principal components obtained by a PCA (principal component analysis). The auditory classifier is trained on low noise levels (SNR = 0 dB), and tested on a wide range of noise intensities (extrapolation paradigm), whereas the visual classifier exploits geometrical parameters describing the shape of the lips, computed by an automatic lip feature extractor [LAL 90]. For pure auditory recognition (Figure 11.17a), the behavior of the system is unsurprising: when the number of components decreases, the identification performance decreases too, whatever the noise level.

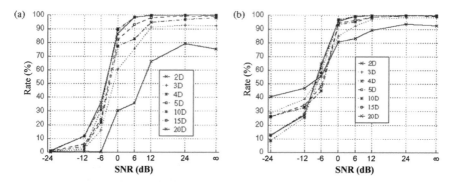

Figure 11.17. *Effect of dimensionality reduction on recognition performance:*
(a) audio, (b) audiovisual

For audiovisual recognition however, system behavior varies with the number of dimensions in an unanticipated manner (Figure 11.17b). For test conditions matching the training conditions (medium and large SNR), audiovisual identification performance decreases with dimensionality reduction. Up to approximately 10 dimensions, this reduction suppresses redundant information: and audio and audiovisual performances are similar to those obtained without pre-processing (20D). For dimensions between 5 and 10, some information loss takes place: audio performance decreases significantly compared to cases where no pre-processing is done. However, the audiovisual score does not decrease dramatically, due to the presence of visual information. Below 5 dimensions, information loss is drastic, and audiovisual scores decrease significantly. However, for test conditions that do not match training conditions (high noise level, between -6 dB and -24 dB), the dimensionality reduction is always beneficial. The audio classifier makes more errors, but the decisions are more ambiguous and bear less weight in the fusion process: audiovisual performance increases with dimensionality reduction.

In summary, dimensionality reduction leads to auditory decisions that are less and less contrasted. This is illustrated by computation of the audio classifier's average ambiguity as a function of the number of dimensions. Figure 11.18 shows the value of the ambiguity for 500 test items at two different noise levels (-24 dB and without noise). "Ambiguity" characterizes the contrast between the main *a posteriori* probabilities yielded by the classifier. Here, we have chosen to define it as follows:

$$Ambiguity = \frac{P_A(i_2|\mathbf{x}_A)}{P_A(i_1|\mathbf{x}_A)}$$

$$with \quad i_1 = \arg\max_j P_A(j|\mathbf{x}_A) \quad and \quad i_2 = \arg\max_{k \neq i_1} P_A(k|\mathbf{x}_A)$$

[11.5]

where i_1 is the first choice and i_2 is the second choice of the audio classifier. It can be shown that Ambiguity decreases with dimensionality reduction. For the noise-free case, this can be an issue, as the decision is in principle correct; which explains the performance degradation. Nevertheless, for high noise levels (-24 dB in our example), this is beneficial as the decisions are generally erroneous and a strong auditory ambiguity places more weight on the visual classifier so audiovisual scores increase and a better generalization is observed in mismatched conditions. A better generalization may not mean a higher performance level; rather it implies a better estimation of reliability for decision. According to this figure, the situation is far from optimal: for instance, at -24 dB, audio recognition performance is 0%, even though the ambiguity is not very high. This becomes a major problem beyond 4 dimensions, where ambiguity is almost zero: the audio classifier makes an incorrect decision with a large confidence level. This erroneous decision is very detrimental to fusion; hence audiovisual performance becomes similar to auditory performance.

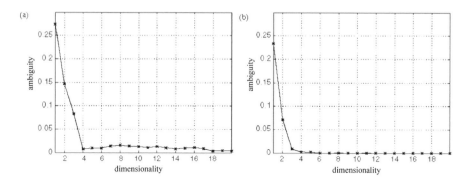

Figure 11.18. *Ambiguity values as a function of the number of dimensions: (a) for a SNR equal to -24 dB, (b) for the noise-free case*

Thus, data pre-processing in fusion processes is essential. Reducing the dimensionality of the audio data drives the classifier into more balanced decision-making, enabling the visual modality to take more importance in the fusion process. This is very useful in extrapolated noise conditions, where audiovisual scores increase significantly. If the dimensionality reduction is carried out "intelligently", i.e. without deteriorating the noise-free data structure, while enhancing the modifications that take place in noisy conditions, the classifier remains efficient in matched conditions. It also makes even more reliable decisions in mismatched conditions, enabling a better fusion. This is observed in [KRO 97] with Linear Discriminant Analysis, and in our own work [TEI 98], which exploits an original multidimensional structure-unfolding algorithm, namely the constrained curvilinear component analysis.

Finally, one more step consists of de-noising auditory data using the visual stream by exploiting the internal consistency of audiovisual data This is a path we explored a few years ago by exploiting filtering algorithms [GIR 96, 01] and, more recently, methods inspired from blind source separation techniques [SOD 04, RIV 07]. This is undoubtedly a direction that will lead to new developments in this domain, as we have already underlined.

11.5. Conclusions

Needless to say at this stage, the multimodal nature of speech provides a fascinating research paradigm. For theoreticians and experimenters, it offers full scope for working on the interaction between sensory modalities and the relationship between perception and action. It also provides a roadmap for the study of auditory processing, and determines some checkpoints between the speech signal and the linguistic code. All in all, it constitutes a favored ground for developing theories of speech perception, phonology and memory, in other words, everything that would be included in a "speech module"... if it exists!

For engineers, multimodality leads to more robust voice interface for man-machine dialog: speech recognition and speech synthesis (see Chapter 3). Multimodality is also bound to be advantageous in the telecom sector, offering new methods for jointly processing audio and video contents, de-noising, compression and coding [GIR 01, GIR 04].

As a challenging research topic this area is full of applicative potential in the domain of hearing aids and language learning. In fact, multimodality is a perfect illustration of the target focus of this book, lying between cognitive models and communicating machines... multimodal machines, of course.

11.6. References

[ABO 07] ABOUTABIT N., BEAUTEMPS D., CLARKE J., BESACIER L., "A HMM recognition of consonant-vowel syllables from lip contours: the Cued Speech case", *Proc. InterSpeech*, p. 646-649, Antwerp. Belgium, 2007.

[ABR 95] ABRY C., LALLOUACHE T., "Le MEM: un modèle d'anticipation paramétrable par locuteur. Données sur l'arrondissement en français", *Bulletin de la Communication Parlée*, 3, p. 85-99, 1995.

[ABR 96] ABRY C., LALLOUACHE M.-T., CATHIARD M.-A., "How can coarticulation models account for speech sensitivity to Audiovisual desynchronization?", in D. STORK and M. HENNECKE (ed.), *Speechreading by Humans* and *Machines*, NATO ASI Series F: Computer and Systems Sciences, vol. 150, p. 247-255, Springer-Verlag, Berlin, 1996.

[ABR 97] ABRY C., SCHWARTZ J.L., "La parole cognitive" in M. GORDON and H. PAUGAM-MOISY (ed.), *Sciences Cognitives – Diversité des approches*, p. 103-114, Hermès, Paris, 1997.

[ADJ 96] ADJOUDANI A., BENOÎT C., "On the integration of auditory and visual parameters in an HMM-based ASR" in D.G. STORK and M.E. HENNECKE (ed.), *Speechreading by Man* and *Machine: Models, Systems* and *Applications*, p. 461-472, NATO ASI Series, Springer, 1996.

[ADJ 97] ADJOUDANI A., Reconnaissance automatique de la parole audiovisuelle. Stratégies d'intégration et réalisation du LIPTRACK, labiomètre en temps réel, PhD thesis, Institut National Polytechnique de Grenoble, Signal-Image-Parole, 1997.

[ALE 92] ALEGRIA J., LEYBAERT J., CHARLIER B., HAGE C., "On the origin of phonological representations in the deaf: Hearing lips and hands" in J. ALEGRIA, D. HOLENDER, J. JUNCA DE MORAIS and M. RADEAU (ed.), *Analytic Approaches to Human Cognition*, p. 107-132, Elsevier Science Publishers, Amsterdam 1992.

[ALI 96] ALISSALI M., DELÉGLISE P., ROGOZAN A., "Asynchronous integration of visual information in automatic speech recognition system", *Proc. ICSLP'96*, p. 34-37, 1996.

[AND 96] ANDRÉ-OBRECHT R., JACOB B., SENAC C., "Words on lips: How to merge acoustic and articulatory informations to automatic speech recognition", *EUSIPCO'96*, p. 16311634, 1996.

[AND 97] ANDRÉ-OBRECHT R., JACOB B., PARLANGEAU N., "Audiovisual speech recognition and segmental master-slave HMM", *Proc. AVSP'97*, p. 49-52, 1997.

[ATT 04] ATTINA V., BEAUTEMPS D., CATHIARD M.A., ODISIO M., "A pilot study of temporal organization in Cued Speech production of French syllables: rules for a Cued Speech synthesizer", *Speech Communication*, 44, p. 197-214, 2004.

[BAR 98] BARKER J., BERTHOMMIER F., SCHWARTZ J.L., "Is primitive AV coherence an aid to segment the scene?", *Proc. AVSP'98*, p. 103-108, 1998.

[BEN 94] BENOÎT C., MOHAMADI T., KANDEL S.D., "Effects of phonetic context on audiovisual intelligibility of French", *J. Speech* and *Hearing Research*, 37, p. 1195-1203, 1994.

[BEN 96] BENOÎT C., Intelligibilité audiovisuelle de la parole, pour l'homme et pour la machine, Habilitation à diriger des recherches, Institut National Polytechnique de Grenoble, 1996.

[BER 89] BERNSTEIN L.E., EBERHARD S.P., DEMOREST M.E., "Single-channel vibrotactile supplements to visual perception of intonation and stress", *J. Acoust. Soc. Am.*, 85, p. 397405, 1989.

[BER 96] BERNSTEIN L.E., BENOÎT C., "For speech perception by humans or machines, three senses are better than one", *Proc. ICSLP'96*, p. 1477-1480, 1996.

[BER 98] BERNSTEIN L.E., DEMOREST M.E., TUCKER P.E., "What makes a good speechreader? First you have to find one", in R. CAMPBELL, B. DODD and D. BURNHAM (ed.), *Hearing by Eye, II. Perspectives* and *Directions in Research on Audiovisual Aspects of Language Processing*, p. 211-228, Psychology Press, Hove (UK) 1998.

[BER 04] BERNSTEIN L.E., AUER E.T., MOORE J.K., "Audiovisual speech binding: Convergence or association", in G.A. CALVERT, C. SPENCE and B.A. STEIN (ed.) The *Handbook of Multisensory Processes*, p. 203-223, MIT Press, Cambridge, 2004.

[BLO 96] BLOCH I., "Information combination operators for data fusion: a comparative review with classification", *IEEE Trans. Systems, Man* and *Cybernetics*, 26, p. 52-67, 1996.

[BOE 00] BOË L.J., VALLEE N., SCHWARTZ J.L., "Les tendances des structures phonologiques: le poids de la forme sur la substance", in P. ESCUDIER and J.L. SCHWARTZ (ed.), *La parole, des modèles cognitifs aux machines communicantes – I. Fondements*, Hermes, Paris, 2000.

[BRE 86] BREEUWER M., PLOMP R., "Speechreading supplemented with auditorily presented speech parameters", *J. Acoust. Soc. Am.*, 79, p. 481-499, 1986.

[BRO 96] BROOKE N.M., "Using the visual component in automatic speech recognition", *Proc. ICLSP'96*, p. 1656-1659, 1996.

[BRO 98] BROOKE N.M., "Computational aspects of visual speech: machine that can read and simulate talking face", in R. CAMPBELL, B. DODD and D. BURNHAM (ed.), *Hearing by Eye II*, p. 109-122, Psychology Press, Hove (UK), 1998.

[BUR 98] BURNHAM D., LAU S., "The effect of tonal information on auditory reliance in the McGurk effect", *Proc. AVSP'98*, p. 37-42, 1998

[CAL 82] CAMPBELL R., DODD. B., "Some suffix effects on lipread lists", *Canadian Journal of Psychology*, 36, p. 508-514, 1982.

[CAL 03a] CALVERT G.A., CAMPBELL R., "Reading speech from still and moving faces: the neural substrates of visible speech", *J. Cognitive Neuroscience*, 15, p. 57-70, 2003.

[CAL 03b] CALLAN D.E., JONES J.A., MUNHALL K., CALLAN A.M., KROOS C., VATIKIOTIS-BATESON E., "Neural processes underlying perceptual enhancement by visual speech gestures", *Neuroreport*, 14, p. 2213-2218, 2003.

[CAM 98] CAMPBELL R., "How brains see speech: The cortical localisation of speech reading in hearing people", in R. CAMPBELL, B. DODD and D. BURNHAM (ed.), *Hearing by Eye, II. Perspectives* and *Directions in Research on Audiovisual Aspects of Language Processing*, Psychology Press, Hove (UK), p. 177-194, 1998.

[CAT 88] CATHIARD M.A., "La perception visuelle de la parole: aperçu de l'état des connaissances", *Bulletin de l'Institut de Phonetique de Grenoble*, 17-18, p. 109-193, 1988.

[CAT 91] CATHIARD M.-A., TIBERGHIEN G., TSEVA A., LALLOUACHE M.-T., ESCUDIER P., "Visual perception of anticipatory rounding during acoustic pauses: a cross-language study", *Proceedings of the XIIth International Congress of Phonetic Sciences*, 4, p. 50-53, 1991.

[CAT 94] CATHIARD M.A., La perception visuelle de l'anticipation des gestes vocaliques: cohérence des événements audibles et visibles dans le flux de la parole, PhD Thesis, Pierre Mendès France University, Grenoble, 1994.

[CAT 95] CATHIARD M.-A., LALLOUACHE M.-T., MOHAMADI T., ABRY C., "Configurational *vs.* temporal coherence in audiovisual speech perception", *Proceedings of the XIIIth International Congress of Phonetic Sciences*, 3, p. 218-221, 1995.

[CLA 90] CLARK J.J., YUILLE A.L., *Data Fusion for Sensory Information Processing Systems*, Kluwer Academic Publishers, Boston, 1990.

[COR 67] CORNETT R.O., "Cued Speech", *American Annals of the Deaf*, 112, pp. 3-13, 1967.

[DAS 94] DASARATHY B.V., *Decision Fusion*, IEEE Computer Society Press, Los Alamitos, California, 1994.

[DAV 98] DAVIS C., KIM J., "Repeating and remembering foreign language words: does seeing help?", *Proc. AVSP'98*, pp. 121-125, 1998.

[DEG 92] DE GELDER B., VROOMEN J., "Abstract versus modality-specific memory representations in processing auditory and visual speech", *Memory* and *Cognition*, 20, p. 533-538, 1992.

[DIX 80] DIXON N.F., SPITZ L., "The detection of audiovisual desynchrony", *Perception*,9, pp. 719-721, 1980.

[DRI 96] DRIVER J., "Enhancement of selective listening by illusory mislocation of speech sounds due to lip-reading", *Nature*, 381, p. 66-68, 1996.

[DUP 05] DUPONT S., MÉNARD L., "Vowel production and perception in French blind and sighted adults", *J. Acoust. Soc. Am.*, 117, p. 2622, 2005.

[EBE 90] EBERHARDT S.P., BERNSTEIN L.E., DEMOREST M.E., GOLDSTEIN M.H., "Speechreading sentences with single-channel vibrotactile presentation of fundamental frequency", *J. Acoust. Soc. Am.*, 88, p. 1274-1285, 1990.

[ERB 69] ERBER N.P., "Interaction of audition and vision in the recognition of oral speech stimuli", *J. Speech* and *Hearing Research*, 12, pp. 423-425, 1969.

[ESC 80] ESCUDIER P., BENOÎT C., LALLOUACHE T., "Indentification visuelle de stimuli associés à l'opposition /i/-/y/: Etude statique", *1er Congrès Français d'Acoustique*, 1, pp. 541-544, 1980.

[FAR 76] FARWELL R.M., "Speechreading: a research review", *American Annals of the Deaf*, February, p. 19-30, 1976.

[FOW 86] FOWLER C.A., "An event approach to the study of speech perception from a directrealist perspective", *J. Phonetics*, 14, p. 3-28, 1986.

[FOW 91] FOWLER C.A., DEKLE D.J., "Listening with eye and hand: Crossmodal contributions to speech perception", *J. Experimental Psychology: Human Perception* and *Performance*, 17, p. 816-828, 1991.

[GIB 66] GIBSON J.J., *The Senses Considered as Perceptual Systems*, Houghton Mifflin Co, Boston, 1966.

[GIR 96] GIRIN L., FENG G., SCHWARTZ J.-L., "Débruitage de parole par un filtrage utilisant l'image du locuteur: une étude de faisabilité", *Traitement du Signal*, 13, p. 319-334, 1996.

[GIR 01] GIRIN L., SCHWARTZ, J.-L., FENG G., "Audiovisual enhancement of speech in noise", *J. Acoust. Soc. Am.*, 109, p. 3007-3020, 2001.

[GIR 04] GIRIN L., "Joint matrix quantization of face parameters and LPC coefficients for low bit rate audiovisual speech coding", *IEEE Transactions on Speech* and *Audio Processing*, 12, p. 265-276, 2004.

[GRA 85] GRANT K.W., ARDELL L.H., KUHL P.K., SPARKS D.W., "The contribution of fundamental frequency, amplitude envelope, and voicing duration cues to speechreading in normal-hearing subjects", *J. Acoust. Soc. Am.*, 77, p. 671-677, 1985.

[GRE 96] GREEN K.P., "Studies of the McGurk effect: Implications for theories of speech perception", *Proc. ICSLP'96*, Philadelphia, USA, p. 1652-1655, 1996.

[GRE 98] GREEN K.P., "The use of auditory and visual information during phonetic processing: implications for theories of speech perception", in R. CAMPBELL, B. DODD and D. BURNHAM (ed.), *Hearing by Eye, II. Perspectives* and *Directions in Research on Audiovisual Aspects of Language Processing*, pp. 3-25, Psychology Press, Hove (UK), 1998.

[HAT 93] HATWELL Y., "Transferts intermodaux et intégration intermodale", in M. RICHELLE, J. REGUIN and M. ROBERT (ed.), *Traité de Psychologie Expérimentale*, p. 543-584, Presses Universitaires de France, Paris, 1993.

[JAC 96] JACOB B., SENAC C., "Un modèle maître-esclave pour la fusion des données acoustiques et articulatoires en reconnaissance automatique de la parole", *Actes des XX Journées d'Etude sur la Parole (JEP'96)*, pp. 363-366, 1996.

[JOU 96a] JOURLIN P., "Asynchronie dans les systèmes de reconnaissance de la parole basés sur les HMM", *Actes des XXIèmes Journées d'Etude sur la Parole (JEP'96)*, pp. 351-354, 1996.

[JOU 96b] JOURLIN P., "Handling desynchronization phenomena with HMM in connected speech", *Proc. EUSIPCO'96*, p. 133-136, 1996.

[KNU 78] KNUDSEN E.I., KONISHI M., "A neural map of auditory space in the owl", *Science*, 200, pp. 795-797, 1978.

[KRO 97] KRONE G., TALLE B., WICHERT A., PALM G., "Neural architecture for sensor fusion in speech recognition", *Proc. AVSP'97*, p. 57-60, 1997.

[KUH 82] KUHL P.K., MELTZOFF A.N., "The bimodal perception of speech in infancy", *Science*, 218, pp. 1138-1141, 1982.

[KUH 94] KUHL P.K., TSUZAKI M., TOHKURA Y., MELTZOFF A.N., "Human processing of auditory-visual information on speech perception: potential for multimodal human-machine interfaces", *Proc. ICLSP'94*, p. 539-542, 1994.

[LAL 90] LALLOUACHE M.T., "Un poste 'visage-parole'. Acquisition et traitement de contours labiaux", *Actes des XVIIIèmes Journées d'Etude sur la Parole (JEP'90)*, p. 282-286, 1990.

[LEG 96] LEGOFF B., BENOÎT C., "A text-to-audiovisual speech synthesizer for French", *Proc. ICSLP'96*, p. 2163-2166, 1996.

[LEY 98] LEYBAERT J., ALEGRIA J., HAGE C., CHARLIER B., "The effect of exposure to phonetically augmented lipspeech in the prelingual deaf", in R. CAMPBELL, B.

DODD and D. BURNHAM (ed.), *Hearing by Eye, II. Perspectives* and *Directions in Research on Audiovisual Aspects of Language Processing*, p. 283-301, Psychology Press, Hove (UK), 1998.

[LIB 85] LIBERMAN A., MATTINGLY I., "The motor theory of speech perception revised", *Cognition*, 21, p. 1-36, 1985.

[LIS 92] LISKER L., ROSSI M., "Auditory and visual cueing of the [±rounded] feature of vowels", *Language* and *Speech*, 35, p. 391-417, 1992.

[LUE 98] LUETTIN J., DUPONT S., "Continuous Audiovisual speech recognition", *Proc. 5th European Conference on Computer Vision*, 1998.

[MAS 77] MASSARO D.W., WARNER D.S., "Dividing attention between auditory and visual perception", *Perception* and *Psychophysics*, 21, p. 569-574, 1977.

[MAC 87] MACLEOD A., SUMMEFIELD Q., "Quantifying the contribution of vision to speech perception in noise", *British Journal of Audiology*, 21, p. 131-141, 1987.

[MAS 87] MASSARO D.W., *Speech Perception by Ear* and *Eye: A Paradigm for Psychological Inquiry*. Laurence Erlbaum Associates, London, 1987.

[MAS 93] MASSARO D.W., COHEN M.M., GESI A., HEREDIA R., TSUZAKI M., "Bimodal speech perception: an examination across languages", *J. Phonetics*, 21, p. 445-478, 1993.

[MCG 76] MCGURK H., MACDONALD J., "Hearing lips and seeing voices", *Nature*, 264, p. 746-748, 1976.

[MEI 96] MEIER U., HÜRST W., DUCHNOWSKI P., "Adaptive bimodal sensor fusion for automatic speechreading", *Proc. ICASSP'96*, p. 833-836, 1996.

[MEL 83] MELTZOFF A.N., MOORE K.M., "Newborn infants imitate facial gestures", *Child Development*, 54, p. 702-709, 1983.

[MIL 51] MILLER G.A., HEISE G.A., LICHTEN W., "The intelligibility of speech as a function of the context of the test materials", *J. Exp. Psychol.*, 41, p. 329-335, 1951.

[MIL 87] MILLS A.E., "The development of phonology in the blind child", in B. DODD and R. CAMPBELL (ed.), *Hearing by Eye: the Psychology of Lipreading*, p. 145-161, Lawrence Erlbaum Associates, 1987.

[MON 94] MONTACIE C., CARATY M.J., ANDRE-OBRECHT R., BOË L.J., DELEGLISE P., EL-BEZE M., HERLIN I., JOURLIN P., LALLOUACHE T., LEROY B., MELONI H., "Application Multimodales pour Interfaces et Bornes Evoluées (AMIBE)", in H. MELONI (ed.), *Fondements et Perspectives en Traitement Automatique de la Parole*, p. 155-164, Université d'Avignon et des pays de Vaucluse, 1994.

[MOV 96] MOVELLAN J.R., MINEIRO P., "Modularity and catastrophic fusion: a Bayesian approach with applications to audiovisual speech recognition" *CogSci UCSD Technical Report 97.01*, 1996.

[MUL 88] MULFORD R., "First words of the blind child", in M.D. SMITH and J.L. LOCKE (ed.), *The Emergent Lexicon: The Child's Development of a Linguistic Vocabulary*, p. 293338, Academic Press, New York, 1988.

[NAK 97] NAKAMURA S., NAGAI R., SHIKANO K., "Adaptative determination of audio and visual weights for automatic speech recognition", *Proc. Eurospeech'97*, p. 1623-1626, 1997.

[NEE 56] NEELY K.K., "Effects of visual factors on the intelligibility of speech", *J. Acoust. Soc. Am.*, 28, p. 1275-1277, 1956.

[NOR 77] NORTON S.J., SCHULTZ M.C., REED C.M., BRAIDA L.D., DURLACH N.I., RABINOWITZ W.M., CHOMSKY C., "Analytic study of the Tadoma method: background and preliminary results", *J. Speech Hear. Res.*, 20, p. 574-595, 1977.

[PET 84] PETAJAN E.D., Automatic lipreading to enhance speech recognition, PhD Thesis, University of Illinois, 1984.

[PET 87] PETAJAN E.D., BISCHOFF B.J., BODOFF D.A., BROOKE, N.M., "An improved automatic lipreading system to enhance speech," *Technical Report TM 11251-871012-11*, AT&T Bell Laboratories, 1987.

[POT 98] POTAMIANOS G., GRAF H.P., "Discriminative training of HMM stream exponents for Audiovisual speech recognition", *Proc. ICASSP'98*, 6, p. 3733-3736, 1998.

[POT 03] POTAMIANOS G., NETI C., DELIGNE, "Joint Audiovisual speech processing for recognition and enhancement", *Proc. AVSP'03*, p. 95-104, St. Jorioz, France, 2003.

[RAD 94] RADEAU M., "Auditory-visual spatial interaction and modularity", *Cahiers de Psychologie Cognitive*, 13, p. 3-51, 1994.

[REE 82a] REED C.M., DOHERTY M.J., BRAIDA L.D., DURLACH N.I., "Analytic study of the Tadoma method: further experiments with inexperienced observers", *J. Speech Hear. Res.*, 25, p. 216-223, 1982.

[REE 82b] REED C.M., DURLACH N.I., BRAIDA L.D., "Research on tactile communication of speech: A review", *Am. Speech Lang. Hear. Assoc. (ASHA)*, Mono no 20, p. 1-23, 1982.

[REI 87] REISBERG D., MCLEAN J., GOLFIELD A., "Easy to hear but hard to understand: a lipreading advantage with intact auditory stimuli", in B. DODD and R. CAMPBELL (ed.), *Hearing by Eye: The Psychology of Lipreading*, p. 97-113. Lawrence Erlbaum Associates, London, 1987.

[RIV 07] RIVET B., GIRIN L., JUTTEN C., "Mixing audiovisual speech processing and blind source separation for the extraction of speech signals from convolutive mixtures", *IEEE Transactions on Audio, Speech* and *Language Processing*, 15, p. 96-108, 2007.

[ROB 95a] ROBERT-RIBES J., Modèles d'intégration audiovisuelle de signaux linguistiques: de la perception humaine à la reconnaissance automatique des voyelles, PhD Thesis, Institut National Polytechnique de Grenoble, Signal-Image-Parole, 1995.

[ROB 95b] ROBERT-RIBES J., SCHWARTZ J.L., ESCUDIER P., "A comparison of models for fusion of the auditory and visual sensors in speech perception", *Artificial Intelligence Review Journal*, 9, p. 323-346, 1995.

[ROB 98] ROBERT-RIBES J., SCHWARTZ J.L., LALLOUACHE T., ESCUDIER P., "Complementarity and synergy in bimodal speech: auditory, visual and audiovisual identification of French oral vowels in noise", *Journal. Acoust. Soc. Am.*, 103, p. 3677-3689, 1998.

[ROG 96] ROGOZAN A., DELEGLISE P., ALISSALI M., "Intégration asynchrone des informations auditives et visuelles dans un système de reconnaissance de la parole", *Actes XXIèmes Journées d'Etude sur la Parole (JEP'96)*, p. 359-362, 1996.

[ROG 97] ROGOZAN A., DELÉGLISE P., ALISSALI M., "Adaptative determination of audio and visual weights for automatic speech recognition", *Proc. AVSP'97*, p. 61-64, 1997.

[ROS 81] ROSEN S., FOURCIN A.J., MOORE, B., "Voice pitch as an aid to lipreading", *Nature*, 291, p. 150-152, 1981.

[SAM 05] SAMS M., MÖTTÖNEN R., SIHVONEN T. "Seeing and hearing others and oneself talk", *Cognitive Brain Research*, 23, p. 429-435, 2005.

[SAV 88] SAVAGE-RUMBAUGH S., SEVCIK R.A., HOPKINS W.D., "Symbolic cross-modal transfer in two species of chimpanzees", *Child Development* 59, p. 617-625, 1988.

[SCH 97] SCHWARTZ J.L., BOË L.J., VALLÉE N., ABRY C., "Major trends in vowel system inventories", *Journal of Phonetics*, 25, p. 233-254, 1997.

[SCH 98] SCHWARTZ J.L., ROBERT-RIBES J., ESCUDIER P., "Ten years after Summerfield ... a taxonomy of models for audiovisual fusion in speech perception", in R. CAMPBELL, B. DODD AND D. BURNHAM (ed.), *Hearing by Eye, II. Perspectives* and *Directions in Research on Audiovisual Aspects of Language Processing*, p. 85-108, Psychology Press, Hove (UK), 1998.

[SCH 04] SCHWARTZ J.L., BERTHOMMIER F., SAVARIAUX C., "Seeing to hear better: Evidence for early Audiovisual interactions in speech identification", *Cognition*, 93, p. B69–B78, 2004.

[SCH 06] SCHWARTZ J.L., "Bayesian model selection: the 0/0 problem in the fuzzy-logical model of perception", *J. Acoust. Soc. Am.*, 120, p. 1795-1798, 2006.

[SEK 93] SEKIYAMA K., TOHKURA Y., "Inter-language differences in the influence of visual cues in speech perception", *J. Phonetics*, 21, p. 427-444, 1993.

[SEK 96] SEKIYAMA K., "Cultural and linguistic factors influencing audiovisual integration of speech information: The McGurk effect in Chinese subjects", *Perception* and *Psychophysics*, 59, p. 73, 1996.

[SER 00] SERNICLAES W., "Perception de la parole", in P. ESCUDIER and J.L. SCHWARTZ (ed.), *La parole, des modèles cognitifs aux machines communicantes -I. Fondements*, Hermès, Paris, 2000.

[SHA 76] SHAFER G., *A Mathematical Theory of Evidence*, Princeton University Press, New Jersey, 1976.

[SIL 94] SILSBEE P.L., "Sensory Integration in Audiovisual Automatic Speech Recognition", *Proc.28th Annual Asilomar Conference on Signals, Systems,* and *Computers*, 1, p. 561-565, 1994.

[SIL 96a] SILSBEE P.L., BOVIK A.C., "Computer lipreading for improved accuracy in automatic speech recognition", *IEEE Trans. on Speech* and *Audio Processing*, 4, p. 337-351, 1996.

[SIL 96b] SILSBEE P.L., SU Q., "Audiovisual sensory integration using Hidden Markov Models", in D.G. STORK and M.E. HENNECKE (ed.), *Speechreading by Man* and *Machine: Models, Systems* and *Applications*, p. 489-496, NATO ASI Series, Springer, 1996.

[SKI 07] SKIPPER J.I., VAN WASSENHOVE V., NUSBAUM H.C., SMALL S.L., "Hearing lips and seeing voices: how cortical areas supporting speech production mediate audiovisual speech perception", *Cerebral Cortex*, 17, p. 2387-2399, 2007.

[SOD 04] SODOYER D., GIRIN L., JUTTEN C., SCHWARTZ, J.L., "Further experiments on Audiovisual speech source separation", *Speech Communication*, 44, p. 113-125, 2004.

[STE 93] STEIN E.B., MEREDITH M.A., *The Merging of Senses*, Cambridge, The MIT Press, London, 1993.

[STO 88] STOEL-GAMMON C., "Prelinguistic vocalizations of hearing-impaired and normally hearing subjects: a comparison of consonantal inventories", *J. Speech Hear. Dis.*, 53, p. 302-315, 1988. [STO 96] STORK D.G., HENNECKE M.E. (ed.), *Speech reading by Man* and *Machine: Models, Systems* and *Applications*, Springer, NATO ASI Series, 1996.

[SUM 54] SUMBY W.H., POLLACK I., "Visual contribution to speech intelligibility in noise", *J. Acoust. Soc. Am.*, 26, p. 212-215, 1954.

[SUM 84] SUMMERFIELD Q., MCGRATH M., "Detection and resolution of Audiovisual incompatibility in the perception of vowels", *Quarterly J. Experimental Psychology: Human Experimental Psychology*, 36, p. 51-74, 1984.

[SUM 87] SUMMERFIELD Q., "Some preliminaries to a comprehensive account of Audiovisual speech perception", in B. DODD and R. CAMPBELL (ed.), *Hearing by Eye: The Psychology of Lipreading*, p. 3-51, Lawrence Erlbaum Associates, London, 1987.

[TEI 95] TEISSIER P., "Fusion audiovisuelle avec prise en compte de l'environnement", *DEA Signal-Image-Parole*, Institut National Polytechnique de Grenoble, 1995.

[TEI 98] TEISSIER P., SCHWARTZ J.L., GUÉRIN-DUGUÉ A., "Models for Audiovisual fusion in a noisy recognition task", *Journal of VLSI Signal Processing Systems, Special Issue on Multimédia Signal Processing*, 20, p. 25-44, 1998.

[TEI 99a] TEISSIER P., Fusion de capteurs avec contrôle du contexte: application à la reconnaissance audiovisuelle de parole dans le bruit, PhD Thesis, Institut National Polytechnique de Grenoble, Signal-Image-Parole, 1999.

[TEI 99b] TEISSIER P., ROBERT-RIBES J., SCHWARTZ J.L., GUÉRIN-DUGUÉ A., "Comparing models for audiovisual fusion in a noisy-vowel recognition task", *IEEE Transaction on Speech* and *Audio Processing*, 7, p. 629-642, 1999.

[TII 04] TIIPPANA K., ANDERSEN T. A., SAMS M., "Visual attention modulates audiovisual speech perception", *European Journal of Cognitive Psychology*, 16, p. 457-472, 2004.

[VAR 98] VARNIER J., Un synthétiseur audiovisuel 3D du Langage Parlé Complété (LPC), Projet de troisième année, ENSERG, Institut National Polytechnique de Grenoble, 1998.

[VIH 85] VIHMAN M.M., MACKEN M.A., MILLER R., SIMMONS H., MILLER J., "From babbling to speech: A re-assessment of the continuity issue", *Language*, 61, p. 397-445, 1985.

[WAT 03] WATKINS K. E., STRAFELLA A. P., PAUS T., "Seeing and hearing speech excites the motor system involved in speech production", *Neuropsychologia*, 41, pp. 989–994. 2003.

[WEL 89] WELCH R.B., "A comparison of speech perception and spatial localization", *Behavorial* and *Brain Sciences*, 12, pp. 776-777, 1989.

[WEL 96] WELCH R.B., WARREN D.H., "Intersensory interactions", in K.R. BOFF, L. KAUFMAN and J.P. THOMAS (ed.), *Handbook of Perception* and *Human Performance, Volume I: Sensory Processes* and *Perception*, pp. 25-1-25-36, Wiley, New York, 1996.

[XU 92] XU L., KRZYZAK A., SUEN C.Y., "Methods of combining multiple classifiers and their applications to handwriting recognition", *IEEE Trans. Systems, Man* and *Cybernetics*, 22, pp. 418-435, 1992.

Chapter 12

Speech and Human-Computer Communication

12.1. Introduction

The past decades have witnessed impressive advances in the development of automatic speech processing. The VIP100 (the first recognition system to be commercialized in 1972 by Threshold Technology Inc.) could only recognize 100 isolated words after a long and tedious training phase and occupied a volume of approximately one cubic meter. Nowadays, adaptive systems may be integrated into a portable computer and are able to recognize virtually unlimited vocabularies.

At the end of the 1970s, when microchips appeared on the market, the first results obtained in automatic speech processing were so promising that they triggered very ambitious predictions: a journalist claimed that typewriters with keyboards would totally disappear by 1986 and that they would be entirely replaced by vocal interfaces. We are still far from achieving this today, but a number of evolutions have taken place, one of the milestones being the presentation by Philips of the first (almost) real-time continuous speech recognition system in German, at the Eurospeech conference in Berlin in 1993 [EUR 93]. Since then, sharp competition has been observed between research groups and companies, leading to a significant drop in product prices and a noticeable improvement in their performance and usability. Voice technologies are now not only used by professionals, but also by the general public: voice servers are one example.

Chapter written by Wolfgang MINKER and Françoise NÉEL.

With the advent of the Internet, anyone from anywhere in the world may access any remote database, whilst communicating with several other users, and here, speech offers a particularly natural and efficient mode of communication. It is therefore worth examining the current position of voice technologies in this context, either as the unique mode of communication, or as a complement to other communication modes.

In the past decades, the awareness of the specificity of speech has increased, and its role has somehow been clarified in complex multimodal interactions, where several perception and production modes interact. Many application domains can be identified, in which the use of speech, either seconded by gestures and eye movements or subordinated to them, aims at responding to the user's expectations in an optimal way.

The aim of this chapter is to identify the favored role that speech holds in natural interactions with machines, with a particular focus on speech recognition (speech synthesis will only be mentioned in the context of specific applications). Section 12.2 recalls the current use, which, as a result of several factors, is favorable to the use of voice technologies in an ever-growing number and variety of applications in everyday life. Section 12.3 underlines the advantages and the limitations of the speech mode and voice technologies, and proposes a multidimensional typology to help analyze the characteristics of products available on the market. Some of the most essential application domains are then presented in section 12.4. Focus is placed on home, industry and office applications where only speech is used, such as quality control systems, automatic dictation, training and automatic translation. In these domains, numerous products are widely available. In section 12.5, domains where speech may be combined with other communication modes are mentioned. In these cases, the systems presented are prototypes or proofs of concept. Finally, it is important to note that this chapter does *not* address telecommunication applications [GAG 98], as a specific chapter in this book is dedicated to them.

12.2. Context

Several factors have simultaneously favored more widespread use of voice technologies in new interfaces. Among them, it is worth mentioning:
-    the development of micro-electronics;
-    the expansion of information and communication technologies, and the ever growing interconnection of computer systems;
-    the coordination of research efforts and the improvement of automatic speech processing systems.

12.2.1. *The development of micro-electronics*

Rapid developments in micro-electronics have led to an increase in the power of microchips and memory capacity (approximately doubling every second year, according to Moore's law). As a result, signal and speech-processing algorithms (recognition and synthesis), which are greedy in terms of computation costs and memory requirements, can now operate on even basic equipments (a SoundBlaster card, for example) and do not require any specialized hardware. Small vocabulary systems such as those designed by Dragon System, IBM and Philips can operate on a PC and even on a PDA (Personal Digital Assistant). They can therefore be used by millions of people [HUN 96]. Some applications are even integrated into text processing software (such as *Word Perfect Office 2002* by Corel integrating *Dragon NaturallySpeaking 5.0* for a price of approximately €150, or Lotus in *Smart Suite Millennium* integrating *ViaVoice* by IBM).

The fact that signal and speech processing software has become commonplace has caused a spectacular drop in the price of these products: The *Speech Server Series* proposed by IBM in 1993 for 170,000 FF (€26,000) for 4 workstations was replaced in 1995 by a fully software-based version of large-vocabulary speech recognition (*IBM Voice Type*) for 8,000 FF (€1,200). Then, after 1997, several new versions emerged on the market, with prices ranging (in 2000) from 38 FF (€6) (*Simply Speaking 3.0*) to 1,500 FF (€230)! Prices of products proposed by Dragon Systems and Philips have followed a similar downward trend.

Moreover, miniaturization has favored nomadic devices, and has enabled speech recognition to be integrated into mobile phones (Nokia, Philips, etc.). More recently, the expansion of satellite communication technologies and of HPC (handled PC) terminals, has led to small and compact handhelds, which combine telephone functionalities, Internet access and *GPS (global positioning satellite)* systems to determine the location of a person or a vehicle. A future step could be a total dilution of the interface in the environment: as indicated by the theme *"The Disappearing Computer"* in a call for proposals by the European Commission in 2000, audio sensors and processors could become incorporated in clothes *("wearable computer")* as, for example, in the *Navigator* system proposed by Carnegie Mellon University (CMU) [SMA 94], or even in the walls of houses.

12.2.2. *The expansion of information and communication technologies and increasing interconnection of computer systems*

In the past decades, information and communication technologies, particularly telecommunication and computer networks, have grown exponentially. The Web enables professionals and the general public to access, anywhere, anytime, a considerable volume of data (texts, images, videos, speech) spread worldwide and accessible through fixed and mobile telephone networks. This development has benefited directly from progress made in signal processing (coding, compression and multiplexing) which optimizes the transmission (via optical fibers or satellites).

Worldwide, the number of computers connected to the Internet was estimated at 934 million at the end of 2004, including more than 25 million in France, whereas it was approximately equal to 40 million worldwide in 1995 (and hardly 100,000 in France). Estimates for 2007 are of 1.35 billion worldwide. France is bridging the gap with other countries, in particular, relative to Germany (42 million Internet users in 2004) and the UK. The number of Internet users can only increase as prices become lower and ADSL (asymmetric digital subscriber line) connections, yielding 100 times faster transmission rates on conventional telephone lines, becoming more common.

NUMERIS, the digital telephone network only represented 10% of telephone lines in 1993, has undergone continuous expansion, and this has stimulated the development of specialized services (RNIS, Réseau Numérique à Intégration de Services). Conventional landline telephone networks are currently replaced by high-speed Internet protocol (IP) networks.

The standardization of the Itineris/GSM (Global System for Mobile Communications) radio-frequency network, replaced by the GPRS (General Packet Radio Service), gave almost complete coverage of the French territory, before the 3rd generation technology, i.e. the UMTS (Universal Mobile Telecommunication System) was in full operation. Based on very fast transmission rates (from 384 kbps to 2 mbps), the UMTS system offers the possibility of deploying a wide range of multimedia services. The growth of mobile phones is tremendous, as a result of the deployment of low orbit satellite networks (Galileo project, etc.) [CHO 95]: the number of subscribers in France reached 17 million at the end of 1999, and had exceeded 38 million in January 2003, i.e. more than one inhabitant out of two owned one (compared to 8 million in September 1998). The total number of mobile phones in the world is expected to reach 10 billion units by 2010 (compared to 400 million at the end of 1999).

In fact, a network with total interconnection is gradually forming, between workstations, telephones, personal computers, television sets, etc. New systems such as "Net Telephones" tend to integrate multiple functionalities in the framework of the WAP (Wireless Application Protocol) protocol, which offers the user the possibility of connecting to the Web from a mobile phone, thanks to the interoperability of the GSM standard with the Internet.

In this context, speech recognition technology is in expansion on many handheld devices such as mobile phones or PDAs, with their minimal keyboards (for instance, the QteK S100 Smartphone, with the VoiceCommand software by Microsoft).

12.2.3. *The coordination of research efforts and the improvement of automatic speech processing systems*

Significant performance improvements in speech systems have been made since the 1980s: indeed, in 1985, systems were just about able to recognize one speaker uttering isolated words, after a training process requiring the pronunciation of several hundreds of words. In this first decade of the 21st century, systems are now able to recognize virtually any person speaking in continuous mode and can use a vocabulary of several hundreds of thousands of words (or even more). They are also able to carry out a dialog with a vocabulary of several thousand words on the telephone network, on a specific topic (train or plane time-table enquiry, for instance). This progress is not only due to scientific steps (the corresponding algorithms were already known in the 1970s), but also to the organization of systematic evaluation campaigns. These were feasible thanks to progress in micro-electronics , making computational resources and large databases available. However, the campaigns themselves resulted from a greater awareness from specialists of the importance of the *evaluation paradigm*. This paradigm was initially defined in American programs, and became central in many national and European projects.

In the USA, the DARPA program (Defense Advanced Research Project Agency) on *Speech and Natural Language* was launched in 1984. Its goal was to set up evaluation campaigns on a regular basis to encourage the objective comparison of different systems and methods. The results obtained during these campaigns have shown continuous improvements in recognition systems over the years and enhanced versatility in the corresponding software [PAL 90a].

Such initiatives rely not only on a prior consensus on the use of common methods and protocols, but also on availability of shared resources and tools which make it possible to measure any progress in a rather formal way. In fact, the most commonly used methods, either stochastic (hidden Markov models) or based on neural networks [MAR 90a] require large corpora of signals and texts for training the systems [NEE 96]. In the context of the DARPA program, efforts have therefore been put on the collection of large databases that cover various applications domains: the *WJS (Wall Street Journal)* and *ATIS (Air Travel Information Services)* corpora are pioneer examples.

The distribution of such resources has also been facilitated by the creation in the US of a specific institution, the LDC (Linguistic Data Consortium), which aims at improving coordination in the collection and the dissemination of (speech and text corpora) via the Internet and on CD-ROMs.

In Europe, in the framework of the *ESPRIT* and *LRE* programs, many projects were launched by the European Community on the topic of evaluation: in 1988, the *ESPRIT SAM/SAM-A* (multilingual speech input/output standardisation, assessment and methodology) project [SAM 92], which grouped more than 30 academic and industrial partners, was one of the first European projects to define standards for speech recognition, speaker verification and speech synthesis systems. This project also yielded the first set of CD-ROMs containing a multilingual corpus (*EUROM0, EUROM1*) which was based on the conclusions of the GDR-PRC CHM Pôle Parole (French Research Group of the Coordinated Research Program on Human-Computer Communication), and was inspired from the first CD-ROMs recorded in France (BDSONS) [DOL 88].

Some other projects followed, including:

–   *LRE RELATOR* for the definition of a data collection and distribution structure for multilingual linguistic resources. The outcome of the project was the creation in February 1995 of the ELRA association (European Language Resources Association), the European equivalent of the LDC;

–   *LRE SQALE* [STE 95] for the adaptation of the ARPA-LVCSR (Large Vocabulary Continuous Speech Recognition) algorithm [PAL 95a] in a multilingual context (American-English, British-English, French and German) [YOU 97];

–   *LRE EAGLES* (Expert Advisory Group on Language Engineering Standards) for setting up a handbook in linguistic resources development and voice system assessment [GIB 97]. The *ISLE* project (2000-2002), which was embedding multimodal interactions, was a continuation of *EAGLES*;

–   *LRE Eurocoscoda*, for setting up representative corpora of spoken language for European languages, in relation to the international program *COCOSDA (Coordinating Committee on Speech Databases and Speech Input/Output Systems Assessment)* [MAR 93a].

In France, on a national level, a similar goal was pursued, with the creation of the *BDSONS* database by the GRECO-CP (Spoken Communication Group) and then, as early as 1991, by the GDR-PRC CHM, the LIMSI-CNRS (Laboratoire d'Informatique pour la Mécanique et les Sciences de l'Ingénieur of the CNRS) and the OFIL (Observatoire des Industries de la Langue), with the creation of a large size corpus of spoken French, the *BREF* database [GAU 90] [LAM 91].

In French-speaking countries, the AUF (Agence Universitaire de la Francophonie, formerly AUPELF-UREF (Association des Universités Partiellement ou Entièrement de Langue Française – Université des Réseaux d'Expression Française) was organized from 1995 onwards, under the umbrella of the *FRANCIL* network *(Réseau FRANCophone de l'Ingénierie de la Langue)*. Several initiatives to evaluate speech recognition, dialog and speech synthesis systems for the French language (Actions de Recherche Concertées (or ARC, for Concerted Research Actions) and Actions de Recherche Partagées (or ARP, for Shared Research Actions) on spoken and written Language Engineering [CHI 99a]) also began. The goal of these actions was also to disseminate large size corpora to the French-speaking research community at large.

More recently, the EVALDA project, funded by the French Ministry of Research within the framework of their "TechnoLangue" program, has aimed at setting up a perennial infrastructure dedicated to the evaluation of language technologies in French [EVA 05]. The project was to set up various components which can be used for evaluation purposes, related for instance to the organization, the logistics, the linguistic resources, the evaluation protocols, the methodologies and the metrics. Moreover, EVALDA aimed at gathering the key actors of the domain (into scientific committees, mailing lists, internet forums, experts, partners, etc.). Along with the opportunity of cross-comparing experimental results, this effort also contributes to the emergence of collaborations between partners and to new evaluation campaigns.

In the framework of EVALDA, the goal of the sub-project MEDIA (dedicated to the methodology of automatic evaluation of comprehension of dialog in and out of context) was to define and test a methodology for the assessment of the understanding capability of spoken language dialog systems [DEV 04]. An evaluation paradigm was set up, which is founded on the definition and use of test banks built from real corpora and from a semantic representation of common metrics. This paradigm aims at enabling both in-context and out-of-context understanding capabilities for dialog systems. An evaluation campaign of various systems from different participants was conducted, so as to validate the paradigm and the representations on a common enquiry task.

These evaluations allow us to compare the system responses with reference corpora, but only measure one aspect of the system's performance (for example, its recognition or understanding error rate). They cannot be considered sufficient to guarantee a correct level of acceptance from the users. It is also necessary to carry out systematic tests in *operational situations*, such as those achieved by telecommunication operators, who design, develop and distribute products. For many years, the CNET (now FTR&D) [SOR 94], AT&T, NTT, British Telecom, Deutsche Telekom, the CSELT, Infovox-Telia, Unisource, and Telefonica have been evaluating the help of real users, interactive systems that integrate speech recognition and synthesis. These on-site operational evaluations have provided accurate diagnostics on the limitations of voice technologies and have also improved perception of real user needs, resulting in a better adaptation of the proposed services to the public. Experimentation in *close-to-real conditions* is inevitably more expensive, and obviously involves a number of limitations (the real system does not yet exist or at least it does not implement all foreseen functionalities). However, this approach, even with simplified prototypes under construction, provides invaluable information that helps to identify priorities and focus research perspectives.

12.3. Specificities of speech

All the factors mentioned above highlight the importance of including speech in the new universe of communication. In fact, automatic processing has now reached a certain level of maturity. It is interesting to identify what has been brought about by speech signal processing techniques in the field of human-computer communication, and also to examine what they have induced in related domains, such as mediated communication and the automatic indexing of audio-visual documents. For this purpose, it is necessary to recall the advantages of speech over other perception and production modes, without ignoring it's drawbacks.

12.3.1. *Advantages of speech as a communication mode*

Taking the human model as a reference, the advantages of speech seem decisive, at least at first sight, and include:

– (i) *naturalness*: speech is the most natural mode of communication from person to person. It is acquired since early childhood, as opposed to typing on a keyboard, for instance;

– (ii) *speed/efficiency:* several ergonomic studies have shown that the speaking rate in spontaneous speech is about 200 words per minute, as compared with the rate of 60 words per minutes reached by a trained typist [LEA 80].

The efficiency of speech does not only stem from its higher rate of information transmission, but also in the fact it is easy to use it simultaneously with other means of communication. Speech leaves the user free of their movements. Speech is therefore particularly well adapted to applications in which the user has to carry out parallel tasks or is in charge of controlling complex processes that require a sustained use of gestures or vision [CHA 79b].

It should also be noted that, from a cognitive point of view, the combination of speech with visual information results in an improved memorization capability, and enables one to underline the most salient elements. This property is frequently needed in interactive learning.

However, while speech shows an advantageous rate for the emission of a message by a computer or by a human, this advantage does not hold with human perception. Speech is a sequential and one-dimensional process. Therefore, listening to a voice message (at a rate of approximately 200 words per minute) requires some attention (because the information is non-permanent), and necessitates more time for the user to access the message than it does to read a page on a screen (for which a rate of 700 words per minutes can be achieved). As a consequence, these experimental results proscribe long generated spoken messages: it is preferable to display them on a screen as soon as they exceed a certain length, or to synthesize a condensed version by voice. This demonstrates that one communication mode cannot always be substituted directly by another one, without some prior adaptation of the message's form, even if the semantic content remains the same;

– (iii) *scope:* speech provides immediate access to information without having to go through a full hierarchy of menus: pronouncing a single word can thus replace up to ten elementary commands corresponding to a sequence of functional keystrokes and mouse clicks [BEL 97].

Speech also enables access to a non-visible object (on a screen for instance). It can be used to designate and to name objects, and also to achieve complex semantic exchanges and to manipulate abstract notions, which can contribute to modifying and enriching the dialog with the computer. Supremacy of speech is obvious for global commands (for instance, color modifications); it is not so much the case for incremental commands (step-wise movement of a object on a screen for instance), for which any designation mode (mouse, joystick) is more efficient.

12.3.2. *Limitations of speech as a communication mode*

The aforementioned advantages must be balanced by a consideration of the inherent variability of the speech communication mode, which stems from various sources: from the speaker (physical condition, cognitive state), from the environment, from the transmission channel or medium and from the language itself (ambiguities). On top of these sources of variability, technological limitations impose various additional constraints with respect to recording conditions and operational usage:

– *Speaker variability:* two distinct utterances of the same message by the same person are not strictly identical. The signal varies as a function of the physical and emotional state of the speaker (stress, tiredness, hoarseness), but also depending on the speaking style (formal or casual). Decoding is all the more complex as the signal is not affected linearly when the speaking rate is slowed down or accelerated: in fast speaking mode, stable portions of vowels are more drastically modified than consonantal portions [LIE 77]. Naturally, inter-speaker variability (male, female or child voice) is even more considerable.

It is of course necessary to model this variability not only in speech recognition, but also in speech synthesis: when a computer talks, human perception is expecting some form of variability (which does not prevent comprehension in the least). If this variability does not exist in the synthetic voice (for instance, if the prosody is too repetitive), the artificial character of the speech is not easily accepted by the user.

– *Acoustic variability:* depending on its characteristics (and, in particular, its directivity), the microphone captures a number of surrounding noises, together with the speech signal. Non-stationary noises are of course the most disturbing ones for the recognition system: on an airplane, for instance, the breathing noise of a pilot in his oxygen mask can be more disturbing than the noise generated by the engine of the plane.

– *Transmission variability*: besides bandwidth limitations (300-3400 Hz), analog telephone lines create various distortions (dampening, echo) that are not identical across channels. Their replacement by digital lines (NUMERIS network) facilitates telephone speech recognition. On the other hand, speech recognition on mobile phones (*Itineris/GSM* network) raises even more problems than analog lines, because of the low bit-rate (13 kbps), radio transmission and digital error correction. This should evolve favorably with the *GPRS* standard.

– *Linguistic variability:* it is essential to model accurately the various levels of knowledge in speech (lexical, syntactic, semantic and pragmatic) that contribute to the interpretation of the spoken message. The speech signal is continuous, without delimiters (pauses) between words, which means that, given the lexical richness of the language, its segmentation into words is not necessarily unique. Moreover, speech is generated in real time. *Spontaneous speech* is therefore studded with many hesitations, restarts, self-corrections, etc. which are not always compatible with the list of words in the lexicon, or with the expected syntax usually based on written texts [MOR 88a][NEE 99].

– *Recording conditions*: in order to be efficient, the sensor (the microphone) is often intrusive. It may require the user to wear a headband for holding it, which is unacceptable for the general public. A recording set-up based on several coupled microphones may be necessary so as to keep performance at a reasonable level.

Two types of constraints act on the *operational usage*:

– *context detection:* speech is a natural mode of communication. It is therefore essential for the computer to be able to differentiate between command words related to the application and comments made by the user to a third party. This problem of context analysis and detection, which also occurs for gesture recognition, is far from being solved;

– *non-confidentiality and isolation*: even if headphones are used for the audio feedback, speaking aloud to the computer cannot guarantee confidentiality of the information. It may also generate disturbances to the other users (as is also the case with mobile phones). Moreover, wearing headphones isolates the user from the external world.

12.3.3. *Multidimensional analysis of commercial speech recognition products*

There is still a long way to go before a recognition system will be able to handle all the variability and the richness of the speech signal at large, even though the human perception system is able to deal with it. Products currently available represent a compromise between the following criteria:

– *Speaking mode*: speech recognition can operate in *continuous mode* or on *isolated words*. In the latter case, the speaker is forced to mark a short pause between each word, rather than speaking continuously. Isolated word systems are more and more frequently replaced by continuous speech recognition systems. Some systems achieve *word spotting* in the speech stream, so as to offer more spontaneity to the user [JOU 91] [GAG 93].

– *Speaking style*: either the speaker is constrained to speak with a reading style and a regular voice, or he/she can speak spontaneously to the system [ESK 92].

– *Lexicon and language model*: this criterion is directly related to the previous case. Spontaneous speech does not follow the grammar rules of written language, as opposed to read speech. The language model, which is based on a lexicon, also depends on a given semantic domain. Current dictation systems are able to adjust their language model automatically from text corpora of the same domain.

–   *Speaker-dependence level/training:* speech recognition systems are either speaker-*independent* or speaker-*dependent*. In the latter case, a prior training phase is necessary for any new speaker. The first commercial products were based on a complete re-training by the user, who had to utter the whole vocabulary, sometimes several times. Such a constraint is not acceptable, if the vocabulary exceeds several hundreds of words. For large vocabulary speech recognition, this phase is replaced by an *adaptation* phase.

–   *Adaptation*: in order to update speaker-specific acoustic models in the recognition system, the user is requested to utter a set of sentences that convey a sample of the various phonemes of the language, so that the acoustic references can be adjusted to his/her voice. Such an adaptation phase can take 30 to 60 minutes for dictation systems with virtually unlimited vocabulary.

These systems are now equipped with *on-line* adaptation, i.e. acoustic and linguistic models are modified during the course of the system use. Thus, often-dramatic recognition rates (below 50% correct words) improve gradually and rapidly with each new sentence. Recent versions do not require lengthy adaptation phases and they provide satisfactory performances from the very first session.

–   *Vocabulary size*: the vocabulary size can be small (10-50 words), medium (100-500 words) or large (over 1,000 words, up to hundreds of thousands). Many applications only require a medium vocabulary. However, for automatic dictation, the vocabulary may exceed 100,000 entries; when necessary, it is feasible to partition the vocabulary in several sub-vocabularies. In this case, the size of the *active* vocabulary (i.e. the vocabulary known to the system during recognition) is the really relevant quantity. For dictations systems, this vocabulary is generally in the order of 60,000 words for a total vocabulary of 200,000 words.

–   *Local versus remote recognition*: recognition can take place either directly on the workstation or on the user's device, or even remotely across a telecommunication network (with a reduced bandwidth, for some modes of transmission). Models must be adapted accordingly.

–   *Robustness to noise*: in order to take into account the characteristics of the user's environment, some systems include a pre-processing phase (noise cancellation) for extracting speech from the background, or use noise models (non-stationary ones being more difficult to deal with), which are recognized jointly with the speech signal.

–   *Recording conditions:* the microphone may either be *directive* and *close to the mouth* (which requires the user to wear a headset or a hair-band, unless the microphone is integrated in a telephone handset), or it can be standing on a desk or incorporated into a dashboard. In some noisy conditions, (car, terminal, kiosk, etc) a microphone array enables a better speaker localization and focalization, and therefore a more efficient extraction of speech from the background noise. Another important characteristic is whether an *open microphone* (a system which functions continuously) is used or a *push-to-talk microphone* (with a manual command for triggering the microphone when the user talks to the computer).

–   *Types of users: general public vs. professionals.* In these two broad categories, a further distinction must be made, between on the one hand the novice and expert user, depending on his/her knowledge on the application domain, and on the other hand, between the occasional and regular user, the latter being used to the way the systems operates. These various situations can induce very different behaviors.

–   *Response time:* it is desirable that the response time is below one second, except where the use of a dictaphone is compatible with a delayed processing.

–   *Recognition performance:* the recognition rate must generally be higher than 95% at the word level. Of course, this threshold depends on the ultimate goal of the system in its applicative context. For language identification, for instance, a recognition rate of 50% only can still be useful.

According to the vocabulary size, three categories of systems currently on the market can be identified, corresponding to particular combinations of the above criteria:

–   *Command or dialog systems:* these are based on a small vocabulary (from a few tens to a few hundreds of words). They are speaker-dependent (and require a training phase) and they are robust to noise, i.e. usable in a local mode and in difficult environments.

–   *Voice servers:* these work in a speaker-independent mode, through the telephone network. They function with a vocabulary ranging from 10-50 words up to a few thousand, uttered in isolation or detected by word-spotting.

–   *Dictation systems:* these are speaker-independent systems, with on-line adaptation. They offer the recognition of virtually unlimited vocabularies (several hundreds of words) in a quiet environment, but they impose on the user a training phase of several tens of minutes.

An example from the first category is the integrated recognition system by AEG-Daimler-Benz in top-of-the-range cars, which provides recognition for approximately 300 words for interacting with an onboard computer. Other systems, with a more limited vocabulary, are used daily for quality control and maintenance tasks. In the second category, several systems have been tested for specific tasks (train timetable information, for instance) in the US and in Europe (Philips, CSELT, CNET, etc.). Finally, in the third category, the corresponding systems are particularly popular in office automation, for which several continuous speech dictation systems have been commercialized by IBM, Dragon Systems, Lernout & Hauspie, Philips and more recently by Scansoft-Nuance.

Potential users of these systems are either professionals or the general public, and they determine the acceptable constraints, as their specific demands vary in nature. The *professional* user can tolerate particular constraints (lengthy training phase, poor quality synthesis) but will in general need a rather extended vocabulary (from a few hundreds to a few thousand words). On the other hand, the general public will be eager to benefit from a high quality synthesis and will generally be discouraged by any training phase prior to the use of the application or by constrained recording conditions. They are likely to accept a keyword-based interaction if the system is well-designed for guiding the dialog: in these cases, a few tens of words of vocabulary would probably be sufficient.

12.4. Application domains with voice-only interaction

As already underlined in this chapter, speech offers the advantage of being usable whilst performing other tasks, and enables considerable liberty of movements and free vision. Users can move freely and manipulate objects (for sorting them, routing parcels or handling stocks, for instance) or enter data (bank, insurance, control). While paying attention to a complex process, they can describe visual information (dental examination, x-ray reports, microscope observations, inspection of printed circuits). They can also give directions to a remote robot operating in a hostile environment (weightless conditions, submarine, contaminated atmosphere) [CHO 95].

For speech used as a single mode, several application sectors may be identified: the industry sector with command systems and data acquisition devices, the home automation sector and the office automation sector with dictation systems, professional or personal training and, more recently, automatic translation. In most cases, the products have been on the market for many years and are widely used. In the last two sectors (training and translation), systems are still under study or at a prototype level.

12.4.1. *Inspection, control and data acquisition*

In this case, the *hands-free/eye-free* functionality provided by a voice interface is essential. Many systems are operational in the US, particularly in the military sector, for the maintenance of airplanes: in this case, a portable system is used, and it is sometimes integrated into *clothes* [CHI 96b].

In France, as early as 1988, the *Voice Scribe* system by Dragon was integrated by Auralog and used by the SNECMA to control the quality of airplane engines.

At the end of the 1990s, a portable system called *Talkman* by Vocollect was evaluated in the framework of the research project *SAVOIE*, and released by the SNCF (the national train company in France) for nomadic operators, for collecting information (from approximately 1000 distinct measurement points) on parts of train wagons so as to decide on maintenance requirements. The users were handling various tools and instruments, and were sometimes intervening in dangerous situations (under the wagons, on the roofs, standing on ladders, for example). The evaluations demonstrated a clear preference for voice input as compared to a keyboard. A speaker-dependent system is trained using a vocabulary of 150 words. has also shown a good performance level when used in an industrial environment, with a significant level of noise, according to the maintenance workshops of SNCF.

Again in the train domain, the *MGPT* system for management of works protection is used on a daily basis by SNCF agents who request by voice, in a language constrained by security rules, the authorization to execute works on railway lines. The recognition system, developed by Vecsys, has been in service since January 1997 on the Paris-Lille line, and is being deployed in other regions (the South and the East of France).

For inspection of the stages of the *ARIANE* rocket, the company EADS has been using a recognition system called *MARIE*, also developed by Vecsys. This system lists by voice each checkpoint of the inspection procedure, and the operator can simultaneously detect abnormalities visually and describe them orally.

In France, a third system also developed by Vecsys should be mentioned, in the domain of anatomo-cytopathology: for some medical examinations, a report is simply produced from a list of bible-codes referring to a codified dictionary of lesions.

12.4.2. *Home automation: electronic home assistant*

In the domain of home automation, several voice control systems have already been put on the market, especially in the USA (such as *HAL* by Cocoonz or *Butler-in-a-Box* by Mastervoice). These are used for piloting simple commands on a number of devices including heating regulation and a shutter's orientation. The recognition systems can be coupled to a telephone and some cameras, and should offer, in the future, multiple assistance services in everyday life.

Many of these systems can also bring some help to people with motor handicaps, for example, *device control*. In France, the *NEMO system* (formerly *TETRAVOX*), designed by Vecsys and commercialized by Proteor, offers tetraplegics the possibility of controlling objects around them by voice (television, shutters, telephones, etc.).

The European project *TIDE (Telematics for the Integration of Disabled and Elderly) HOME,* in which the LIMSI-CNRS and Vecsys participated [SHA 98], aimed at designing a control platform for a wide variety of home devices. This platform was mainly dedicated to elderly or handicapped people, in order to assist them in controlling the wide range of functions offered by these devices.

12.4.3. *Office automation: dictation and speech-to-text systems*

Beyond motor handicap (for people suffering from osteoarthritis or repetitive strain injury) linked to repetitive tasks, one of the initial reference applications for large vocabulary dictation systems dealt with the medical sector. Indeed, typing medical reports (especially x-ray reports) can easily be automated, given that these reports are highly structured and use a very specific vocabulary. Another reason for using an automatic approach lies in the fact that introducing a voice input, while simplifying the task for the secretaries, does not involve any modification in the habits of the doctors, as the use of dictaphones with recognition is becoming more common.

Currently, similar needs have been identified in the legal sector, as well as in banking and insurance sectors, where the generation of reports contributes significantly to the volume of activity.

In these sectors, several of the above-mentioned criteria and modalities of use (corrections, feedback, etc.) should be detailed:

–   *Language model*: the measure commonly used to characterize the complexity of a language is called the *perplexity* (or dynamic branching factor).

This measure is based on communication theory, and indicates the average number of possible candidate words after a word has been recognized. It relies on a language model in the absence of which all the words would be equally probable. The perplexity of professional languages, such as those used in medical reports, is generally lower than 60 words, and is quite often close to 10. On the contrary, literature texts (novels, narrations, etc.) can reach perplexities of several hundreds of words [BAK 93].

–   *Lexicon and language model customization/application independence*: a personal dictionary of a few thousand words to be consulted as a priority is sometimes proposed for grouping together those words and expressions that are the most commonly used by a speaker. For the *DragonDictate* system, the additional vocabulary is virtually unlimited, thanks to a dynamic management of the *active* and *passive* vocabularies. And the procedures for adding new vocabularies are made as easy as possible. Some systems require a complete reconfiguration of the vocabulary when the application changes. Dragon Systems was the first provider to propose the *DragonDictate* system for isolated words, independently from any particular application, in 1990 [HUN 98]. However, if the language model is to be focused on a specific domain (law, bank, medicine), this model can be adapted quickly using text files from the domain. *NaturallySpeaking*, a subsequent system developed by Dragon, offered the user the ability to create multiple language models: one for reports in a particular domain, another one for home correspondence, and a third one for producing meeting minutes for an association, for example.

–   *Speaking mode*: generally, these systems provide two modes of use: one using key-words or *macros*, the other *free dictation*. In the *macros* mode, the pronunciation of a word can trigger the typing of a whole sentence. This mode is adapted to codified languages which follow a pre-established scheme: thus, in a recognition system for creating X-ray reports, the pronunciation of the words *heart, lungs, normal* will be sufficient to get the system to write *the heart and the lungs show a normal aspect*. Dragon Systems proposed an additional mode, the *command mode*, for which the commands for text creation and management were themselves input by voice.

–   *Correction*: in general, the recognition system is integrated into a word processor, which enables a fast and easy interactive correction of badly recognized words, using a menu which list alternative candidates.

–   *Feedback*: most of the time, the feed-back is visual: recognized words are displayed on the screen as soon as they are recognized. The latest versions also implement auditory spell checking by voice synthesis.

Since the end of the 1980s, there has been growing activity in this domain, and several products have been put on the market. Kurzweil AI was the first company to commercialize an application, as early as 1986: the *VoiceRAD* system (5,000 words in isolation), which was dedicated to the redaction of X-ray reports. This system used a dictaphone and, optionally, text-to-speech synthesis and voice mail. Later, several other American companies proposed similar products on the basis of research carried out since the 1970s, for instance Dragon Systems [BAK 75] [BAK 93] and IBM [JEL 76].

The years 1992-93 were particularly eventful: first, in November 1992 IBM announced the commercialization of the *Speech Server Series* (approximately 24,000 words), building up from the *TANGORA* prototype which had been designed in the 1970s by the group of F. Jelinek, in the T.J. Watson Research Center of IBM [JEL 87]. Then, in September 1993, the first prototype of a continuous speech recognition system in German was presented by Philips (Aachen) at the Eurospeech conference. The system was designed specifically for the medical domain. It did not work in truly real-time and required a special ASIC card, but was perfectly suited to the profession's habits, as it was based on the doctor using a dictaphone as usual.

More recent systems offer regular upgrades with improved performance in terms of recognition rate and user constraints. In particular:

–   recognition is working for continuous speech, i.e. without requiring any pause between words;

–   training time is reduced to a few tens of minutes;

–   an interactive correction software is integrated;

–   vocabulary is unlimited in size, or easily modified;

–   no specific hardware is required, all the products being based on software;

–   recognition is integrated in standard operating systems (typically Windows);

–   the system is domain-independent: for continuous speech recognition Dragon Systems (in April 1997) and IBM (a few months later) proposed domain-independent recognizers (in English), respectively *Dragon NaturallySpeaking*^{TM} and *ViaVoice*^{TM}. This feature has been now extended to all the versions in other languages.

An increasing number of languages are handled: Dragon Systems presented the French version in 1996 and, by 2000, was distributing versions of the same application in German, Italian, Spanish, Sweden, American-English, British-English and Arabic. The IBM system [DER 93] was made available in French, Italian, German, British English, and American English and, since 1995; Arabic, Japanese, Korean and Chinese (Mandarin) have been added.

Finally, in a number of cases (IBM, Microsoft), it is possible to download free versions from websites. These versions are available with a set of development tools, which facilitates their integration in applications others than dictation.

12.4.4. *Training*

Children and adults are attracted to speech-bestowed games like talking dolls, board games, video games and educational toys. Computer-aided learning (and particularly language laboratories) is integrating more and more audio functionalities (sound effects, synthetic sounds) and is evolving towards increased interactivity: computed aided systems for foreign language learning, which facilitate the acquisition of correct pronunciations and the mastering of both vocabulary and syntax, benefit from the contribution of voice technology, which also provide a playful dimension.

12.4.4.1. *Foreign language learning*

The SNCF Eurostar drivers have been using a computer-aided learning system developed by Cap-Gemini-Innovation and Vecsys, to enable them to familiarize themselves with emergency procedures that require dialog in English with ground staff. The system helps them improve their pronunciation while evaluating their ability to understand various English accents and to adapt to adverse conditions.

Auralog has also carried out a number of experiments in secondary schools in France and they distribute the *AURALANG* self-learning system on CD-Roms, which helps the user to improve his/her pronunciation of a foreign language, using a speech recognition system.

Several European projects have been focusing on the improvement of prosody (intonation, accentuation) and of the pronunciation of specific phonemes (the *ESPRIT SPELL* project [HIL 93b]), the *ILAM* project (with Auralog, the British Council, the Goethe Institute, Dragon Systems and AFPA (Association Nationale pour la Formation Professionnelle des Adultes)).

Similar projects have been pursued in the framework of the ARP (shared research actions) of the *FRANCIL* network, in particular at the University of Geneva in Switzerland [HAM 97] and at the University of Ottawa in Canada [CHA 97a].

12.4.4.2. *Teaching to read with computers*

In the USA, the *LISTEN* project integrates the *Sphinx-II* speech recognition algorithm developed at CMU. The goal is to guide children in their practice of reading and in their pronunciation [MOS 99]. The system acts as a teacher

("*Reading Tutor*") and interacts with the child when pronunciation errors or hesitations are detected. The prototype has been tested in several elementary classes and the results of comparative studies have shown the benefit of the approach versus conventional methods, especially regarding the fast acquisition of the basics of reading and comprehension.

A similar project, integrating the $AURIX^{TM}$ recognition system was developed in the UK at the beginning of the 1990s, jointly by the DRA (Defense Research Agency), HWCC and HAWTEC (Hereford and Worcester Training and Enterprise Council), in the framework of the program *STAR (Speech Training Aid Research)*.

12.4.4.3. *Training of air traffic controllers*

At the turn of the 1980s, a definite interest has been witnessed, for systems enabling air traffic controllers to acquire better knowledge of the *phraseology* of their specific non-ambiguous operative language, which they commonly use when communicating with pilots. The contribution of voice technologies was to place the controller in conditions close to operational situations. Several similar projects were carried out in France, in England, in Germany and in the USA. An industrial prototype was designed in France in 1993 by CENA (Centre d'Etudes de Navigation Aérienne) [MAR 93b], LIMSI-CNRS [MAT 91], Stéria Ingénierie & Télécom, Sextant-Avionique and Vecsys; this prototype was then evaluated by controllers in the Orly and Charles De Gaulle airports, as well as by trainees at ENAC (Ecole Nationale de l'Aviation Civile) in Toulouse. In England, at the end of the 1990s, Airliner put approximately 50 work positions in operation using Vecsys boards for a similar application.

In this latter domain, very active studies are still taking place in Europe and the US to combine speech with other modes of communication, not only in the context of training air traffic controllers, but also to improve their working equipment in operational situations.

12.4.4.4. *Assistance to disabled people*

Several European programs (*TIDE*, etc.) have taken into account the various types of disabilities by which the population is affected, together with the corresponding number of people. In 1996, 12 million partially-sighted were counted in Europe, including one million blind people, 81 million hearing-impaired, including one million deaf people, approximately 30 million people with a motor handicap affecting their upper members and 50 million with a disability of their lower members. Unfortunately, these figures are expected to increase with the aging population.

By offering the possibility to control a number of devices, voice technologies can contribute to bring handicapped people to a certain level of autonomy and to enable them to be better inserted in their work and family environments. Speech can replace the defective sense: when sight is impaired, speech can both replace the computer-mouse for designating an icon and also be used to describe graphics (icons, schemas, images), for example:

– *Pronunciation aids:* voice pathologies may be detected and quantified. Voice re-education by speech therapists shares common points with learning of a foreign language. IBM has for several years been developing an interactive system called *Speech Viewer III*, which enables hearing-impaired children to better control their pitch and the intensity of their voice.

– *Voice prostheses*: speech synthesis may be used as a voice prosthesis, so as to replace the impaired voice of physically disabled people or, more generally, for people suffering from muscular dystrophy (for instance, the Infovox-Telia system).

– *Hearing aids*: hearing impaired people also benefit from progress made in, micro-electronics , especially hearing-aid devices which are better adapted to their impairment, for instance cochlear implants (in France: Avicenne Hospital in Bobigny and Saint-Antoine Hospital in Paris).

– *Document access and editing:* the *Méditor* prototype, which integrates a speech recognition board from Vecsys and a synthesis software developed by Elan, was developed in 1994 by LIMSI-CNRS in partnership with INSERM-CREARE and INJA (National Institute of Young Blind Persons). It aimed at facilitating access and editing tasks on documents: voice synthesis was used to indicate typographic attributes of the text, that were not visible in Braille (font type, bold characters, underlined text portions, capital letters, color, etc.) and to speak out footnotes. Voice commands were used to easily modify the text (insertion, suppression and modification of the typographic attributes) [BEL 97] (Figure 13.1). The extension of such a system for learning French and for reading Webpages for blind people has been considered [FAR 99]. Similar projects were ongoing at IRIT, such as the *SMART* project [VIG 95] and, more recently the *IODE* project, which facilitates voice access by blind people to structured documents: tests in 1996 have shown that the adjunction of speech helped the users in accomplishing their task faster.

– *Assisting blind people to move in their environment*: the objective of the TELEFACT project (partnership between LIMSI-CNRS and the Aimé Cotton laboratory) has been to provide to visually impaired people a system to assist them when moving in their environment and which offers a maximum level of security [JAC 04]. Various forms of information presentation have been studied, among these auditory icons, speech synthesis, tactile feedback, effort feedback and the combination of several such modalities. A laser device and an infrared scanner are able to detect reliably obstacles over a distance up to 15 meters.

Figure 12.1. *The Méditor application [BEL 97]. The blind student can use a Braille terminal, a speech recognition system and a speech synthesis system, to consult and edit documents*

12.4.5. *Automatic translation*

Longer term projects relate to *automatic translation*: a very ambitious project was launched by ATR in Japan, as early as 1986 [FUJ 87]: the goal was to achieve a *simultaneous translation dialog* system allowing a person to have a spontaneous conversation on the phone with an interlocutor speaking a different language: this person's message would be automatically translated with a voice that would preserve all the characteristics of the original speaker [BLA 95a].

The *ASURA* prototype [MOR 93b] integrates some of these functionalities and provides automatic translation from Japanese to English and German.

A joint project connected *ASURA* to *JANUS* [WAI 96], another prototype developed in the framework of collaboration between the University of Karlsruhe in Germany and CMU in the US. The latter is able to translate from English and German to German, English and Japanese.

The *C-STAR-I/II* project [C-ST 00] gathered around ATR the following partners: CMU (US), University of Karlsruhe (Germany), ETRI (Korea), CLIPS/IMAG-Grenoble (France) and IRST (Italy). A demonstrator for automatic translation integrating speech recognition and speech synthesis was presented in 1999.

The German national program *Verbmobil* for face-to-face speech translation [REI 95] started in 1993 and terminated in 2000. It federated almost 30 partners from academia and industry: the goal was to have a system which would be fully substitutable to a speaker, when the latter is unable to translate his message into English or even requires to spell words in his/her own language [HIL 95a]. In the latter case, whichever language is used (Japanese or German), the target language is always English. The recognition system should be able to work with spontaneous language, a vocabulary of approximately 2500 words, and a headset microphone. As opposed to dictation or direct translation, speech understanding and dialog management modules were also required [ALE 96] [MAI 96].

The awareness, at the European level, that linguistic engineering industries constitute an economical challenge which can contribute to warrant a multi-lingual communication environment is very favorable to the development of such applications. New programs are dedicated to the set up of infrastructures for the dissemination of linguistic resources and the development of advanced linguistic tools for aided oral and written translation.

12.5. Application domains with multimodal interaction

When looking at somebody speaking, speech appears as intimately linked to the other modes of communication, in particular hand gestures, face expressions, etc. Human communication calls for a wide range of perception and production modes, even if speech generally remains dominant: the correct understanding of a message depends for instance on the consistency between the visual and auditory perception [MOO 94] [BEN 96b] [BER 96a]. This is why, for over a decade, research has focused on the identification of speech specificities in order to integrate it optimally with other communication modes and thus to provide human-computer interactions with a greater naturalness.

Along these lines, we develop in the next sections the application profiles of interactive terminals, computer-aided graphical design, on-board applications (plane, car, etc.), human-human communication and automatic indexing of audio-visual documents. It should be underlined that these applications are nowadays dedicated to the general public.

Moreover, some of the applications already mentioned in section 12.4 may be extended to a multimodal context in the future: for example, for *environment control*, speech recognition associated to position and movement (hand and eye gestures) helps greatly specifying some commands; also, in the case of *training* applications, a multimodal input is already frequently used (touch and voice for the *Meditor* prototype, for instance). In addition, a multimodal output (in the

TELETACT project, for instance) is certainly going to be developed in the future, to help the interaction be more attractive.

12.5.1. *Interactive terminals*

Research on multimodal interactive terminals aims at offering the users, services comparable to those already accessible on voice servers. In particular, these services concern train, plane or boat timetable enquiries and reservations. Many such systems have been tested, for instance those stemming from the European projects *Railtel* [BEN 96a] [BIL 97] and *ARISE (Automatic Railway Information System for Europe)* [BLA 98b] (RECITAL system in France). Presently, interactive terminals only use the tactile modality, but the adjunction of speech would greatly speed up and increase the naturalness of the interaction. The size of the accessible vocabulary (especially as to what concerns the names of the stations) can easily be increased so as to avoid tedious lists to search through, when using a touch screen.

In the framework of the *ESPRIT MASK (Multimodal Multimedia Service Kiosk)* project [BEN 95] [GAU 96] which gathered the LIMSI-CNRS, the SNCF, MORS and UCL (University College London), the proposed human-computer interface combined voice and touch with a multimedia presentation of the result at the output. An initial prototype (Figure 12.2) was used in September 1995 to record a database in French [LAM 95] comparable to the *ATIS* database (air traffic information in English). Tests with the general public (over 200 people) have shown the interest of adding speech to the traditional touch screen: it makes it indeed possible to specify the request by a single message containing at once all the relevant information [TEM 99]. Similar findings where reported with the *ARISE* voice servers (Philips, CSELT, etc.), the dialog being typically three times faster (40 seconds *versus* two minutes) with speech instead of *DTMF (Dual Tone MultiFrequency)* commands [BLO 98].

Figure 12.2. *Prototype of the MASK kiosk terminal [LAM 98]. This prototype was intended to allow a user to obtain train traffic information and to reserve a ticket, using a touch screen and a voice interface (recognition and synthesis)*

The experimental *GEORAL* system [SIR 95], combining touch and voice, was developed at IRISA, in close collaboration with CNET, to provide tourism information for the Tregor area in Brittany.

The *WAXHOLM* prototype, for boat timetable enquiries was developed in Sweden by KTH [BER 95c] and was based on a multimodal interaction which included speech and graphics. Text-to-speech synthesis was complemented by talking face syntheses, which had lip movements synchronized with the vocal message and eye movements pointing towards the relevant window on the screen.

The *AMIBE* (Applications Multimodales pour Interfaces et Bornes Evoluées) project [MON 94] was developed within the framework of the GDR-PRC CHM working group. It aimed at adding a user verification module to both voice and face movement modalities, including the detection of face contour. Lip movement tracking was here reinforcing speech recognition under noisy conditions.

The DARPA *Communicator* program was introduced as a follow-up to the *ATIS* project, encompassing multimodal interaction and relying on a multi-agent architecture [WAL 99]. The goal was to propose a new generation of conversational interfaces in which speech is assisted with pointing gestures at the input and graphics (maps) offering a complement to speech synthesis at the output. This project was intended to offer access to databases from heterogeneous devices (PDAs, telephones, computers, etc.) and to facilitate simultaneous communication between users. The available information was more elaborated than in the case of ATIS and included not only plane timetables but also tourist information and weather forecasts. Several research groups participated to this program, among which CMU, MIT, BBN, IBM, Microsoft and SRI. Several demonstrators have been developed, which integrated the *Sphinx-II* speech recognition system and the *Phoenix* semantic analyser from CMU [RUD 99] or the *Galaxy* system from MIT [GOD 94].

12.5.2. *Computer-aided graphic design*

Speech yields the advantage of allowing designers to dedicate themselves to their primary task, which is to define graphically the shape of 3D objects, while facilitating communication with the computer for operations of lower importance (choice of mathematical functions, object manipulation such as rotation or zoom, file handling, etc.): the designer's attention can thus concentrate on the graphical zone, which avoids frequent head or eye movements between this zone and the one which is dedicated to menus and commands. Moreover, the integration of speech in the designing software allows the user a direct and fast access to complex functionalities that would otherwise necessitate the navigation in an obscure forest of menus. A multimodal prototype called *MIX3D*, including speech, gesture and vision, has thus been developed at LIMSI-CNRS [BOU 95]. An extension of this system allowed for interaction in virtual 3D environment.

12.5.3. *On-board applications*

On-board cars or planes, the driver or pilot's tasks are complex, and the dashboard is usually small and crowded. Speech offers an additional means of interaction with the computer, without preventing the accomplishment of simpler tasks that require visual attention. In the car, new communication infrastructures by satellite for high-rate transmission of images and sounds, the European mobile phone standard *GSM* and the *GPS* (global positioning satellite) system for vehicles provide the driver with dynamic knowledge of the traffic and weather conditions. They help the driver to plan their itinerary, they guide them with voice messages and, when the car is stopped, they give access to tourist databases and a variety of services.

12.5.3.1. *Aided navigation on-board cars*

These systems rely on the deployment of radio-transmission networks such as the *RDS-TMC (Radio Data System-Traffic Message Channel)* and more recently the *GTTS (Global Transport Telematic System*, which integrates *GPS*, *GSM* and Internet). They have been the focus of several national and European projects. The *CARIN* project [CAR 95], developed by Philips and Lernout & Hauspie was intended to respond to the needs of both the professionals (truck drivers) and the general public. A high quality multilingual synthesis [VAN 97] in French, English, German and Dutch delivered, through the onboard radio, information concerning the road status. As the message was transmitted in a codified form, it was easy to generate the information in a given language. The system had been designed so as to guarantee language independence, thanks to a microchip card which enabled the selection of the information in the driver's language, whichever country the truck was circulating in. A regular *TMC* broadcasting service for the German language was set up in 1997 and it should be extended to other countries and other languages. The system used text-to-speech synthesis for message generation, so as to guarantee minimal memory requirements and some flexibility in the perspective of future modifications. The naturalness of the voice was preserved using a technique called prosodic transplantation [VAN 94].

Several other projects in the same domain should be mentioned: in France, the *CARMINAT* project involving Renault, Peugeot, Philips, TDF and Sagem; in Italy, the *Route Planner* system by Tecmobility Magneti Marelli [PAL2 95]; in England, the information query system for the Inmarsat-C satellite, studied by Marconi [ABB

93]; in Sweden, the *RTI* project in which Infovox-Telia and KTH were partners [GRA 93].

In most of these projects, only speech synthesis was implemented. However, one of the most crucial functionalities for ensuring the driver's safety is the ability to command the various equipments by voice. Therefore, speech recognition systems integrated in the dashboard have been proposed: for instance, the *AudioNav* prototype for the recognition of key-words and short sentences developed by Lernout & Hauspie, in the framework of the *VODIS* project with the participation of Renault and PSA; the recognition system by DaimlerChrysler, currently available on top-of-the-range cars. The SENECa (Speech Control modules for Entertainment, Navigation and communication Equipment in Cars) spoken language dialog system demonstrator [MIN 03], resulting from an industrial research cooperation within the framework of an EU-HLT project, provides speech-based access to a wide range of entertainment, navigation and communication applications.

The first evaluations in simulated situations have shown a mitigated interest for speech in this context: the driver's safety being crucial, it is imperative for voice interaction to be reduced both ways to short synthetic messages. Moreover, speech could favorably be complemented by gesture detection during driving. In this direction, some linguistic and cognitive studies have been undertaken on multimodal communication (speech and gesture), from recordings of dialogs between pilots and their co-pilots onboard a car, so as to model the knowledge required and to identify a typology of the relevant concepts [BRI 96]. More generally speaking, psycholinguistic studies are under way in order to investigate on the relation between the mental image of a complex urban route in a human brain and its linguistic description, with the ultimate goal to improve remote navigation-aid systems [DAN 98].

Above all, when the vehicle is stopped, an enriched multimodal interaction with the onboard computer is particularly relevant for accessing tourist information through complex functions.

12.5.3.2. *Avionics*

Since the beginning of the 1980s, parallel studies by Sextant-Avionique (now Thalès) in France (*TOP-VOICE* system) [PAS 93], and by Marconi & Smith Industries in England and in Germany (*CASSY* system) [GER 93] have aimed at making speech recognition systems more robust to noise (plane engine, oxygen masks, etc.) and to the effects of acceleration on voice. Similarly to speech recognition onboard cars, additional studies on recording conditions (type and place of the microphones in the oxygen mask, for instance) remain necessary. Despite many experiments, no system is operational yet, even though speech recognition is still planned on the Rafale and may be combined with a touch-based interaction [ROU 04].

12.5.4. *Human-human communication facilitation*

In the framework of the *ESPRIT LTR I3 (Intelligent Information Interfaces)* program, one of the projects called *Magic Lounge* (Figure 12.3) aimed at offering to *non-professional* users a collective multimodal communication space [BER 98b]. One of the reference activities, in the initial concept, was to allow people in different locations to prepare a joint trip. The project was co-ordinated by the DFKI in Saarbrücken (Germany), the other participants being LIMSI-CNRS (France), the University of Compiègne (France), Siemens (Germany), the NIS (Denmark) and the "Bank of Ideas of the Smaller Danish Isles", a Danish association, the members of which were forming a potential user group. Participants were meant to communicate with one another with heterogeneous equipments (mobile phones, PDAs or high definition multimedia devices, for example), while having access to a variety of local and Internet services.

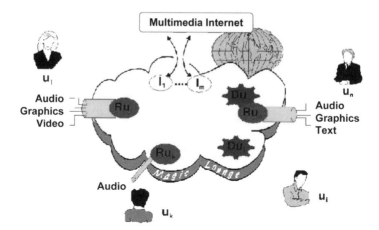

Figure 12.3. *Functionalities of the Magic Lounge demonstrator [BER 98b]. Users ($u_{1,...,n}$) who are spread geographically, can communicate with their equipment, while having access to a wide range of information sources ($I_{1,...,m}$) and possibly, be replaced by a representative ($Ru_{1,...,n}$) or choose a delegate($Du_{1,...,n}$) if they could not attend the meeting*

For the final phase of the project, the study was focused on the automatic or semi-automatic structuring of a collective memory built during the successive meetings [RIS 98], the objective being to provide information to absent or late participants, and summarize the decisions made. A longer-term objective was to automatically extract the relevant information from the exchanges, which requires an adequate combination of spontaneous speech recognition methods, dialog tracking techniques and automatic summarization approaches. Another perspective could be to integrate the studies concerning the oral and gestural querying of graphic databases and to plan a data conversion step from one mode to the other so that a partner which has, for instance, a mobile phone, can obtain a linguistic description of the graphic displayed on the HD screen of another participant. Socio-economical aspects raised by these new communication modes are the focus of specific research (collaboration between LIMSI-CNRS and the University of Compiègne).

SmartKom [WAL 05] has been steering a long-term research effort (1999-2003) funded by the German Federal Ministry for Education and Research (BMBF). Following Verbmobil, SmartKom has been a consortium of mainly industrial and academic research institutions in Germany. The main goal of the SmartKom project has been to provide non-expert users with an intuitive and intelligent interface for everyday computer-based applications in different scenarios of use. Within this context, the co-coordinated use of different input modalities, like speech and language, gestures and mimics as well as speech synthesis and graphical output has been considered to be one of the central aspects. Intuitiveness has been achieved by enabling natural conversational interaction with a lifelike virtual character (called Smartakus) that embodies the system's functionality. One of the major scientific goals of SmartKom has been to explore new computational methods for the integration and mutual disambiguation of multimodal input and output on a semantic and pragmatic level [WAL 05].

Similar to SmartKom, the purpose of the INTERACT project (interactive systems laboratories at Karlsruhe and Carnegie Mellon University) [INT 05] is to enhance human-computer communication by the processing and combination of multiple communication modalities known to be helpful in human communicative situations. One of the project goals is to derive a better model of where a person is in a room, who he/she might be talking to, and what he/she is saying despite the presence of jamming speakers and sounds in the room (the cocktail party effect). The project consortium is also working to interpret the joint meaning of gestures and handwriting in conjunction with speech, so that computer applications ("assistants") can carry out actions more robustly and naturally and in more flexible ways. Several human-computer interaction tasks are explored to see how automatic gesture, speech and handwriting recognition, face and eye tracking, lip-reading and sound source localization can all help to make human-computer interaction easier and more natural.

12.5.5. *Automatic indexing of audio-visual documents*

A closely related area in which automatic speech processing is expected to provide more efficient access to the information contained in audio-visual documents is addressed in the DARPA *Broadcast News Transcription and Understanding TDT* (*Topic Detection and Tracking*) program. This area is very active at present and it has given rise to several projects and achievements [MAY 99]: the European project *OLIVE* [JON 99] [GAU 00], the *MAESTRO* system by SRI [SRI 00], the *Rough'n'Ready* system by BBN [KUB 00], etc. These projects have a cross-disciplinary dimension as they call for speech processing methods, for artificial intelligence techniques (automatic summarization, semantic knowledge extraction, etc) and also for image processing methods.

12.6. Conclusions

It is difficult to predict the position of speech in tomorrow's human-computer communications. One may wonder why gadgets with speech recognition and synthesis functionalities, such as the *Voice Master VX-2* watch which was put on the market as early as 1987, never met the expected success. In the 1980s, an idealistic objective seemed to be the substitution of every human-computer communication mode by speech, excluding all other options. Today, the perimeter in which speech can bring a real advantage is better understood, as are its limitations. It is becoming increasingly clear that speech is complementary to other perception and production modes (gesture, touch, etc), and this necessitates a radically different system design so as to make them easy to integrate.

Currently, voice interfaces are used in professional environments rather than by the general public. However, voice technologies are gradually entering other areas in everyday life, especially *interactive voice information servers*, *onboard car navigation tools* and *learning aids*. Besides these expanding domains, *automatic dictation* of written documents has been for now quite some time a very active sector, all the more as systems can now handle an increasing number of languages.

Human-computer communication stands in a wider context of *digital nomadism* for which a generalized multi-agent architecture would offer comparable services to the user, whether he is in a car, in the street, at the office or at home. This trend was already discernible in programs and projects such as *Verbmobil*, *Magic Lounge*, *SmartKom*, *Communicator*, etc. A rich multimodal *interaction including speech* is becoming more and more necessary to answer the multiplicity of situations. May be, the Man of the Future will control directly or remotely his environment with a *pocket telecomputer* integrated to his *clothes,* combining virtual reality, gesture detection, pointing functionalities and speech recognition, to access information databases spread out all over the world. Voice interfaces are all the more necessary as the size of the equipment shrinks. Speech is also an asset since it allows some user-customization of the interface as a function of individual preferences.

Another area closely related to human-computer communication, and which appears to be very promising, is the immediate re-transcription of public meetings, debates and conferences (all of them including spontaneous speech). This task will help structure the collective memory of remote interactions and indexing automatically multimedia documents so as to facilitate their consultation. This domain has recently triggered a number of American and European projects.

However, the wide use of voice technologies will only become effective if the performance of the systems reach an acceptable level for the general public, in terms of reliability but also in terms of usability. As of what concerns speech, the current effort that has been undertaken on the international level to define and disseminate assessment methodologies and tools for linguistic systems, but also, more recently, towards multimodal interfaces (*Communicator* project), represents a first step. This effort must be accompanied by socio-economical studies, in order to more precisely understand the final user needs in real-life conditions.

12.7. References

[ABB 93] ABBOTT M. and BATEMAN C., "Applications of speech technologies in vehicles", *Joint ESCA-NATO/RSG.10 Tutorial and Research Workshop on Applications of Speech Technology*, Lautrach, September 1993.

[ALE 96] ALEXANDERSSON J., "Some Ideas for the Automatic Acquisition of Dialogue Structure", Eleventh Twente Workshop on Language Technology (TWLT 11): *Dialogue Management in Natural Language Systems,* LUPERFOY S. NIJHOLT A. and VELDHUIJZEN VAN ZANTEN G. (Eds), June 1996.

[BAK 75] BAKER J.K., "The DRAGON system – an overview", *IEEE Transactions on ASSP,* Vol. 23, No. 1, pp. 24-29, 1975.

[BAK 93] BAKER J.M., "Using Speech Recognition for Dictation and Other Large Vocabulary Applications", *Joint ESCA-NATO/RSG.10 Tutorial and Research Workshop on Applications of Speech Technology*, Lautrach, September 1993.

[BEL 97] BELLIK Y., "Multimodal Text Editor Interface Including Speech for the Blind", *Speech Communication,* Vol. 23, No. 4, pp. 319-332, 1997.

[BEN 95] BENNACEF S., Modélisation du Dialogue Oral Homme-Machine dans le cadre des applications de demande d'informations, PhD Thesis, University of Paris XI, July 1995.

[BEN 96a] BENNACEF S., DEVILLERS L., ROSSET S., and LAMEL L., "Dialog in the RailTel telephone-based system", in *International Conference on Speech and Language Processing*, Philadelphia, October 1996.

[BEN 96b] BENOÎT C., MOHAMADI T. and KANDEL S.D., "Effects of phonetic context on audio-visual intelligibility of French", *Journal of Speech and Hearing Research*, No. 37, pp.1195-1203, 1996.

[BER 96a] BERNSTEIN L.E. and BENOÎT C., "For speech perception by humans or machines, three senses are better than one", *ICSLP'96*, Philadelphia, October 1996.

[BER 98b] BERNSEN N.O., RIST T., MARTIN J-C., HAUCK C., BOULLIER D., BRIFFAULT X., DYBKJAER L., HENRY C., MASSOODIAN M., NÉEL F., PROFITLICH H.J., ANDRÉ E., SCHWEITZER J. and VAPILLON J., "Magic Lounge: a thematic inhabited information space with "intelligent" communication services", *International Conference Nîmes on Complex Systems, Intelligent Systems & Interfaces*, Nîmes, May 1998.

[BER 95c] BERTENSTAM J., BESKOW J., BLOMBERG M., CARLSON R., ELENIUS K., GRANSTRÖM B., GUSTAFSON J., HUNNICUTT S., HÖGBERG J., LINDELL R., NEOVIUS L. NORD L., DE SERPA-LEITAO A. and STRÖM N., "The Waxholm system – a progress report", *ESCA Workshop on Spoken Dialogue Systems*, Vigso, May 1995.

[BIL 97] BILLI R. and LAMEL L., "RailTel: railway telephone services", *Speech Communication*, Vol. 23, No. 1-2, pp. 63-65, 1997.

[BLA 95a] BLACK A.W. and CAMPBELL N., "Predicting the intonation of discourse segments from examples in dialogue speech", *ESCA Workshop on Spoken Dialogue Systems*, Vigso, May 1995.

[BLA 98b] BLASBAND M., "Speech Recognition in Practice", *International Conference Nîmes on Complex Systems, Intelligent Systems & Interfaces*, Nîmes, May 1998.

[BLO 98] BLOOTHOOFT G. "Spoken dialogue systems", *Elsnews* 7.3, July 1998.

[BOU 95] BOURDOT P., KRUS M. and GHERBI R., "Gestion de périphériques non-standards pour des interfaces multimodales sous Unix/X11: Application à un modeleur tridimensionnel", 4èmes *Journées Internationales Montpellier 95 sur l'Interface des Mondes Réels et Virtuels*, June 1995.

[BRI 96] BRIFFAULT X. and DENIS M., "Multimodal Interactions Between Drivers and Co-Drivers: An Analysis of On-board Navigational Dialogues", *12th European Conference on Artificial Intelligence, 2nd Workshop on Representation and Processing of Spatial Expressions*, Budapest, 1996.

[C-ST 00] http://www.c-star.org/

[CAR 95] CARDEILHAC F. and PALISSON F., "Système de Navigation CARIN", *SIA/FIEV/EQUIP'AUTO*, Paris, October 1995.

[CHA 97a] CHANIER T., DUQUETTE L., LAURIER M. and POTHIER M., "Stratégies d'Apprentissage et Evaluation dans des Environnements Multimédias d'Aide à l'Apprentissage de Français", *1ères JST FRANCIL* 1997, Avignon, April 1997.

[CHA 79b] CHAPANIS A., "Interactive Communication: A few research answers for a technological explosion", in *Nouvelles Tendances de la Communication Homme-Machine*, AMIRCHAHY M.A. AND NEEL D. (Eds), INRIA, Orsay, April 1979.

[CHI 99a] CHIBOUT K., MARIANI J.J. , MASSON N. and NEEL F. (Eds), *Ressources et evaluation en ingénierie de la langue*, Collection Champs linguistiques (Duculot) et Collection Universités francophones (AUF), December 1999.

[CHI 96b] CHINNOCK C , CALKINS D., COVIN C., FRIEL K., JENKINS M. and NEWMAN E., "Hands-free mobile computing: a new paradigm", *5èmes Journées Internationales Montpellier 96, L'Interface des Mondes Réels & Virtuels*, Montpellier, May 1996.

[CHO 95] CHOLLET G., "Les domaines d'application des technologies vocales", Ecole thématique *Fondements et Perspectives en Traitement Automatique de la Parole*, Marseille-Luminy, July 1995.

[DAN 98] DANIEL M-P. and DENIS M., "Spatial descriptions as navigational aids: a cognitive analysis of route directions", *Kognitionswissenschaft*, No. 7, pp. 45-52, 1998.

[DER 93] DEROUAULT A.M., KEPPEL E., FUSI S., MARCADET J.C. and JANKE E., "The IBM Speech Server Series and its applications in Europe", *Joint ESCA-NATO/RSG.10 Tutorial and Research Workshop on Applications of Speech Technology*, Lautrach, September 1993.

[DEV 04] DEVILLERS L., MAYNARD H., ROSSET S., PAROUBEK P., MCTAIT K., MOSTEFA D., CHOUKRI K., BOUSQUET C., CHARNAY L., VIGOUROUX N., BÉCHET F., ROMARY L., ANTOINE J.Y., VILLANEAU J., VERGNES M. and GOULIAN J., "The French MEDIA/EVALDA Project: the evaluation of the understanding capability of spoken language dialogue systems" in *LREC*, Lisbon, May 2004.

[DOL 88] DOLMAZON J-M., CERVANTES O. and SERIGNAT J-F., "Les bases de données de sons du français", *Premières Journées Nationales du GRECO-PRC Communication Homme-Machine, Parole, Langage Naturel et Vision*, Paris, November 1988.

[ESK 92] ESKÉNAZI M., "Changing styles: speakers' strategies in read speech and careful and casual spontaneous speech", *ICSLP'92*, Banff, 1992.

[EUR 93] EUROSPEECH'93, Berlin, 1993.

[EVA 05] http://www.elda.org/rubrique25.html

[FAR 99] FARHAT S. and BELLIK Y., "SeeWeb: dynamic improvement of the accessibility of HTML documents for blind persons", *INTERACT99*, Edinburgh, August/September 1999.

[FUJ 87] FUJISAKI H., "Overview of the Japanese national project on advanced man-machine interface through spoken language", *European Conference on Speech Technology*, Edinburgh, September 1987.

[GAG 93] GAGNOULET C. and SORIN C., "CNET speech recognition and text-to-speech for telecommunications applications", *Joint ESCA-NATO/RSG.10 Tutorial and Research Workshop on Applications of Speech Technology*, Lautrach, September 1993.

[GAG 98] GAGNOULET C., "Parole et Télécommunications" Ecole d'Eté *"La Parole, des Modèles Cognitifs aux Machines Communicantes"*, Grenoble, September 1998.

[GAU 90] GAUVAIN J.L., LAMEL L. and ESKÉNAZI M., "Design considerations and text selection for BREF, a large French read-speech corpus", *ICSLP'90*, Kobe, November 1990.

[GAU 96] GAUVAIN J.L. and TEMEM J.-N., "Le système de dialogue oral du projet MASK", *5èmes Journées Internationales Montpellier 96, L'Interface des Mondes Réels & Virtuels,* Montpellier, May 1996.

[GAU 00] GAUVAIN J.L., LAMEL L. and ADDA G.,"Transcribing broadcast news for audio and video indexing", *Communications of the ACM*, Vol. 43, No. 2, pp. 64-70, 2000.

[GER 93] GERLACH M. and ONKEN R., "Speech input/output as interface devices for communication between aircraft pilots and the pilot assistant system "Cassy", Joint ESCA-NATO/RSG.10 *Tutorial and Research Workshop on Applications of Speech Technology*, Lautrach, September 1993.

[GIB 97] GIBBON D., MOORE R. and WINSKI R. (Eds), *Handbook of Standards and Resources for Spoken Language Systems*, Walter de Gruyter, Berlin/New York, 1997.

[GOD 94] GODDEAU D., BRILL E., GLASS J., PAO C., PHILLIPS M., POLIFRONI J., SENEFF S. and ZUE V., "Galaxy: a human-Language interface to on-line travel information", *ICSLP'94*, Yokohama, 1994.

[GRA 93] GRANSTRÖM B., BLOMBERG M., ELENIUS K., ROXSTRÖM A., NORDHOLM S., NORDEBO S., CLAESSON I., WAERNULF B. and EKE K., "An experimental voice-based information provider for vehicle drivers", *Joint ESCA-NATO/RSG.10 Tutorial and Research Workshop on Applications of Speech Technology,* Lautrach, September 1993.

[HAM 97] HAMEL M-J. and WEHRLI E. "Outils de TALN en EIAO, le projet SAFRAN", *1ères JST FRANCIL 1997*, Avignon, April 1997.

[HIL 95a] HILD H. and WAIBEL A., "Integrating spelling into spoken dialogue recognition", *EUROSPEECH'95,* Madrid, September 1995.

[HIL 93b] HILLER S., ROONEY E., JACK M. and LEFÈVRE J.P., "SPELL: A pronunciation training device based on speech technology", *Joint ESCA-NATO/RSG.10 Tutorial and Research Workshop on Applications of Speech Technology*, Lautrach, September 1993.

[HUN 96] HUNT M.J., "Reconnaissance de parole pour le poste de travail", *Les Entretiens de la Technologie,* Paris, March 1996.

[HUN 98] HUNT M.J., "Practical automatic dictation systems", *The ELRA Newsletter*, February 1998.

[INT 05] http://www.is.cs.cmu.edu/js/

[JAC 04] JACQUET C., BELLIK Y., BOURDA Y., "A context-aware locomotion assistance device for the blind", HCI 2004, *The 18th British HCI Group Annual Conference, Leeds Metropolitan University, UK* 6-10 September 2004.

[JEL 76] JELINEK F., "Continuous speech recognition by statistical methods*", IEEE,* Vol. 64, No. 4, pp. 532-556, 1976.

[JEL 87] JELINEK F. *et al.*, "Experiments with the Tangora 20,000 word speech recognizer", *IEEE ICASSP*-87, Dallas, 1987.

[JON 99] DE JONG F., GAUVAIN J.L., DEN HARTOG J. and NETTER K., "Olive: speech based video retrieval", *CBMI'99,* Toulouse, October 1999.

[JOU 91] JOUVET D., BARTKOVA K. and MONNÉ J., "On the modelisation of allophones in an HMM-based speech recognition system", *EUROSPEECH'91*, Genoa, September 1991.

[KUB 00] KUBALA F., COLBATH S., LIU D., SRIVASTAVA A. and MAKHOUL J., "Integrated technologies for indexing: spoken language", *Communications of the ACM,* Vol. 43, No. 2, pp. 48-56, 2000.

[LAM 91] LAMEL L., GAUVAIN J.L. and ESKÉNAZI M., "BREF, a large vocabulary spoken corpus for French", *EUROSPEECH'91,* Genova, September 1991.

[LAM 95] LAMEL L., ROSSET S., BENNACEF S., BONNEAU-MAYNARD H., DEVILLERS L. and GAUVAIN J.L., "Development of spoken language corpora for travel information.", *EUROSPEECH'95*, Madrid, September 1995.

[LAM 98] LAMEL L., BENNACEF S., GAUVAIN J.L., DARTIGUES H., TEMEM J.N., "User evaluation of the MASKkiosk", *ICSLP'98*, Sydney, 1998.

[LEA 80] LEA W.A., *Trends in Speech Recognition*, Prentice Hall, 1980.

[LIE 77] LIENARD J.S., *Les Processus de la Communication Parlée*, Masson, 1977.

[MAI 96] MAIER E., "Context construction as subtask of dialogue processing – the VERBMOBIL case", *Eleventh Twente Workshop on Language Technology (TWLT 11): Dialogue Management in Natural Language Systems*, LUPERFOY S., NIJHOLT A. and VELDHUIJZEN VAN ZANTEN G. (eds), June 1996.

[MAR 90] MARIANI J., "Reconnaissance automatique de la parole: progrès et tendances", *Traitement du Signal,* Vol. 7, No. 4, pp. 239-266, 1990.

[MAR 93a] MARIANI J.J., "Overview of the Cocosda initiative", *Workshop of the International Coordinating Committee on Speech Databases and Speech I/O System Assessment, Berlin, 93,* September 1993.

[MAR 93b] MARQUE F., BENNACEF S., NÉEL F. and TRINH F., "PAROLE: a vocal dialogue system for air traffic control training", *Joint ESCA-NATO/RSG.10 Tutorial and Research Workshop on Applications of Speech Technology 93*: Lautrach, September 1993.

[MAT 91] MATROUF K. and NÉEL F., "Use of upper level knowledge to improve human-machine interaction", *2nd Venaco ESCA ETR Workshop on The Structure of Multimodal Dialogue, Acquafredda di Maratea, Italy 91*, September 1991.

[MAY 99] MAYBURY M., "multimedia interaction for the new millenium", *EUROSPEECH'99*, Budapest, September 1999.

[MIN 03] MINKER W., HAIBER U., HEISTERKAMP P., and SCHEIBLE S. (2003). "Intelligent dialog overcomes speech technology limitations: the SENECA example", in *Proceedings of International Conference on Intelligent User Interfaces (IUI)*, pp 267–269, Miami, Florida, USA.

[MON 94] MONTACIE C., "Projet AMIBE: Applications Multimodales pour Interfaces et bornes Evoluées", GDR No. 39, rapport d'activité 1994.

[MOO 94] MOORE R., "Twenty things we still don't know about speech", CRIM/FORWISS *Workshop on Progress and Prospects of Speech Research and Technology, 94,* Munich, September 1994.

[MOR 88] MOREL A.M. *et al.*, *Analyse Linguistique d'un Corpus de Dialogues Homme-Machine: Tome I. Premier corpus: Centre de Renseignements SNCF à Paris*, Ed: University of Paris III – Sorbonne Nouvelle, 1988.

[MOR 93] MORIMOTO T., TAKEZAWA T., YATO F., SAGAYAMA S., TASHIRO T., NAGATA M. and KUREMATSU A., "ATR's speech translation system: ASURA", *EUROSPEECH'93,* Berlin, September 1993.

[MOS 99] MOSTOW J. and AIST G., "Authoring New Material in a Reading Tutor that Listens", *Sixteenth National Conference on Artificial Intelligence (AAAI-99),* Orlando, FL, July 1999, in the refereed *Intelligent Systems Demonstration* track. Also presented at *37th Annual Meeting of the Association for Computational Linguistics* (ACL'99), College Park, MD, June, 1999.

[NEE 96] NEEL F., CHOLLET G., LAMEL L., MINKER W. and CONSTANTINESCU A., "Reconnaissance et Compréhension de la Parole: Evaluation et Applications", in MELONI H. (Ed.), *Fondements et Perspectives en Traitement Automatique de la Parole, publications AUPELF-UREF*, 1996.

[NEE 99] NÉEL F. and MINKER W., "Multimodal speech systems", in PONTING K.M. (Ed.): *Computational Models of Speech Pattern Processing*, pp. 404-430, Springer-Verlag, Berlin, 1999.

[PAL 90] PALLETT D.S., "Issues in spoken language system performance assessment in the United States", *International Symposium on International Coordination and Standardization of Speech Databases and Assessment Techniques for Speech Input/Output,* Kobe, November 1990.

[PAL 95a] PALLETT D.S. *et al.*, "1994 Benchmark Tests for the DARPA Spoken Language Program", *DARPA Workshop Spoken Language System Tech*, 1995.

[PAL 95b] PALLME M. and MAGGIA G., "Route Planner, un système de navigation flexible", *SIA/FIEV/EQUIP'AUTO,* Paris, October 1995.

[PAS 93] PASTOR D. and GULLI C., "D.I.V.A. (DIalogue Vocal pour Aéronef) Performances in simulated aircraft cockpit", *Joint ESCA-NATO/RSG.10 Tutorial and Research Workshop on Applications of Speech Technology*, Lautrach, September 1993.

[REI 95] REITHINGER N., MAIER E. and ALEXANDERSSON J., "Treatment of incomplete dialogues in a speech-to-speech translation system", *ESCA Workshop on Spoken Dialogue Systems,* Vigso, May 1995.

[RIS 98] RIST T., ZIMMERMANN D., MARTIN J-C., NÉEL F. and VAPILLON J., "Virtual meeting places with intelligent memory functions", 7th *Le Travail Humain Workshop on La conception de mémoires collectives/Designing collective memories*, Paris, September 1998.

[ROU 04] ROUSSEAU C. BELLIK Y., VERMIER F. and BAZALGETTE D., "Architecture framework for output multimodal systems design", *OZCHI 2004*, University of Wollongong, November 2004.

[RUD 99] RUDNICKY A.I., THAYER E., CONSTANTINIDES P., TCHOU C., SHERN R., LENZO K., XU W. and OH A., "Creating natural dialogs in the Carnegie Mellon Communicator System", *EUROSPEECH'99*, Budapest, September 1999.

[SAM 92] SAM, Multi-lingual speech input/output assessment methodology and standardisation, Final Report, June 1992.

[SHA 98] SHAO J., TAZINE N.-E., LAMEL L., PROUTS B. and SCHRÖTER S., "An open system architecture for a multimedia and multimodal user interface", *3rd TIDE Congress*, Helsinki, June 1998.

[SIR 95] SIROUX J., GUYOMARD M., JOLLY Y., MULTON F. and REMONDEAU C., "Speech and tactile-based GEORAL system", *EUROSPEECH'95*, Madrid, September 1995.

[SMA 94] SMAILAGIC A. and SIEWIOREK D.P., "The CMU mobile computers: a new generation of computer systems", *IEEE COMPCON 94*, IEEE Computer Society Press, February 1994.

[SOR 94] SORIN C., "Operational and experimental French telecommunication services using CNET speech recognition and text-to-speech synthesis", IVTTA'94, *Second IEEE Workshop on Interactive Voice Technology for Telecommunications Applications*, Kyoto, Japan, September 1994.

[SRI 00] SRI TEAM, "MAESTRO: Conductor of Multimedia Analysis Technologies", *Communications of the ACM*, vol. 43, No. 2, pp. 57-63, 2000.

[STE 95] STEENEKEN H.J.M. and VAN LEEUWEN D.A., "Multi-lingual assessment of speaker independent large vocabulary speech-recognition systems: the SQALE project", *EUROSPEECH'95*, Madrid, September 1995.

[TEM 99] TEMEM J.N., LAMEL L. and GAUVAIN J.L., "The MASK demonstrator: an emerging technology for user-friendly passengers kiosk", *World Congress on Railway Research*, Toulouse, October 1999.

[VAN 94] VAN COILE B., VAN TICHELEN L., VORSTERMANS A., JANG J.W. and STAESSEN M., "ProTran: a prosody transplantation tool for text-to-speech applications", *ICSLP'94*, Yokohama, September 1994.

[VAN 97] VAN COILE B., RÜHL H.W., VOGTEN L., THOONE M., GOSS S., DELAEY D., MOONS E., TERKEN J.M.B., DE PIJPER J.R., KUGLER M., KAUFHOLZ P., KRÜGER R., LEYS S. and WILLEMS S., "Speech synthesis for the new pan-European traffic message control system RDS-TMC", *Speech Communication*, Vol. 23, No. 4, December 1997.

[VIG 95] VIGOUROUX N., SEILER F.P., ORIOLA B. and TRUILLET P., "SMART: system for multimodal and multilingual access, reading and retrieval for electronic documents", *2nd TIDE Congress*, PLACENCIA PORRERO I., PUIG DE LA BELLACASA R. (Eds), IOS Press, April, Paris.

[WAI 96] WAIBEL A., "Interactive translation of conversational speech", *Computer* 27 (7), pp. 41-48, 1996.

[WAL 99] WALKER M.A., HIRSCHMANN L. and ABERDEEN J., "Evaluation for DARPA COMMUNICATOR spoken dialogue systems," in *LREC*, 1999.

[WAL 05] WAHLSTER W., (Ed.). *SmartKom: Foundations of Multi-modal Dialogue Systems*, Springer, Berlin/Heidelberg, Germany, 2005.

[YOU 97] YOUNG S., ADDA-DECKER M., AUBERT X., DUGAST C., GAUVAIN J.L., KERSHAW D.J., LAMEL L., LEEUWEN D.A., PYE D., ROBINSON A.J., STEENEKEN H.J.M. and WOODLAND P.C., "Multilingual large vocabulary speech recognition: the European SQALE project", *Computer Speech and Language,* Vol. 11, pp. 73-89, 1997.

Chapter 13

Voice Services in the Telecom Sector

13.1. Introduction

Human beings have a natural preference for communicating by speech. It is therefore natural that they also favor speech for remote or off-line communication. Even though the volume of data transferred in telecommunication networks grows every day, especially as the Internet and the related services expand, speech still represents a significant proportion of the traffic and the economic stake for telecom operators. This is the main reason telecommunication operators consider automatic speech-processing techniques critical – they are technologies they *must* master. Over the past few years, voice services have become more and more widespread for three main reasons: the technologies have come of age, call centers are developing at a rapid pace and access to information needs to be diversified.

In this chapter, we will mainly focus on worldwide commercial voice services that rely on advanced speech technologies. We will attempt to present their economical and ergonomic benefits, and also the associated constraining factors and limitations. In fact, the telecommunication sector is probably one of the most constraining areas for automatic speech processing, because of the multiple variations and alterations which affect the speech signal – the various transmission channels and terminals. Moreover, the variety of targeted users (the general public as well as professional users) causes additional complexity that cannot be underestimated when designing "advanced" voice services with the aim of a successful application.

Chapter written by Laurent COURTOIS, Patrick BRISARD and Christian GAGNOULET.

After a brief overview of the main voice technologies used in the telecom area, and a presentation of their relative importance, we focus more specifically on the one that has the strongest development potential and is gradually becoming the preferred access mode for telephone voice services: automatic speech recognition.

13.2. Automatic speech processing and telecommunications

When considering the four main domains of automatic speech processing, namely speech coding, speech recognition, text-to-speech synthesis and speaker verification, the first (i.e. speech coding) is by far the most universally used in telecommunication networks. Only in the past few years have the other technologies gradually been deployed in some services, in particular voice command.

DataMonitor, a market analysis firm [DAT 04], has shown that the income generated by the whole chain of voice technologies and services grew at an annual rate of 20.3% in the period 2001 to 2005.

A projection of these figures for 2008 shows a steady and significant growth, even though it is well below the somewhat exaggerated curves forecast by analysts at the end of the 1990s. These figures need to be compared to the telephone service market (which mainly exploits speech coding techniques), which reaches several billion Euros in France. It is worth noting that in these areas, market studies have often produced predictions totally invalidated by the facts, but today, the observed growth of voice technologies is steady, and the market has at last reached maturity.

Several companies are strengthening their positions in the European market for speech recognition: Nuance (the market leader), Telisma (a spin-off from France Telecom) and Loquendo (subsidiary of Telecom Italia).

13.3. Speech coding in the telecommunication sector

The voice technologies of widest interest in the telecommunication sector are those connected to speech coding. Indeed, the telephone service, the most basic telecom service, takes immediate advantage, via the digital networks, of each progress made in speech compression techniques. It is obvious that the massive deployment of mobile networks would not have been conceivable in such proportions without research and the industrial development of economically viable techniques for low rate speech compression. They have indeed circumvented a major bottleneck: handling the communication traffic under the constraints imposed by the radio-frequency transmission bandwidth.

It is worth noting that the irreversible evolution of speech coding techniques towards lower rates has been greatly facilitated by the tolerance of the human ear to degradations in speech signals. In fact, the final user has often in the past, been rather lenient as regards tolerating the speech quality of voice services. This has led some service providers selling answering machines and messaging applications with a much lower quality than that used by telecommunication networks. Today however, the user is becoming more and more accustomed to high quality audio environments (FM, CD, MP3, Home Cinema, etc.). This offers a competitive advantage to voice messaging services which are embedded into the network, as they can provide the user with a constant audio quality far better than that available on most local or PABX messaging systems. However, the bit-rate reduction has yielded an additional factor of complexity for speech recognition and speaker verification tasks in telecom services.

Beyond basic telephone service that use compression techniques transparent to the client, coding techniques are also used in speech storage applications within the network. These are either voice messaging services, which store user- specific private information inside the network, or simpler services for the distribution of pre-recorded information to a group of users. These so-called *voice services* are usually interactive, and provide information to the user in the form of a spoken message (banking information, weather forecast, etc.).

At every level in telecommunication networks, speech is coded, compressed, transmitted, stored, restituted and distributed to every terminal. Speech compression is the dominant speech processing technique used.

13.4. Voice command in telecom services

13.4.1. *Advantages and limitations of voice command*

Telecommunication services are becoming more widespread, varied and complex, but the clients expect them to become increasingly simple to use. This essential need for simplicity has increased operator's interest in speech recognition.

In a context of both optimistic market studies and more or less conclusive experimentation, only major technical difficulties have hindered wider deployment of voice command services. Since the beginning of the 1990s, this is no longer a dream, nor is it a limited experimentation field: some services successfully deployed today are based entirely on speech recognition technology.

The main factors responsible for the success of voice services in the telecom sector are:

–   that more reliable, more robust technologies offered by providers which have now reached a real level of industrial maturity;

–   the widespread adoption of dedicated standards for voice technologies, whether they concern application development (VoiceXML language) or the integration of speech recognition and synthesis engines with the MRCP protocol[1];

–   the enhanced visibility of the values and benefits yielded by voice applications, due to an increased appreciation from marketing and project units of what can be expected from voice technologies: namely that they are neither a panacea nor a repellent.

However, the technical difficulties to be overcome when using speech recognition in a telecom environment are numerous. They result largely from the wide variability of channels (from radio transmission used for mobiles, up to voice transmission on the Internet (VoIP), including the analog line of the landline network and all types of terminals with uncontrolled microphone qualities). Paradoxically, the degradations resulting from signal compression generally remain minor compared to those induced by transmission itself, even for low bitrates, as in GSM.

These difficulties require experience for a satisfactory implementation, which leads corporate clients to outsource their voice services to dedicated companies.

Another limitation to a more widespread use of voice command in telecom services stems from the users themselves and their attitudes towards speech recognition systems. In this respect, a noticeable difference exists between services for the general public and those dedicated to professional users. In the latter case, the user is motivated by the benefits brought by the service, and its regular use makes them an "expert" (they know how to be understood by the service and how to overcome residual difficulties). When used by the general public, the system is faced with users who may never have heard about speech recognition, and whose first contact with the service may be critical. It may turn out to be fatal if the system's ergonomics are poorly designed, or if the communication environment created around the service is not adapted to the target. It is therefore more difficult to develop a voice service for the general public than for professional users. It is also essential to accompany the launch of such a service with repeated information campaigns to increase public awareness and inform them of the advantages of voice command.

Nowadays, speech recognition systems have become much more robust and service providers, product manufacturers and software editors have focused their efforts on ensuing robustness, not simply on the achievement of an ideal "recognition rate" which does not mean much *per se*, given the impact of other external factors. From the client's point of view, the behavior of a system when confronted with

1 MRCP: multimedia resource control protocol.

unusual conditions (noise, unexpected spontaneous speech, speaker's hesitations) is much more important than recognition "errors" (substitutions) which are now marginal for a high quality system [JOU 96].

The ergonomic aspects of voice command services are also key factors in the success of a service. They are fundamentally distinct from those of interactive services based on phone DTMF keystrokes. The recent arrival on the market of speech recognition products able to interpret commands formulated spontaneously by the speaker, with sufficiently wide vocabularies to offer flexibility (use of synonyms) is bound to help the market take off, and the conditions are now ripe for voice command to become natural to everyone. Voice technology providers have understood this: they do not limit their range of products to "raw" technologies; they also develop "bricks" to construct complete services with carefully planned ergonomic designs validated by clients in large-scale experiments.

13.4.2. *Major trends*

13.4.2.1. *Natural language*

Negative feedback from unsatisfied clients using voice services based on DTMF is increasing. These services often generate great frustration from users who are misdirected, or hang up because they do not understand the logic of the service.

Thanks to the last generation of continuous speech recognition engines (n-gram technology), it is possible to move towards higher quality user interactions. By calling for a semantic analysis component, the system is able to associate the spoken utterance of a user with a particular service, depending on the type of application, without requiring the user to limit themselves to a predetermined vocabulary restricted to a few isolated words.

Moreover, this new technology overcomes the problems caused by mouth noise, hesitations and fillers, and the overall efficiency of recognition is greatly improved. At last, these technologies can be coupled to larger vocabularies which integrate all possible ways for a user to express their expectations, making it possible to design services which "understand the clients" as opposed to those for which the vocabulary is imposed by the rules and the habits of a profession.

Continuous speech recognition also enables the association of several command words and parameters into one sentence, mainly for frequent use applications such as "Call Grandma at Home".

This type of technology is generally found in call routing, information and communications services. France Telecom, as a real pioneer in Europe, has launched several speech-enabled, self-care services based on natural language interaction: fix-line, self-care (the 3000), call routing voice services for banks and insurance companies.

13.4.2.2. *Embedded speech recognition*

Together with the emergence of embedded operating systems in communicating objects (especially mobile phones), the increase of computing and memory capabilities have now made it possible to integrate fully fledged, speaker-independent, speech recognition software functionalities in mobile devices. Speech recognition engine providers sell their software to a number of mobile device manufacturers that offer voice access to local phonebooks that may contain several hundred names. Samsung, on the other hand, has launched a terminal that provides voice dictation functionalities for texting.

It is easy to see that a large number of communicating objects will soon integrate speech recognition capabilities for local and remote services, themselves coupled to resources centralized in a network.

13.4.3. *Major voice command services*

Two distinct motivations stimulate a telecom operator to integrate voice command into its services. The first is to lower the cost of running operator-based services, by automating the most basic requests from clients, leaving human operators the task of handling only the more complex questions. A second reason is the simplicity that speech recognition offers for interfacing to complex systems, yielding a higher use of the services and a corresponding increase in turnover. In both cases, the use of voice command is of economic interest, and most operators consider it as a strategic move in the future development of their services.

13.4.4. *Call center automation (operator assistance)*

From telecom to financial services, from transport information to mail order, simplicity and efficiency are keywords for call routing, to ensure the client is rapidly directed towards an appropriate human operator. A survey carried out by Forrester [FOR 05] indicates that in 2005 21% of Call Centers in North America and 19% in Europe were using speech recognition functionalities.

A major objective for voice service providers is to better understand the caller's request and to route them towards the correct destination for their needs. Expansion of such services by being given a long list of telephone numbers (difficult to remember for the clients and heavy to use with a DTMF interface) are motivations for the development of more natural ways of handling the clients' requests on the phone.

There is a growing awareness of the fact that call centers are not simply cost generating centers, but can also generate value for the company, as they are part of the marketing process. As a contact point, the call center is also there to improve the company's perception of the users' needs, to gain their confidence and to bring a genuine value without becoming financially prohibitive.

Thanks to voice technologies, the automation of call centers relieves human operators from repetitive and tedious tasks, and allows them to focus on sales activities, problem solving and user-specific information provision.

The benefit of these services is therefore economical. When a user calls an operator, some time units dedicated to conversation with them can be saved by automation some of the request, yielding very substantial gains. The first such service put in operation was the AABS (automatic alternative billing service) by AT&T. This service allows the user to choose automatically, without any human intervention, the billing mode for their call (call collect, credit card or third party bank account). Today, this service is deployed over the entire North American continent and has been in use for more than 10 years. AT&T claim this automated service is responsible for benefits in the range of 100 million dollars per year. France Telecom also deployed a similar service a few years ago: PCV France (dial 3006).

Among the many lessons learned from such large scale public services (several million calls per day), the most obvious is the need for very robust systems. For instance, in order to simply recognize the words "yes" and "no" and their synonyms, Nortel's system calls for approximately 200 distinct models. Similarly, these services have shed light on the need to inform the general public about voice command services, and to provide information about their deployment. Moreover, it appeared as very relevant to enrich progressively the speech recognition models from voice data collected in real use conditions.

Another way of reducing the costs of operator-based services worth mentioning is the *partial* automation of phone directory services (for instance, phone enquiries by France Telecom). These services are more complex than those formerly mentioned: the recognition system must identify and distinguish words belonging to a large inventory of City and people names, which requires technological solutions that can deal with hundred of thousands of words. Several services of this type have been deployed in Italy and in Sweden since 2001, also in Belgium (service "Lisa") and

Germany. In France, several years ago, France Telecom launched the fully automated directory assistant of phone directory enquiries for the general public. The user must pronounce the name of the city and then the family name. Once the number has been found by the system, it is delivered to the user by speech synthesis.

Portals for access to services or companies, such as front desk services proposed by France Telecom, have the advantage of offering a wide range of services to the user through a single number, and can route them to the proper service by simply uttering its name.

Company telephone switchboards follow the same principles. A voice server directs the caller towards their interlocutor in the company without going through a hostess, but simply by uttering the name of the requested service or person, and matching it in the company's directory. Many improved variants and extensions are possible on these principles.

13.4.5. *Personal voice phonebook*

Speech recognition is not only useful to reduce the costs of employing telecom operators. It also offers new uses and functionalities to clients. A typical example of such a service is the *personal voice phonebook*.

Here, speech recognition is used to associate a phone number to a spoken name corresponding to a contact, providing a natural and fast way to establish communication, compared to keystroke. This range of service is especially useful when the client's hands and/or eyes are busy with a parallel task that makes conventional dialing difficult or even dangerous, for instance whilst driving. For mobile phones, the service is often coupled with a voice dialing system (preferably handling continuous speech), so that all calls can be placed in a hands-free mode.

When speech recognition is implemented inside the user's device, the system is usually speaker-dependent (the names in the phonebook are specific to each user). However, technological progress makes it possible to deploy speaker-independent services, in "flexible" modes, which allow the automatic real-time construction of a recognition model. The flexible mode enables the immediate access of any entry by simply adding it on to (or modifying) the current phonebook. When combined with continuous speech recognition technology, this allows direct expression of a composite contact request, such as: "*Pierre Martin* on his *mobile*". Additionally, the system can integrate dialogue interruption functionalities to allow even faster usage by experienced clients.

These systems all offer the user a single natural interface, which goes far beyond the simple phonebook functionality. Other functions giving access to additional telephone services have been tested in the USA by both Sprint and Bell Atlantic, for example, call-back or call waiting. In 2004, France Telecom opened a commercial voice service for the general public giving access to "My Contact List". As well as handling the phonebook via a Web interface, it allows addition of new contacts by voice, and will be upgraded soon to include access through the Web.

13.4.6. *Voice personal telephone assistants*

This term refers to clusters of services that are all accessible by voice command. They include all the services which have been listed above for directory assistance (access to additional telephone services, personal phonebook, voice dialing) plus messaging services (answering machines, voice mail, etc), services dedicated to mobile users (landline-mobile network continuity, etc) and various services close to secretarial work (call filtering, personalized message delivery).

Today, these services are not widely available. The emergence of telephone services over IP, and the architectural simplifications that stem from this should revive this type of application, made popular by WildFire (bought by Orange) as early as 1996, but which did not catch on because of the market's immaturity and the technologies involved.

13.4.7. *Other services based on voice command*

Beyond the aforementioned services that currently dominate the market, *voice command* is also spreading progressively in a wide range of voice services. Messaging services are generally quite complicated for the clients, because of the large number of available functions. They thus offer a very favorable context for voice command, by bringing naturalness and simplicity of use for navigation of the various services. Orange now provides its clients with voice command functionalities for voice mail services. Similarly, *online services* (such as Audiotel in France) benefit greatly from speech recognition, and it is now easier to navigate in their complex tree-structure with speech than with a keyboard.

Information delivered by companies to their clients, or to the general public by the media is undoubtedly going to become more and more regularly accessible by voice, using speech synthesis and recognition techniques. Voice indexing technologies also open interesting prospects, as they offer speech content-based search functionalities.

All these services should gradually benefit from new, more natural interaction modes with the user, relying on spontaneous speech recognition and large enough vocabularies, so the user does not feel constrained by an over-rigid dialogue structure.

13.5. Speaker verification in telecom services

Today, this technology is still restricted to very specific contexts and services, such as home curfew control, or "Offender Tagging". Soon, it should benefit from progress in the speech recognition field. As for voice command, the telecom environment is slightly more complex, as a reliable authentication requires accurate characterization of the speaker, while the variety of networks and terminals affect the voice features of a given speaker. If the user always accesses the service from the same environment (for instance, in incarceration applications), the problem is much simpler.

Depending on the service, it can be necessary to tune the false rejection rate (wrong rejection of a valid user) and the false acceptance rate (acceptance of an impostor) differently. Various verification modalities can be considered: key words (chosen by the user or by the system), or in a totally transparent manner, by coupling with voice command.

Voice authentication is also used in access control services to local data networks and to the telephone service (caller identification). For example, this function allows privileged access to specific messages for some users, in the context of voice mail services.

Only a few services fully exploit speaker verification technologies. The operators and public authorities keep testing the technologies, most of which come either from voice recognition software providers like Nuance, or specialized companies like Persay. The main services that use voice authentication are involved in preventing telephone fraud on mobile networks and call cards. The Foncard service by Sprint is also an example of how speech recognition and speaker verification can utilize the same utterance spoken by a card bearer.

13.6. Text-to-speech synthesis in telecommunication systems

Text-to-speech synthesis has been operational for quite some time in telecommunication services, even though margins for progress remain in terms of naturalness. However, as opposed to voice command (interest in which is universal), speech synthesis seems restricted in application to services requiring a bridge between written and oral language. Speech synthesis allows the voice delivery of written messages, of textual data from large datasets and of information that evolves frequently over time (directories, weather forecast, etc.).

At the beginning of the 2000s, the speech synthesis industry was able to market a new generation of synthesis engines based on longer speech units, and have managed to considerably increase the number of services using it. However, progress needs to be made before speech synthesis can be used in a front-desk application, where the natural voices of professional speakers are still required.

Mobile services will probably be those that most exploit speech synthesis technology in the future. The ability of a mobile user to access his written messages (faxes, E-mails, etc.) from any telephone is essential, and the service is sufficiently important to the client that they will generally tolerate some defects in the synthetic speech. In France, France Telecom launched such a service in 1999. The coupling between automatic text indexing techniques and speech synthesis may also circumvent some bottlenecks that exist in the automatic reading of written messages, for instance, the sometimes incomprehensible task of e-mail reading.

Another service which exploits text-to-speech synthesis, is the *reverse directory service*, which delivers directory information related to a particular phone number. France Telecom has opened the "QuiDonc" (Reverse directory) service, which can verbalize directory registrations. These services have already met with significant commercial success.

13.7. Conclusions

This overview of various implementations of advanced voice processing techniques in telecom services is far from exhaustive. Many experiments carried out worldwide by various operators have not been mentioned, including some in Europe [SOR 96]. We have limited this presentation to services deployed at a certain scale, which are more indicative of the real growth of this market. Today, many factors contribute to the growth forecast for voice services in the telecom sector: the increased robustness of speech recognition systems, a greater maturity of the industrial solutions, a better user-experience generated (including larger vocabularies but most of all, the handling of a more spontaneous speaking mode) and finally, more intimate coupling between speech recognition and dialogue.

From the simple voice phonebook to the automatic on-line translation service, the telecom sector is probably bound for rapid deployment of speech processing technologies. It is at least the sector that has enabled validation of the most significant progress in the domain, and is the most rapidly expanding applications sector.

13.8. References

[DAT 04] DATAMONITOR, "Voice Business", Market Update, *The Emergence of Packaged Applications*, April 2004.

[FOR 05] FORRESTER: *Contact Center and IVR Spending: Cross-Channel Platforms Offer Growth Potential*, August 2005.

[JOU 96] JOUVET D., 1996: " Robustesse et flexibilité en reconnaissance automatique de la parole ", *Echo des Recherches*, no. 165, pp. 25-38.

[SOR 96] SORIN C. *et al.*, 1996: "Current and experimental applications of speech technology for telecom services in Europe". *IVTTA '96*, Basking Ridge, NJ, USA, pp. 1-6.

List of Authors

Martine ADDA-DECKER
LIMSI-CNRS
Orsay
France

Brigitte BIGI
LIG
CNRS
Grenoble
France

Frédéric BIMBOT
IRISA
CNRS-INRIA
Rennes
France

Olivier BOËFFARD
IRISA-ENSSAT
University of Rennes 1
Lannion
France

Patrick BRISARD
Formerly France Télécom R&D
Lannion
France

Laurent COURTOIS
Orange Labs R&D
Lannion
France

Christophe D'ALESSANDRO
LIMSI-CNRS
Orsay
France

Alain DE CHEVEIGNE
Laboratoire de Psychologie de la Perception
CNRS - ENS - Paris Descartes University
Paris
France

Renato DE MORI
LIA
Université d'Avignon et des Pays du Vaucluse
Avignon
France

Pierre ESCUDIER
Formerly Institut de la communication parlée
CNRS
Grenoble
France

Gang FENG
GIPSA-lab
CNRS - Grenoble University - Grenoble INP
Grenoble
France

Christian GAGNOULET
France Télécom R&D
Lannion
France

Jean-Luc GAUVAIN
LIMSI-CNRS & IMMI-CNRS
Orsay
France

Laurent GIRIN
GIPSA-lab
CNRS - Grenoble University - Grenoble INP
Grenoble
France

Thierry GUIARD-MARIGNY
SIP CONSEIL
MEYLAN
France

Jean-Paul HATON
LORIA-INRIA
Henri-Poincaré University
Nancy
France

Lori LAMEL
LIMSI-CNRS
Orsay
France

Joseph MARIANI
LIMSI-CNRS & IMMI-CNRS
Orsay
France

Wolfgang MINKER
Institute of Information Technology
Ulm University
Germany

Françoise NEEL
Formerly LIMSI-CNRS
Orsay
France

Jean-Luc SCHWARTZ
GIPSA-lab
CNRS – Grenoble University
Grenoble
France

Pascal TEISSIER
Formerly Institut de la communication parlée
INPG
Grenoble
France

Index

global system for mobile
   communications, 64, 420, 442, 458
GlobalPhone, 293, 294
glottis, 3, 6
GMM acoustic distributions, 257
GMM, see Gaussian mixture models
goat, 347
GPRS, see general packet radio
   service
GPS, see Global Positioning Satellite
   system
grammar, 214
grammatical constraints, 241
grapheme, 101
grapheme-to-phoneme conversion,
   108, 259
grapheme-to-phoneme conversion
   rules, 259
grapheme-to-phoneme transcription,
   105
graphemic sequences, 102
GSM, see global system for mobile
   communications

H

half total error rate, 346
hand-free/eye-free functionality, 431
harmonic coding, 49
harmonic plus noise model, 120
harmonicity, 192, 195, 201
head-set microphone, 262
hearing impaired listener, 172
hearing-aid, 437
Hermitian symmetry, 44
hesitations, 245
hidden Markov model, 110, 204, 216,
   247, 249, 254, 255, 302, 332, 397,
   421
   acoustic, 304
   gender-dependent, 304
   phone-based, 302, 303
HMM topology, 334
HMM, see hidden Markov model
Holdsworth's model, 195

homograph heterophone, 109
homomorphic analysis, 361
HTER, see half total error rate
HTML, see hyper-text markup
   language
hyperplane, 301
hyper-text markup language, 143
hyphens, 242

I

I3, see intelligent information
   interfaces
ICASSP, 294, 396
ICSLP, 294, 396
identification score, 335
IETF, see Internet Engineering Task
   Force
IID, see independent and identically
   distributed
illusion of continuity, 193
IMELDA, 361
impostor, 464
   score distribution, 341
   speaker, 344
independent and identically
   distributed, 232
index 323
indexing of audio data, 261
InfoVox, 115
INMARSAT, see International
   Maritime Satellite
insertions, 262
intelligent information interfaces, 444
intelligibility, 146, 172, 282
   mutual, 282
intensity, 14
interactive terminals, 440
interaural delay, 198
interlocutor, 321
International Maritime Satellite, 78
international phonetic alphabet, 3, 5,
   283, 305
   symbols, 283